VOICES FROM THE SOIL

Timeless interviews with visionaries who fought for a scientific,
soil-health-based system of farming

Acres U.S.A.
Greeley, Colorado

Dedicated to the voice of Charles Walters,
founder of Acres U.S.A.

Special thanks to the brave editors, farmers and writers who contributed to this series of revolutionary interviews, which started in 1971 and continue to this day in Acres U.S.A. magazine and the Tractor Time podcast.

Voices from the Soil
Copyright © 2023, Acres U.S.A.

Acres U.S.A.
P.O. Box 1690
Greeley, Colorado 80632 U.S.A.
phone 970-392-4464
info@acresusa.com • www.acresusa.com • www.ecofarmingdaily.com

Printed in the United States of America
First Edition, Winter 2022-2023

Cataloging-in-Publication Data
Voices from the soil / by Acres U.S.A.. • 1st ed.
Greeley, CO, ACRES U.S.A., 2023
384 pp., 7"x10"

Acres U.S.A. owns the rights to all interviews and illustrations. Thanks to contributions from Charles Walters, Fred Walters, Chris Walters, Tara Maxwell, Paul Meyer, Ben Trollinger and all the outstanding editors who have contributed to Acres U.S.A., the Voice of Eco-Agriculture. All typos herein are to be circled and mailed to editor@acresusa.com

Cover and Interior Design by Carl Chiocca. Edited by Ryan Slabaugh.

ISBN 9781601731722 (pbk.)
ISBN 9781601731739 (ebook)

1. Agriculture — biological farming. 2. Soil science.
3. Soil plant relationship. 4. Fertilizers and plant nutrition.

II. Title.
S605.5.Z56 2017 631.5184

On Selecting the Voices

At first glance, one might immediately wonder how we selected the interview subjects to include in this book. After all, Acres U.S.A. editors have published more than 550 centerpiece, long-form interviews during the life of the magazine, 74 interviews in the life of our podcast, and selecting only 25 left almost all of them on the cutting room floor. This was beyond a "Greatest Hits" exercise. It felt, at times, doomed to fail. We had to review each interview for timeliness, integrity and historical importance, and balance that with the need for a wide variety of ideas and philosophies. We forged ahead.

In the end, we feel as though this book ended up being a useful tool on two fronts: First, this collection creates a living history at a critical time when more and more of the world is rightfully questioning the efficacy of our global food supply. Secondly, we want to provide a diverse set of individuals for the world to examine, to learn from, and to use as an example in their own advocacy. These voices share a unique talent in explaining the why: specifically, why we need to work with nature when we grow our food, and why we must work together to counterbalance the economic forces working against this concept. One day, we truly hope this goal is null and void, but for today, we see it as the most critical issue facing the planet, or more accurately, its human inhabitants.

So, what is in store ahead? Through these chronologically organized conversations, you can start in 1972 with Dr. William A. Albrecht, before eventually connecting his ideas to one of his last students, Neal Kinsey, and livestock manager Pat Coleby, and then to the ferocious advocacy of Vandana Shiva, and mixed between all those, to the down-home practicality of everyday farmers, ending with our talk with Gabe Brown in 2018. We chose 2018 as the end year because we truly feel Gabe has established a role with his book, *Dirt to Soil,* in helping energize the movement at just the right time. Certainly there are more great interviews to come.

One other point I want to clarify that did occur in our editorial conversations: citizens of the world have been talking and writing about soil health for centuries—millennia, in reality, proven by the rock drawings of harvesting equipment found in the Mideast from 23,000 years ago. Looking back for just our publishing history for voices—the last 50 years—uncomfortably eliminates the majority of voices who came before, from the early 20th-century biodynamic scientists to the voices of migrant workers who for centuries powered the global agriculture economy. And we can't forget the Mayan who modified their jungle ecosystems to include sustainable cornfields, and the Maasai with their own sustainable desert systems.

Vacancies aside, we also selected these interviews, because the stories the subjects tell go worlds beyond their own. You will find stories from Bill Mollison about the work of Vietnamese farmers teaching Americans after the Vietnam War, and you will read about the legacy of black farming that has empowered Leah Penniman's Soul Fire Farm in New York. You will read thoughts about globalization, the World Trade Organization, political laziness and a misled public, and you will read a timeless essay about living with fear from Wendell Berry. This is a rare collection of people and ideas that establish the strength, the vision and the confidence behind these voices of and from the soil.

We hope you enjoy the conversations that follow and you find they what we intended: inspiration, education and courage to speak up when you see a chance to make a better world.

A few notes for the reader before we get started:
- All of these interviews were published previously in the pages of Acres U.S.A. magazine between 1971 and today.
- "Missionary zeal," is what Fred Walters, son of founder Charles Walters, co-editor of this book and conductor of many of the interviews here, would call the energy in these interviews. We liked it so we thought we'd share.
- The illustrations included with the essays were not originally published with the interviews—they were commissioned to a professional artist in

2020 specifically for this book.

- This book can be one in a series. If the world finds it useful, we will create other volumes around livestock, nature and soil science.

On behalf of all the publishers, editors and interviewers at Acres U.S.A., thank you for reading.

Sincerely,
The Voices at Acres U.S.A.

Foreword

Anything that can stand strong for a half-century in our fast-changing world does so because it has merit and value. The fact that there was a in-depth interview with an eco-farming thought leader in the very first issue of Acres U.S.A. magazine—and still is today—is nothing short of astounding. It certainly makes the magazine and its intrepid interviews worthy of a deeper look.

Charles Walters was an old-school, small-press journalist and author who pounded words into a manual typewriter by the tens of thousands in order to fill every issue with thought-provoking content and never miss a deadline. He often used the phrase "get the story out" to describe his mission in life. There were amazing discoveries, insights and tales to be told that all too often mainstream agricultural media ignored either for lack of insight or seeking to avoid displeasing the increasingly dominant concentration of power in agriculture in the toxic, industrial realm.

Since he has passed from the scene, we can no longer ask Charles Walters if he began the practice of a long-form interview as a practical way to generate copy for what began as a one-person enterprise or as an insightful tool to share the bigger story in all its color and richness. Perhaps the birth of the Acres U.S.A. Interview was a bit of both.

This was all before the Internet took over the world. News stories have grown shorter and shorter over the past years to the absurd point where many readers don't delve deeper into a complicated story than a headline measured in characters, not words ... and certainly not in tens of thousands of words.

The world is a complex place. And nature is equally complex. The visionaries and innovators who have shared their knowledge in the form of interviews in the pages of Acres U.S.A. these past 50 years simply knew too much to be constrained to a few hundred words. They simply had to be allowed to tell their stories in full beauty and complexity. This depth is the ongoing goal of the Acres U.S.A. Interview.

The resulting historical record is as fresh and revealing today as in the year the tape recorders first rolled. Professor William Albrecht, an early student of soil composition, health and function, once told our founder, "There is only one story. Your job is to find ten thousand ways to tell it."

He was correct. The central truth that rings out through the hundreds of interviews that have appeared in this little journal present striking internal consistency, truth and wisdom. They stand together as a whole, as lone articles, and when coupled side-by-side.

Before agriculture took leave of its collective senses and chased bins and bushels while killing the soil, polluting the water, and lacing the public food supply with toxic, manmade chemicals, the knowledge to grow bountiful harvests of healthy food was well known. The wisdom and fundamental science of sophisticated, modern farming in harmony with natural systems was and is in our hands. Reading these in-depth interviews selected from the first 50 years of Acres U.S.A. will result in an education, confirmation of the soundness of eco-farming, and head-shaking realization that the knowledge we seek is in our hands. Others have worked so that we may benefit.

We're not saying that science and our understanding of it have not advanced—and also agricultural technology—but core truths stand rock-solid as a foundation for now and the future.

A few generations of gurus have hit the lecture circuit teaching soil health and non-toxic agronomy this past half-century. Most add a wrinkle or two, but tend to repeat and expound on the wisdom of others who passed before. And this is how it should be, to stand on the shoulders of elders and not be forced to repeat their mistakes, not to reinvent the proverbial wheel.

For most of Acres U.S.A.'s history, crafting the monthly Interview was very much a family affair. Charles Walters, founder of the magazine, conducted almost every interview for the first 30 or so years. He was an economist and historian who learned agriculture through a stint with *Veterinary Medicine* magazine and founding and editing the *NFO Reporter*, crisscrossing the country during the holding actions of the 1960s. His multi-disciplined training, interests and insight transformed an interview from a mere tool for self-promotion into a worthy entry into the annals of history.

His firstborn son Chris, also a writer/journalist, tag-teamed with his father for many years and then took over the Interview for a couple of decades after Charles Walters' death. A serious reader and historian, he literally spent days preparing for a one-hour taped interview; his depth and sophistication of questioning is apparent in his hundred-plus resultant articles.

I (middle son Fred) took over the helm as publisher and editor twenty-five years into the magazine's life. Coupling my education (science and journalism) with the deeper education gained over a lifetime of dinner table conversations with my father and visits with our authors and conference speakers, I tried to stick with the tone, style, intent and personality set by previous decades.

The juggernaut that is the mighty little Acres U.S.A. interview continues with a new crop of informed, dedicated journalists. And in this era of tweets and ultra-short, dumbed-down news, its role as a carrier of in-depth substance is perhaps more important than ever.

When considering the wealth of information that has flowed through the pages of Acres U.S.A. its first half-century, the value in the many practical articles, the inspiration in farmer profiles and personal stories, and the truth in the "connecting dots" of news shorts all are overwhelming. But for some reason the Interview stands apart. Each conversation with an eco-farming visionary — from decades past and current, names both well known and now all but forgotten — stands as a supporting column of the temple of knowledge.

And because the pages relay the speaker's own words, reading these interviews is like having a personal conversation with the interview subject. Thoughts and words blossom from the pages with a richness, tone and voice that will resonate as visionary, wise and true. Phrases beautiful and enlightening leap from the page.

Become friends with eco-farming visionaries and elders through the volume you hold in your hands. Mine their wisdom. Read, discover and enjoy.

Fred C. Walters
Publisher & Editor, Acres U.S.A. (retired)
February 2021

Interviews

-

"I've Had to Stand Alone"

An Interview with Dr. William Albrecht

Originally Published: May 1971

In late April 2019, a group of people met to honor a scientist in Columbia, Missouri, and a man who helped start Acres U.S.A.: William A. Albrecht, 1888-1974.

Neal Kinsey, a longtime contributor, author and speaker for Acres U.S.A., whose interview is also found in these pages, presented the honor to Dr. Albrecht's family, and he was chosen for a good reason. Mr. Kinsey, and Charles Walters, Acres U.S.A. founder, were two of the last students to work with Dr. Albrecht, a timely connection that turned into a gift for human kind.

The Acres U.S.A. founder, Mr. Walters, served as an editor with Veterinary Medicine magazine in the 1950s and traveled the country for the NFO in the 1960s. This was the era when animal care and farming based not on natural processes but domination by toxic chemistry came into being. In fighting for the survival of family farmers—he entered agriculture as a trained economist —he saw the only way agriculture can be economical was to be ecological. In rounding up the body of knowledge of the day he kept hearing the name "William Albrecht," not from the University of Missouri, but from esteemed scientists and consultants around the world. He picked up the phone and learned Dr. Albrecht was still alive and held office hours, even though long in retirement. Despite the attempt at dissuasion by the University, he hopped into a car and drove two hours down the road to Columbia and met the good professor.

And so began a weekly tradition of tutorials and lessons, which formed the foundation of the magazine and books and conferences of Acres U.S.A., and the fuel for the eco-agriculture movement.

Here is an interview and an introduction from our second issue in 1971 with William A. Albrecht. "Let me help you catch a vision" was the phrase with which Dr Albrecht greeted the Acres U.S.A. interviewer:

WILLIAM A. ALBRECHT. If you want to reduce human medicine or veterinary medicine to a common denominator, you have to remember that when the animal's physiology is deranged, it doesn't make much difference what you call the problem—but it is very probably a mistake in nutrition often founded on the attempt to be economical. I have come to the conclusion that deficiencies are more often at the base of health irregularities than we realize. I have had occasion to test some trace elements. And two M.D.s hooked up with me at one point. So I jumped in at the deep end because if I got into trouble with the medical profession—and that's very easy to do—I'd have two M.D.s to pull me out. I put a student on studying brucellosis contagious abortion, which they call contagion, which it isn't at all. And we proved it with four generations of a herd of 85 milk cows that were labeled to be slaughtered. We fed them trace elements, and we treated the soil with trace elements while we were getting ready to feed the animals the products of the soil. In four years we had 17 female calves that became heifers and raised calves, and their calves were clean according to the veterinary tests. Because, you see, they introduced an artificial microbe they call a *brucellosis abortus* as though it were a grand name. It is nothing but the symptom name given to the microbe. We had 17 heifers mature after we started feeding trace elements, and they give us calves, and those 17 calves were as clean as could be by any test the veterinarian could run on the bloodstream.

ACRES U.S.A. The remedy turned out to be nutrition?

ALBRECHT. All by just feeding trace elements, and the four trace elements we picked were those a Cleveland concern showed were missing in the nervous tissue of the animals which were infected, and were not missing in the animals that were not infected.

ACRES U.S.A. What were the trace elements?

ALBRECHT. The four trace elements that were missing—manganese, some iron—and I put that in parenthesis, because that wasn't necessarily missing, but it is always necessary to have iron, copper, cobalt and zinc.

ACRES U.S.A. Were these findings reported in the professional literature?

ALBRECHT. Oh, yes. It was a volume, but not too many were printed. In 1949 we held a clinic because I got the state medical association on my neck. The doctors down in Springfield had been giving the people inoculations for brucellosis. The women and the men moved over to Dr. Allison, who fed them the trace elements, changed their blood corpuscles from all white pussey ones to red ones. He was feeding these trace elements with a coating so that they didn't open until they got into the alkaline part of the intestines. We showed that Bang's disease was the result of a trace element deficiency.

ACRES U.S.A. Did you ever consider becoming a physician?

ALBRECHT. Yes. I had a good doctor friend who spent his lifetime teaching people about health, and when he died he had 72 percent of his business still on his books. So I got discouraged as a boy. I said, "I am afraid that I don't have enough association with the medical profession to make a go of it." Having been a country boy with a lot of curiosity, interested in the physiology of plants, animals and man, I decided I'd better stay with plants and agriculture. So I took soil fertility and soil microbiology for my major, and they brought me here to put out cultures of legume bacteria, because at that time soybeans were new and there were no cultures.

ACRES U.S.A. Where did you take your training?

ALBRECHT. All at Illinois. Four degrees. A.B., B.S. in agriculture, M.S. and Ph.D. I'm probably more of a plant microbial nutritionist than anything else. In other words you get down to the single cell.

At this point Dr. Albrecht started questioning
the editor of Acres U.S.A.

ALBRECHT. Does the veterinarian know which way to turn? His animal is on that soil and eats plants from that soil. Why doesn't he get down to the basics? Why doesn't he go down to the foundation?

ACRES U.S.A. You can't get them to listen. Despite the evidence, you can't make anyone listen if they don't want to.

ALBRECHT. Now that's one of my disappointments in teaching and writing and studying. They don't use logic to explain what it is all about. They're only commercial-minded. As a boy— before I left country school—I told my mother I'd learned something. There are no hoop snakes. And I said, "Mother, I'm going to study snakes." I got myself an Osage orange cane with a little fork at the bottom. And a cane is longer than most snakes. That snake has to keep at least half of its body down to get the leverage for the other half to strike. It has to have an anchor. When I finished my graduate schoolwork, I had over 200 specimens of various things of that nature preserved and put away on the stockboard nailed on the joists in the basement, all cured. Alcohol only cost 50 cents a quart. And I knew the saloonkeeper. The thing that disgusts me is that your scientists go to technology instead of teaching. They patent everything and make it secret. I don't like that. So I decided that I was going to study and learn. If you analyze what I've done here that they've paid me for, it's nothing but learning what nature did which had never before been recorded.

ACRES U.S.A. What has been the biggest revelation?

ALBRECHT. In agriculture, and soil microbiology, and in medicine, I discovered what the country boy said when he came home to his dad from the college of agriculture. He said, "Dad, they teach so much that ain't so." So I've spent most of my life finding what is so. As I learned, I wrote everything out and studied it out, and put it into manuscript form.

ACRES U.S.A. On the basis of your research, should fertilizers be soluble?

ALBRECHT. No. Fertilizers are made soluble, but it's a damn fool idea. They should be insoluble but available. Most of our botany is solution botany. When it is solution botany, the first rain would take it out. There's a big difference between the laboratory and the farm.

ACRES U.S.A. Is this the reason we have so many farm wells that are too hot to use?

ALBRECHT. There you are. And we live with our own damnable ignorance because we don't sit and think. We copy to make money. And they teach copy stuff in college. And if a student has an idea, he never gets to say, "I have a hunch." And teachers do not encourage students to have a hunch because they want them to memorize what they've said.

ACRES U.S.A. Is that why the "insoluble but available" idea has not been taught in school?

ALBRECHT. I wrote this (a paper titled *Insoluble Yet Available*) and got it published in the British papers. We've put money between scientific study and publication of the results. So we only tell farmers about products that can make money for the companies.

ACRES U.S.A. I go around to these schools—including the University of Missouri—and all they're teaching is this "soluble" business. There are regulations in some states that in effect define fertilizers as products that are N-P-K rated, and make it difficult for farmers to have ready access to humates, natural mineral fertilizers and the like. Some of my associates inform me that you are the best spokesman—with academic standing—in America today.

ALBRECHT. I've had to stand alone.

ACRES U.S.A. I note that you've done a lot of work, and that this work is not being made available to younger generations—farmers under 40.

ALBRECHT. That's the reason I'm happy to see you. As a journalist you can use quotation marks. You can report. I'll just give you a simple principle. A root puts out carbonic acid and treats the rock with that acid and gets its nutrition. And yet we fight soil acidity.

ACRES U.S.A. In other words, this acidity breaks down the rocks?

ALBRECHT. Of course. The only acid you like to drink is carbonic. You don't drink hydrochloric acid.

ACRES U.S.A. Why, then, have these states come to proscribe against acidity with their fertilizer laws?

ALBRECHT. Because what I say doesn't amount to anything in the eyes of these people. They've bought a conventional truth because there is profit in it for a few big firms.

ACRES U.S.A. What you're telling me isn't what they're teaching in this university?

ALBRECHT. That's the sad part. You see what people take is what the horde take. Not what the fellow who sits and thinks takes.

ACRES U.S.A. Would you agree with the aphorism, "People take leave of their senses as a group. They come to their senses individually?"

ALBRECHT. Always, if they have the courage of their convictions.

ACRES U.S.A. Is this reprint, *Insoluble Yet Available*, your anchor piece?

ALBRECHT. Just horse sense, that's all. I tried to put together the observations that mean something. Let's take this matter of the plant's nutrition. When I came here as a microbiologist, they wanted me to grow a culture. And they thought I could grow a bacteria that would make a plant fix nitrogen and be inoculated. And I was here six months before I discovered that was what they believed, and I was terribly disgusted. I said, I'll have to tell those people that when a bull and a cow get together, the cow has to do her part too, not just the bull. All of my research here is merely that conviction that when my cultures do not make nodules on their legumes, I've got a plant that is too sick to carry its half. But I haven't got that across so far.

ACRES U.S.A. Why has your research turned out so differently from results others have had—results, I might add, more pleasing to commercial firms?

ALBRECHT. Well, Professor Miller thought I should grow bacteria that would make the cow have a calf whether she wanted to or not. And I had to politely show the points I wanted to make. Here (at which point Albrecht produced a report titled, *Some Soil Factors in Nitrogen Fixation* *by Legumes*) is increasing calcium saturation of an electrodialyzed clay. I separated the finest part of the clay out in the centrifuge running 32,000 rpm after the clay had been suspended and settled for three weeks. At the bottom that clay plugged up finally, because the clay was too heavy. But we had thinner and thinner, smaller and smaller clay until about halfway up in that centrifuge—you know what the milk separator is?—there we had it as clear as Vaseline. Now we took the upper half of that clay—a clay so fine that it was like transparent Vaseline. We made pounds and pounds of that because we had put it into the electrical field and made it acidic and took all the cations off so it was an acid clay. That was the thing with which we studied plant nutrition. We studied plant nutrition with that clay by putting different elements on in different orders. We grew plants. We studied plants with this fraction of the clay in the soil that holds the positively charged nutrients. And we could mix them and balance them.

ACRES U.S.A. As controlled experiments, I suppose you reduced the variables so you could take up one element at a time. How did you start?
ALBRECHT. We began with calcium because we found that we had to come up here to 65 percent saturation. In other words, you've got to load that clay in that soil with 65 percent of that clay's capacity to hold calcium against rainwater before you can grow a plant with enough calcium to be healthy.

ACRES U.S.A. Will you explain acidity to me?
ALBRECHT. Your acid clay is nothing more than one that doesn't have the positive ions on it—hydrogen, calcium, potassium, magnesium, sodium and the trace elements. I've got to have 65 percent of that clay's capacity loaded with calcium and 15 percent with magnesium. I've got to have four times as much calcium as magnesium. You see why we ought to lime the soil? We ought to lime it to get it up to where it feeds the plant calcium. Not because it fights acidity.

ACRES U.S.A. You can't perform this function with any soluble fertilizer?
ALBRECHT. It's got to be a positively charged element like your calcium, magnesium and so on.

ACRES U.S.A. When were these findings revealed?
ALBRECHT. This paper was given at the International Society of Soil Science the day Hitler invaded Poland, 1939. In this paper I summarized the work of about a dozen graduate students.

ACRES U.S.A. Why this clay method for research?
ALBRECHT. As a result of using this clay method of learning what plants are fed, I learned

about plant nutrition in the soil, not in solutions, as is common laboratory procedure.

ACRES U.S.A. To what extent did the farm press pick up this material and make it available to farmers?

ALBRECHT. Very little. They say it's too complicated. They say, "I don't know anything about it," and out it goes. But here comes something from the chemistry lab that's advertised for sale, and they swallow it hook, bait and all.

ACRES U.S.A. Do research grants influence scientific findings?

ALBRECHT. Let me answer you this way. I have a concept as to how those positively charged elements are held in the soil against water. My problem is to get a vision, and my graduate students helping me a leg at a time let me catch the vision. I say let's put it into the common man's language of the Creator's business of creation. Not commercialism. The moment you throw money into this thing for a boy to study, you're on the wrong track.

ACRES U.S.A. A couple of years ago, I was writing a book on farm bargaining. I came across some information to the effect that the continued application of salt fertilizers is delivering less and less production. In other words, the American farm plant is over the hill and on the way down. Would you care to comment?

ALBRECHT. I have excellent data on the half-life of our soils. You see the soil is like a radioactive element newly created. When this soil was balanced out there in man's absence, and before man took it over, it was virgin soil. It was in equilibrium with the forces of soil development and leaching. If you start with the desert in the West, on the east side of the coast ranges—because water has all been precipitated on the west side—that's the raw rock with a slight weathering. As you come east, then it is heavier rainfall, and you develop the soil into more than a desert. And that American bison lived where conditions were about balanced, and that's a little above 25 inches of rainfall. Because when you go above 25 inches of rainfall you began leaching. But at 25 inches, you're just about balanced. That buffalo was smart. He had mineral rich soil and not mineral-leached soil.

ACRES U.S.A. Yes, but that same soil is now being irrigated, and nature's 25 inches of rainfall is being sidestepped?

ALBRECHT. Yes. And it's been grown with crops that suck only the back teat, we'll say, and remove certain elements more than others. The buffalo didn't go far east and west, but north and south. He went with the winter and summer, back and forth. He went long distances north and south, but he didn't migrate far east and west, because he would have gone to less rainfall and more rainfall.

ACRES U.S.A. This would put the center about the middle of Kansas?

ALBRECHT. That's right. Here in Missouri, we have virgin soil east of Columbia—soil that has never been plowed. The farm across the road hasn't been farmed since it was broken out in the early days. So we studied that soil, and we have the rate of decline under the old-fashioned horse and collar days against that of virgin prairie. How rapidly did this system of farming tear that soil down after 60 years? Well, the 60 years show how fast it went down. On Sanborn Field

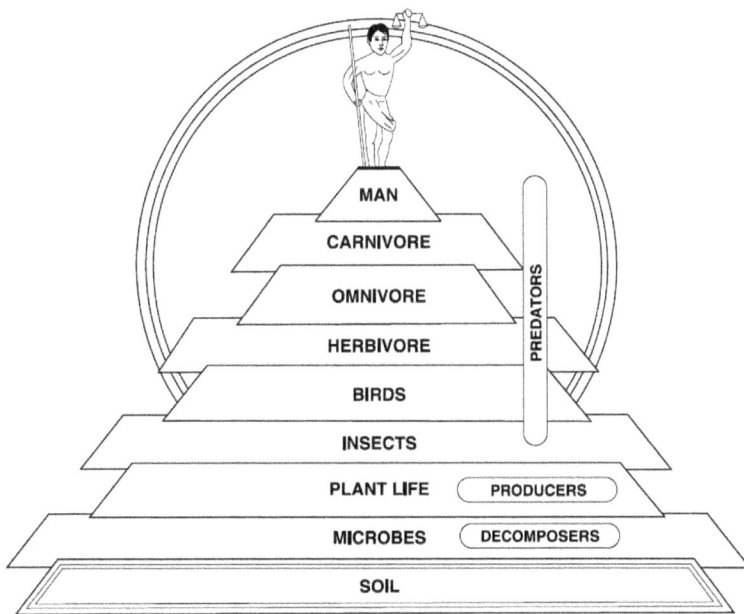

we grew corn continuously with nothing put back. Everything taken off. In 40 years we took 2/5 of the fertility out of the soil. Where we put out corn continuously, in 40 years we exhausted 2/5 of the fertility. I grew wheat continuously for 25 years and used nitrogen fertilizer at 25 pounds per acre. And, including the nitrogen I put back, in 30 years we burned out 50 percent of the soil.

ACRES U.S.A. So you can't farm, say, 100 years on this ground with techniques being used in America?

Above: Dr. Albrecht's view of the ecosystem, as published in his series of books.

ALBRECHT. Not if you take half of the remaining fertility out every 30 years. You see, with salt fertilizers N-P-K rated, you're churning that soil to make the microbial fires burn.

ACRES U.S.A. What can farmers do to farm scientifically and still preserve the topsoil for future generations?

ALBRECHT. Supplement with the first item that is most exhausted. But this isn't as simple as just putting nitrogen back. You have to know that nitrogen is an extremely significant item in the microbial life that is going to live in that soil. You've got to maintain the living soil, and not a dead soil. And the moment you start putting nitrogen on the soil, you burn the carbon out. And you burn out more than you put in.

ACRES U.S.A. How can nitrogen be returned soundly?

ALBRECHT. In my garden I take the leaves of the trees and I compost them. And I keep that carbon high for the nitrogen I put in, and not the nitrogen high to make that carbon burn out and shoot the life out of the soil. I put calcium into that compost—it's all leaves, kind of woody, so it's like wheat straw—ratio of carbon to nitrogen of 100-to-1. But I just put a little nitrogen, and just that so that during the year I get down to where I've got still a lot of carbon. But there is enough nitrogen being used this way. The microbes don't let it leach out because it's always tied up. Carbon ties up the nitrogen.

ACRES U.S.A. Do you do anything else to that compost?

ALBRECHT. Oh, yes, you've got to take care of your phosphorus, calcium, potassium and magnesium. That's nature's way. But it has to be broken up. Now we haven't learned how to appreciate carbon as excess because we've got 3/100ths of a percent in the atmosphere, and what does a plant do mainly? It ties up the active element into that excessive carbon. So you've always got to keep a black soil.

ACRES U.S.A. That's the significance of the black? Carbon?

ALBRECHT. Yes. And you never can go very deep with the black because the air is shut off, and you can't have a deep soil unless it is granulated with calcium. So we had our black soil in the prairie that still had a lot of calcium. And the depth of that black soil merely shows that balance, that accurate balance of good nutrition.

ACRES U.S.A. Since I was a boy in Kansas, we've been growing wheat ...

ALBRECHT. ... High protein wheat.

ACRES U.S.A. ... but the protein is slipping.

ALBRECHT. Oh, yes, I kept the records of Kansas, and in the time I studied it protein dropped dramatically.

ACRES U.S.A. Now they have deep wells, fantastic milo crops, and soil that is starting to leach out.

ALBRECHT. In other words, they're moving it to where the microbial fires are being fanned.

ACRES U.S.A. This is being characterized as efficiency in agriculture, is it not?

ALBRECHT. Efficiency in mining. Once the soil is exhausted, you're on an ash heap. You don't know what's missing except that nothing grows. The mysteries of creation haven't all been put under button pushing technology. And the trace elements are a part of it. I had a letter from a man in Florida. He had read something about my remark that we aren't including an inventory of all the elements that are nutritional. He wrote, "Albrecht, I'm growing citrus down here. I graduated from Purdue." He said, "Tell me what to put on the soil as trace elements. I'd like to try it." I gave him a gunshot. Copper, magnesium, zinc ... I gave him a list. Come Christmas he sent me some fruit. Earlier he said, "We're having trouble keeping our dairy cows out of our citrus grove. Every morning that herd of cows goes through the fence. Now we have a strong fence, but we have one cow that goes right through it anyway. She's in every morning." So I said. "Don't laugh at that critter. She's just a little smarter. She's an A student." Anyway, I received this citrus; I put the fruit in my wine cellar. When we finished that grapefruit bushel, the last two in the bottom had been cracked, but they hadn't spilled any juice. They had split so that the slices separated out. And I checked the shipping date, and it had taken six weeks for them to come to my place. This was interesting, because the fruit hadn't been taken by the green mold, and citrus turns green in a hurry if it isn't properly fertilized. I think the copper he had put on that land at about 5 pounds per acre protected against the green mold. Later he wrote that he had corresponded with some 100 experts. Not one thought we knew what we were doing. But, he said, "I just closed my contract for all of my citrus at a nice markup because I had a better fruit."

ACRES U.S.A. Then the whole business in Florida is a matter of trace element deficiency?

ALBRECHT. Of course. Now why can't our farmers here see that? I've seen that Florida thing for years. They went to seedless fruit because they could grow more seedless fruit without making seeds because you've got to have balanced fertility to make seeds. I just talk straightforward to those people. They're on sand down there, and they have to literally spoon feed their citrus fruit.

ACRES U.S.A. Recently I visited with a man in Minnesota who told me it was costing him twice as much per pound for feedlot grain in Minnesota than it was costing him out West. Is that because the corn is deficient? What about blight? If grapefruit develops green mold because of mineral deficiency, does the same hold for corn blight?

ALBRECHT. I have written several letters telling people that before they fight this corn blight, see what five pounds of copper per acre will do to help corn protect itself. I ate the grapefruit. Mrs. Albrecht ate them. They were not green moldy.

ACRES U.S.A. A lot of farmers are coming to their senses, one at a time. What's the best starting point in considering sound farming, rather than soil mining?

ALBRECHT. We should not start before we include all the potential stages, and see how they fit together. The first stage, when this farmer made his power and his manure, and handled it all himself, he was more nearly natural. We have lost sight of three factors. The microbes in the soil, and on top of the soil, they are the forces that which by decomposition do the recycling. Then we've got the plants that profit by that. And they synthesize sunshine. The microbes can't do that. The microbes have to have synthesized plant foods. My book *Wastebasket of the Earth* outlines this. The microbes in the soil are the decomposers. The plants are synthesizers. All else that grows is a predator on those two. That's why it is so important to treat microbes in the soil with respect, and why it is so important to rebuild soil. Now the German manure system was a tremendous force in rebuilding, but you see that's too much work.

ACRES U.S.A. How can you farm and rebuild that soil if you pursue monoculture or one-crop farming?

ALBRECHT. You can't because you haven't got the manure. You have got to go back to the profile of the soil, maintain the carbon because the carbon reduced is what holds the other things there. When it is not black, she's gone. Wide carbon ratio is the safety of your soil. You can't do that out of a chemical company's paper bag. And this whole business of what's pollution is nothing more than having run the thing lopsided—out of balance.

ACRES U.S.A. Can you give me a precise take on calcium?

ALBRECHT. Calcium granulates your soil and keeps it black. That granulation lets air go deeper. And that lets the microbes burn. Now if you don't have enough air in there, you ferment, and you make alcohol. So when you make that soil anaerobic and don't granulate it, you got too much hooch in it. Oh, that thing is delicate. You can't put nitrogen in that soil without damage.

ACRES U.S.A. Can you comment on natural minerals?

ALBRECHT. I like to use natural minerals. That's what limestone is. That's what rock phosphate is. Your humates. That's when your decomposition is carried under enough air exclusion.

ACRES U.S.A. Humates are not soluble?

ALBRECHT. No. And humus deep down in the soil is anaerobic and tends to be black. You bring it up and cultivate it and oxidize it and you release the things that were reduced.

ACRES U.S.A. It is a sound approach?

ALBRECHT. Oh, yes. Nature builds its own humus down only so far because it gets no air and is preserved down there. Down in Texas you do pretty well because your panhandle of Texas is high calcium. Your Kansas soils are high calcium. But they've been burning the calcium out awful fast.

ACRES U.S.A. What will irrigation do?

ALBRECHT. That saturates the land with water, gives it a fermentation and quick oxidization. Nature does this so gradually. And if there is plenty of calcium there to granulate it, the humus will move down and your roots go down, and you've got a deeper feeding.

ACRES U.S.A. If 25 inches per annum rainfall means perfect balance in the high plains, what effect does irrigation have, especially when carried on to the point where catch basins are used?

ALBRECHT. If you waterlog the field, you're going to be in trouble. It doesn't take many years. Just look what 25 years did on Sanborn Field—burned out 50 percent of our fertility.

ACRES U.S.A. Let's go back to your statement that fertilizers should not be water-soluble.

ALBRECHT. It can't be water-soluble because the preceding rain would have taken it out. The clay humus is a colloid on which the positive ions are held because the clay is negatively charged and holds positive elements. So your calcium is held on the clay. Your hydrogen is held on the clay humus. Your magnesium is held on the clay. Your potassium is held on the clay. You've got the cations—the positive ions—and they have to be balanced for the plant. And I told you, you had to have 65 percent saturation of the calcium, 15 percent of the magnesium, 2 to 5 percent of the potassium ... for your legume plant to take nitrogen from the air, and grow, you've got to have 65 percent saturation on the clay of the calcium, and so on. Now that's a balanced plant diet. But how many plants are fed on that kind of a diet? Nobody talks about a balanced plant diet in terms of positively charged elements because they don't understand it. They only understand mining the soil for a fast profit, with no thought of future generations.

ACRES U.S.A. To have an idea of what's wrong with soil, you'd have to test the soil, wouldn't you?

ALBRECHT. Yes, But how many men doing the testing look for it to be balanced? They only start with the one element, that's least, and dump on an excess and go overboard. So you might talk about the soil in terms of a balanced diet for the plant. But the plant has an advantage. As a root goes down, it is hunting. So you need a deep profile for that plant to be fed in. The plant struggles to survive. Its roots are hunting. The plant does a lot of scratching around. So they're trying to feed this plant and don't realize that as the soil gets dry that root's going for water. When they put water-soluble salts on that soil they unbalance the thing as if you took too much whiskey.

ACRES U.S.A. Some of these points seem complicated. Does this plant physiology have a counterpart for illustration purposes?

ALBRECHT. For about 25 years I've worked on this Epsom salts business. Frequently, after a hernia is repaired, bowels won't move past that hernia. So they give Epsom salts. And if they check they'll see that urine is throwing the protein out of the blood. Protein is wasted because the Epsom salts ruin the membrane in kidneys and keep them from doing their normal work. When you take Epsom salts, that salt replaces the calcium in the wall of your intestines and it throws everything it can because that membrane is no longer normal. It just throws everything from the blood stream till it flushes it and can go back to your bones to get some calcium to rebuild intestine walls. When I gave that to Dr. F.M. Pottenger, he said, "You've got a good theory because if we've got a highly rheumatic person and give him Epsom salts, he's so low in calcium he throws the calcium out so badly that it kills him." Now the medical profession knows that they shouldn't give Epsom salts, but they do. But you see with this hernia, the kidney wasn't functioning when the magnesium went through. The magnesium that the bloodstream had to throw out through the kidney was knocking the kidney. Now here's my theory. Now remember my work. If I didn't have my soil loaded high enough with calcium, the nutrients were going from the plant back to the soil exactly the way they go from an intestine. If I don't have this calcium-saturated soil high enough, the plants throw their fertility back to the clay, instead of from the clay to the plant.

ACRES U.S.A. The plant feeds the soil instead of the soil feeding the plant?

ALBRECHT. The plants will build the fertility up in the soil and the plants will starve to death. Now you see what I mean when I had a different vision of plant nutrition in the soil than solubility. This thing is delicately balanced, but who has a vision of it. If you put chemicals into that soil you've ruined that cell root. These laws of physiology—it doesn't make much difference whether it is a person or a plant. I'm convinced that the Creator knew his business, and man still hasn't learned.

ACRES U.S.A. There is one problem with what you're saying. Hardly any of these farm magazines put this information into layman's language so that farmers can understand.

ALBRECHT. That's the sad part of it. I just want you to do your own thinking. Let me fill you in on why we've been on the wrong track. You can then pass it on to your readers. I just want to say it the same way and have it repeated that way. Don't worry too much if you don't always understand. Keep on studying and it will all come clear.

"We Live on This Earth and It Is Marvelously Made"

An Interiew with Dr. Carey Reams

Originally Published: November 1979

Dr. Carey Reams, 1903-1985, was a native of Orlando, Florida, and spent most of his lifetime as a researcher, agronomist, and student of nutrition for animal and man. During World War II, the Jeep he was driving ran over a land mine in the Philippines. When he awoke some three weeks later, his first words were, "We sure landed easy." He didn't, of course. His spine was severed in one place and crushed in other places. One eye was gone. The physicians told him he would never walk again. This challenge prompted him to dovetail what he knew about agriculture with what was still to be learned about nutrition and repair of the human body. As answers surfaced that were at variance with the pronouncements of the American Medical Association, Reams found himself in court battles that seem to never end. In 1979, Acres U.S.A. found Dr. Reams with a quiet moment on his hands. With a tape recorder running, questions were noted and answers captured for publishing.

ACRES U.S.A. Dr. Reams, your classes of instruction in agronomy often start with Einstein's equation, $E = Mc^2$. And then you go on to explain $Mc^2 = E\ E^2$, E designating heat and E^2 designating electrical energy. In the 40 or so hours that follow, you explain these things—but for now, can you make your approach come clear for our readers?

DR. REAMS. Well, Einstein gave us the mathematical formula for moving an object from A to B. I visited him by appointment at him home many years ago. I asked him, "What I want to know now is how to move the energy from B back to A." He said, "Young man, I'll give you that assignment." So I set to work on it. The only way I could find to do it was either in plating tanks or in the biological field of growth.

ACRES U.S.A. How does the plating tank figure in helping us understand this nutrient movement?

DR. REAMS. It brings you to ionization and electrolytes. A plant doesn't have any brains. It can't go out and get its food. So it has to get it somehow. The ionization method of the line of least resistance of magnetic fields is the way food is carried to the plant.

ACRES U.S.A. Is this what the soil audits call the exchange capacity?

DR. REAMS. No, the exchange capacity is the ratio of various elements during various periods of the growing season, and the effect of temperature and heat and water, and whatever. It affects the electrical, but it isn't necessarily electrical.

ACRES U.S.A. Let's take this thing from the seed up.

DR. REAMS. Yes, and the seed needs carbon to bring water into the seed. Carbon is the governor for water. In turn, organic matter brings carbon to the soil. Carbon holds some 400 percent of its weight in moisture. The carbon level determines stress. More carbon in the seed means more energy in the seed. As the seed swells, friction begins, setting up heat. Then it gives off anionic heat. This moves upward through the shell of the seed. As it moves upward, a vacuum is formed. The cations then fill up the vacuum as they come in with the water if there is nitrogen present. The nitrogen is a carrier. In other words, it attracts the other plant food nutrients to it. It loads itself as it builds itself onto the rootlet.

ACRES U.S.A. Your conceptualization then is that the process is something akin to the planting tank?

DR. REAMS. Exactly. The seed is losing energy. The nitrogen is picking it up. It attaches itself to the rootlet, gets into the plant sap, and then the sap changes it to a part of the plant. It does this in the stump of the plant. That is where it changes it to the frequency of the plant. Synchronizes it. Nitrogen attracts other nutrients the same as a salt in the plating tank. It is an electrolyte. Nitrogen is the one element farmers think of first. It is foolish, of course, to think of one element as being more important than another—and yet nitrogen is very important. Nitrogen is nothing more in biological life than an electrolyte. It functions very much like a metal—that is, it carries an electrical charge. It is the charge that ionizes the water in the soil and makes it possible for minerals to enter the plant. Without nitrogen in the soil the electrical currents won't flow. Also remember, nitrogen can be a cation or an anion, and it can have either a positive or a negative charged nucleus. The anionic form is found in nitrate nitrogen, and the cationic form is found in ammonia. Isotopes in the soil will change to the side of least resistance—to the greatest magnetic attraction. Nitrogen is the sun in every molecule. Take the sun out of a universe—any universe—and you have a zero, a nothing.

ACRES U.S.A. Dr. Reams, can we background your concept of ionization? Can we make what we're visiting about here mote meaningful to an average farmer?

DR. REAMS. Let me put it this way. We live on this Earth and it is marvelously made. About 250 miles southeast of Capetown, South Africa, there is a south magnetic field. And just below

Hudson's Bay in the Eastern Standard Time Zone, there is a north magnetic field. Over the entire crust of the earth there is an electrical charge—from the south magnetic field to the north magnetic field. This is a very important charge. Sometimes this charge runs along just above the surface, and sometimes it runs many hundreds of feet deep. You don't have to be a scientist to prove the fact of this charge. Just take a Boy Scout compass. The same thing that causes an electrical charge to flow north causes the compass to point to the north magnetic field. In the Southern Hemisphere the compass would point to the southern magnetic field. I have an illustration I'll leave with you to illustrate this point.

ACRES U.S.A. We're almost afraid to ask, but what has this to do with a seed sprouting?

DR. REAMS. That seed is a complete little solar system. Deep in that cell there is a trace element called manganese, specific gravity 54. This is very important because manganese is the element of life. Without it there would be no life on this planet as we know it. It is manganese that enables living things to bring forth young, and this holds for the smallest bacteria to the largest sea mammal. The seed has this germ of life in it when you plant it. The seed then becomes a magnet with its little bit of ionized energy. This magnet is positive—never negative. We call it a "cation". Now this moisture arrives and ionization takes place. Ionization is simply the process of taking something apart, ion by ion, or putting it together, ion by ion. As I have pointed out, this is exactly the way it happens in the plating tank with silver, nickel, chromium, copper. Now when this magnetic force passes over the crust of the earth, it brings ionized plant food—like copper to the plate in a tank—into the seed. This supplies the seed with the wherewithal to make its first rootlets. The first rootlets that come out of the plant go straight down because the seed is positive. But as the seed rootlets go down, the positive charges—as they form according to the frequency of this plant—give up ions. This accounts for upward growth.

ACRES U.S.A. Now if the rootlet picks up the nearest nutrient with an electrical charge, does this mean the rootlet will grow in the line of least resistance, and therefore head for magnetic north?

DR. REAMS. Yes. The first rootlet of that taproot will point to magnetic north. You might ask your readers to place a morning glory vine cutting in a glass of water in order to observe roots as they develop. They will grow in a direct northerly line. So when you're setting out a crop, it is best—whenever possible—to plant the rows east and west if the ground is level enough for that. This permits plants to feed in the middle of the row rather than under each other. If you're setting out trees, vines or any plant that lives more than two years, put the side with the most roots to the north.

ACRES U.S.A. In Brazil early this year, it came to the attention of this editor that it took much longer to grow corn than in, say, Iowa. Does the ionization of the earth have something to do with this?

DR. REAMS. Yes. It takes from 9 to 10 months to grow corn in Brazil. In Iowa it takes about 5 months. In Alaska it can take as little as 6 weeks to 2½ months because the ionization of the earth is much greater there. When ionization of the soil is increased, the voltage doesn't go up, but the magnetic field is broadened.

ACRES U.S.A. You used the term frequency and we understood you meant to apply this to plants. Can you explain?

DR. REAMS. All plants have a frequency. Every living thing according to its kind has its own frequency. Now, a frequency is the number of wavelengths per second. These terms are commonly used in the biophysics field. Basically the rules are the same, but the applications are different. Frequency for a human male is .0000024. For the female it is .0000026. The male dog—.000038. A female dog has a frequency of .000040. The numbers for a stud horse are .000049—for a mare, .000046. There is an odd number for citrus—.0009. All other plants have even numbers. A frequency is how much time it takes for one electron to make one complete cycle around one molecule. So frequency is time. Frequency is different from kind to kind, not from plant to plant. Pecans, peaches, prunes, all have the same frequency. Potatoes and tomatoes have the same frequency. All varieties of citrus have the same frequency. The decimal and figures for frequency are determined by an oscilloscope, which is a laboratory instrument used in physics.

ACRES U.S.A. Why is citrus different?

DR. REAMS. I would have to say that is the way God made it. There's a cute-little phenomenon about citrus that is different from all other plants. The male blossom has four petals and the female has either three or five. All wild lemon citrus trees have three petals, but the domestic or hybrids, like the Myers or Californias, have five for the female. If you're watching citrus blossoms you'll notice that almost all blossoms that come out first are female. The females are almost ready to shed by the time the males come out. In May each year in the North Temperature Zone you have a terrific shedding of young fruit off the tree. This means the tree has taken on more young fruit than it can handle, so it drops it on the ground. If you take these drops, dissect them, you'll find most of them deformed. If you find them perfect, then there's not enough plant food.

ACRES U.S.A. How is this important to the grower?

DR. REAMS. Plants do not have to change cells nearly as rapidly as animals. Baby chicks have a base exchange of every cell of their bodies every two days, and they don't have cancer. They're perfectly normal. The higher the frequency, the higher the base exchange. The greater the longevity of a plant, the more important it is to exchange its cells. In pecans, this exchange is about every three years. Citrus, about 18 months. Most plants will not have a base exchange until they start to blossom or fruit. In other words it maintains the same cells to perform the same duties that long. We teach our students these things and dovetail everything with sound fertilization. I really don't think I can go into computation of fertility to satisfy frequency in an interview. It takes a short course.

ACRES U.S.A. Let's look at phosphate.

DR. REAMS. Phosphate is the holding capacity to keep plant nutrient minerals harnessed and hitched and anchored into a certain area to keep them from moving into a wider range. In the Old Testament they used to have to let the farms lie idle every seventh year so that minerals from the forest and air and sea could spread back over them. Also for grass to pick up nutrients from the air. Each blade of grass is an antenna to return nutrients to the soil. Phosphate is the second greatest catalyst there is. Water is the first. That's the way the Creator made it. I don't know any other reason.

ACRES U.S.A. As a catalyst, phosphate shouldn't be used up in the growth process. Why would a farmer develop a phosphate shortage if he ever had a proper level in the first place?

DR. REAMS. It keeps spreading out, otherwise this would not be a problem. In the leaf of the plant in the process of photosynthesis, it joins the carbon the plant takes in with water, and you have $C_6H_{22}0_{12}$, which is a carbohydrate that is stored in the fruit or the wood or the leaf of a tree. In the fall of the year—because of the coolness of the evening and change of temperature—it is squeezed out of this union, and then you have the carbohydrate without the phosphate in it. It remains in the leaf. The leaf drops on the ground of a deciduous tree—or an evergreen, where it takes longer—and you have a continuous evolutionary cycle. But a lot of it is lost to the air, some is cropped off, and in the grain the mineral remains in phosphate form.

ACRES U.S.A. You prefer soft rock phosphate?

DR. REAMS. Yes. Phosphate and other nutrients in soft rock phosphate are colloidal compounds. Com-pound colloids are not water-soluble, but-they stand in suspension in water and give the impression that they are. Compound colloids are so small that a single cubic inch of them will cover 7.5 acres in a solid sheet. They are so fine they fit into the holes between the molecules of water. As a consequence, they are 100 percent available to plants and they won't leach out of the soil. The factor that determines the mineral content in any crop is the phosphate in the soil. The higher the available phosphate, the higher the sugar and mineral content in the produce. The soil has to have a minimum of 400 pounds phosphate per acre, available. It is not possible to get that much down using super-phosphate, triple- super or hard rock phosphate. If you apply enough super-phosphate (0-20-0) to do the job, it will kill the soil so dead you won't be able to grow anything for three years. There is a place for super-phosphate—as a catalyst in order to change the soil from an anionic to a cationic growth condition, and to create energy. Triple- super will leach out. It also creates too much heat. Finally, it causes the soil to become packed and hard. Plant recovery of phosphate applied in this form is very inefficient, possibly 10 to 15 percent utilization.

ACRES U.S.A. What about hard rock?

DR. REAMS. The problem with hard rock phosphate is that it takes 30 to 40 years for it to become available to plants. Only about 3 percent is available in any suitable time frame. In about another 300,000 years another 30 percent would become available.

ACRES U.S.A. What happens when you apply enough calcium and then add enough phosphate to meet your standards?

DR. REAMS. After you apply 1 ton of soft rock phosphate per acre, and 1 to 2 tons of high calcium lime, in that order, a phosphate of calcium is formed. Now a union of the most powerful magnetic forces imaginable in soil chemistry takes place. In the first 14 days after you apply these two elements a battle begins which kills bugs, grubs, bacteria, fungi, weed seeds. You have to let this battle run its course. I always tell farmers not to plant for 14 days after applying soft rock phosphate and lime. But when these two elements finish bonding, rain, flood, sleet, nothing can harm the soil. A glue of sorts is formed. This magnetic union will not allow leaching or erosion.

ACRES U.S.A. In good lay terms, just what does this soft rock phosphate do?

DR. REAMS. It operates like yeast in bread. When the sun strikes the soil, it makes it rise and aerates it. As heat aerates the soil, bacteria penetrate deeper together with oxygen. This has the effect of increasing the depth of topsoil. You see farms with too much compaction. Sodium is the element that cause most of this compaction. Soft rock phosphate counteracts high sodium.

ACRES U.S.A. Dr. Reams, you have called phosphate a cation. Yet the usual soil audits define it as an anion. Can you explain this?

Dr. REAMS. Yes, it is a cation. I don't know where they got the information for anion as expressed in those soil tests. I got mine from the oscilloscope and the direction in which an electron travels. In other words, if it travels counterclockwise, it is a cation. An anion travels clockwise.

ACRES U.S.A. Then calcium would be an anion, not a cation, as is the usual designation in standard soil audits?

DR. REAMS. Yes. Calcium and potassium are anions. And yet conventional soil audits call them cations. I don't see how. The texts I studied many, many years ago had them classified as anions. I don't know why chemistry books changed this. But then there are a lot of strange things these days. For instance, the Dodge automobile has positive called negative and negative called positive on the battery. Just to be different. I have talked to several of the world's outstanding physicists and associated with them and discussed this with them by the hour. On this business of frequency, I wanted another term. They said, "There isn't any." That's the word to use and keep on using it. I'm talking about chemists from Bell Telephone, physicists from Western Electric and so on. I have talked with these people, and they say I'm on the right track.

ACRES U.S.A. Is this to suggest that the avenue science travels is through physics first?

DR. REAMS. That is correct. This holds even for chemistry.

ACRES U.S.A. And so these things we're talking about here are accepted at the physics level, but this intelligence has not permeated down to chemistry, soil science and so on? Is that a correct statement?

DR. REAMS. Yes. It has not gotten to our colleges. In industry what I'm saying is largely accepted.

ACRES U.S.A. How long will it take our colleges to catch on?

DR. REAMS. I have no idea. Perhaps 20 years or more.

ACRES U.S.A. Let's go on to the next element. In Acres U.S.A. we've always called calcium the prince or mother of nutrients. Would you agree with this assessment?

DR. REAMS. I would say it is a key nutrient. But phosphate remains the second key. Water is the first one. Water is a nutrient, definitely so.

ACRES U.S.A. In your classes you show a preference for colloidal phosphate. Why?

DR. REAMS. Colloidal phosphate cannot be tied up or locked up. It has a tendency to float, to come to the top of the soil. Calcium has a tendency to go down. Phosphates and carbons determine your topsoil. Carbon determines holding of water. Phosphates determine holding of other elements and color. Calcium may be high or low or medium. That determines your volume of production. Potassium determines your size of produce.

ACRES U.S.A. What is a brix?

DR. REAMS. A brix is a laboratory measurement used to determine carbohydrates in all plant life. The unit of measurement on a refractometer is a brix, and the thing being measured is specific gravity. The higher the phosphate content of your soil, the higher the sugar content of your crop, whether it's grass, beets, sugar cane, peanuts, turnip greens, whatever. This rule is constant and there is no exception. The higher the sugar content the higher the mineral content. There is no exception to that either. The higher the mineral content the greater the specific gravity of a given bin or bushel. There is another point here. The higher the sugar and mineral content, the lower the freezing point. So there you have it. The P_2O_5 content determines sugar and mineral content. Sugar and mineral content determines production, and immunity to insects and freeze point. Sugar is the greatest enemy bugs and pests have. Insects will not attack plants with a, high reading on the refractometer. Now, for farmers producing alcohol crops, the higher the brix reading on a refractometer, the more alcohol that crop will produce.

ACRES U.S.A. Does this same reading give an index to weed control?

DR. REAMS. That is a general rule. I am not in favor of using herbicides to kill grasses and plants. If you will supply your plant with plenty of nutrients as needed, you won't be bothered much with weeds. The healthy crop will quench them out. Plants are a lot like animals in a barnyard. Take a goose and a horse. You can feed both green grass and they'll live a long time. You can feed them both oats and corn, and they'll live. But feed them both hay, and the goose won't live. That is what you can do for plants. Give the plants you want the vital minerals they need and you'll get rid of the plants you don't want.

ACRES U.S.A. Last year, at the Acres U.S.A. Eco-Ag Conference, your associate Dr. Dan Skow said that muriate of potash should never be used. Since you're not opposed to some other forms of salt fertilizers, why this prohibition on muriate of potash?

DR. REAMS. Muriate of potash contains chlorine, and for this reason it ought to be a federal offense to sell it. KCl is the chemical abbreviation for muriate of potash, which has 50 percent chlorine. The K symbol for potassium comes from the German word "kalium," which means potassium. Now about chlorine: 1,000 milligrams equals 1 gram, 28.35 grams equals 1 ounce, and 16 ounces equals 1 pound. So multiply 28.35 times 1,000 and you get 28,350 milligrams per ounce—this times 16 ounces, or 453,600 milligrams per pound. Now I'll show you how you're killing your bank account. Just imagine applying 500 pounds of potash per acre—well, make it 400 pounds. Your 400 pounds of muriate of potash would contain 90.720,000 milligrams of chlorine per acre. In a quart of water we're supposed to have 2 milligrams. At 5 milligrams it tastes like you're drinking Chlorox. At 8 and 9 the water looks like milk. If you multiply this all out you'll see that with muriate of potash we put 50 times more chlorine in the soil than we use to kill all bacteria in water. What does that do to the soil? It kills it. All you need now is a funeral. Is it any wonder farmers go out of business at the rate of 2,500 a month in the U.S.?

ACRES U.S.A. Dr. Reams, just a few more points to give readers a frame of reference, and to prepare them for your program at the Eco-Ag Conference. Many of our readers have expressed dissatisfaction with Extension soil tests. What is wrong?

DR. REAMS. I recommend farmers run their own tests. They can do this with available kits. The problem is that when you send your soil samples to the Experiment Station, 90 percent of the time the only thing they do with them is drop them in the waste basket. They know the area in which you live and they write back and give the area average. And that area average is made by a spectrometer. A spectrometer is absolutely worthless for biological purposes. For mining it is a wonderful instrument. If you want to know how much gold is in the soil—copper, silver, iron—it's wonderful. But for farming it is worthless because it will tell, you what the total amount of elements are, but not which form. They'll take carbon disulfide and dissolve the phosphate in the soil sample and tell you that you've got enough phosphate to last 40 years. Then they'll run another test using alcohol to test calcium, and according to it you've got enough calcium. What I want to know is this: just what saloon can the plant go to to get alcohol to dissolve its plant food? What laboratory can the plant go to in order to find carbon disulfide to dissolve phosphate? Where can the plants go to buy those extracting solutions?

ACRES U.S.A. Just for the record, what do you think of fluoridation of water supplies?

DR. REAMS. The rule I've given on colloidal substances applies to fluoride. Colloidal fluoride is not poisonous. That is the kind of fluoride that goes into teeth and makes them hard. But this is not the kind of fluoride they put in the city water. Sodium fluoride is a completely different type of compound.

ACRES U.S.A. Obviously, we can't cover everything. But one last question: You have expressed yourself on the value of chicken manure in your lectures. Would you put chicken manure on citrus?

DR. REAMS. Yes, but I'd never dig it in. I'd leave it on top of the ground. Why? Because the boron will ammoniate your trees. It'll never hurt citrus if you leave it on top of the ground. Moreover, if you have proper calcium and phosphate levels, you'll never have to spray the grove for insects. Your chicken manure should be spread from trunk to trunk. By the way, chicken manure is also good for other fruit trees. On deciduous frees it should not be applied until after the crop is off—but leave it on top of the ground. Never, never stir the soil. If you do you cut those rootlets of the deciduous trees and they bleed to death. Don't confuse chicken litter with cage droppings. Cage manure is 100 percent chicken manure with no sawdust or shavings in it.

ACRES U.S.A. Ergs—you mention ergs a lot. What is an erg?

DR. REAMS. Ergs represent energy per second. You measure this with an instrument and it tells you how much plant food there is in a field per second that is available to the plant that second. When a plant is young, it does not need more than 20 to 40 ergs. At fruiting time you need your plant food availability at peak-80 to 200 ergs. In our courses we compute ergs. We teach farmers how to get the energy flow when it is needed. If the pre-season work has gone well, there are a lot of things a farmer can do. We can teach things the Experiment Station can't for one reason. The government furnishes the station and its equipment. The federal government furnishes half and the state furnishes half. But the government doesn't furnish all the salaries. They have to get that money from the fertilizer company, the spray company, the machinery company. To do this they sell their souls. It is a sad, sad situation, and the farmer is holding the sack. Fertilizer laws are not written to protect the farmers. They're written to protect the fertilizer companies.

"You are Now Elevated to the Title of Plant Manager"

An Interview with Don Schriefer

Originally Published: December 1979

In 1979, Don Schriefer, 1927-1998, was working with Advanced Ag in DeMotte, Indiana, as a consultant with many years of experience in making it happen in eco-agriculture. When he sat for an interview, the University of Illinois graduate in agriculture and alumni of the University of Wisconsin talked about his work that went beyond his education in tillage and rooting. He is an ex-schoolmaster, but in 1979, his classrooms were the fields and the short-course rooms where farmers gathered to learn more than what was generally available in agriculture colleges—even today. He went on to write two classics for Acres U.S.A.: *Agriculture in Transition* and *From the Soil Up.*

ACRES U.S.A. Mr. Schriefer, you often talk about innovative practices, particularly in tillage, innovative practices in farming programs. This term innovative, what does it really mean in your consulting service?

SCHRIEFER. To me, an innovator is someone who is creative, logically and intelligently creative. Not a fellow who goes out and blindly probes and experiments without knowledge. An innovator is someone who gets things done. He accomplishes things, but he accomplishes them through experience and knowledge. Being innovative then is being creative. It is an interesting thing. I was just reading the other day that in the research end of agriculture, they are saying that basic technical research is now slowing down. The implications were that they were running out of things to research. I am not too sure that is too bad. But what I am saying now is if we are going to be innovative, we have got to take technical knowledge and practical experience and put them together into a workable program down at farm level. And I would say that it is high time that we started to do that because technical knowledge is the seeking out of raw information, facts, and this type of research is very rarely innovative.

ACRES U.S.A. Is research "innovative" or is harnessing research the real innovation?

SCHRIEFER. The innovation comes in when people take known facts and information that have been learned through research, and then go out and put them to work in the field. The innovator, in my opinion, is not only a creative man but a man that has a lot of spice in his life. Innovation is the spice of life in my book. But it is not a hit and miss thing. That is what I want to make very clear. You do not call yourself an innovator, if you go out there without the background, experience and knowledge and just blindly probe.

ACRES U.S.A. How do your seminars fit into the picture? When you quit school teaching, did you really quit?

SCHRIEFER. No. Education is usually on the top on my mind. We do a lot of it. I just came back yesterday from Michigan. We had a hard day up there. Our state supervisor in Michigan and myself put on a five-hour seminar and we drove six or seven hours to get to O'Hare to catch a plane. But we put on educational seminars around the country and we have sponsors. ... Let me give you the picture. We went up there Thursday because the day was bright and sunny—about 72 to 75 degrees in Michigan. All the farmers were out doing field work or harvesting. Also, we charge $50 per head to come into these five-hour seminars. We can't go all over the country anymore just talking for free because our business demands are too heavy. You have to draw the line. So we charged $50 a head. And these farmers flock in there on this nice, sunny day and they actually filled the room. Two farmers couldn't make it, but they sent their spouses, which I thought was pretty nice. They sat there, and they listened, and they took notes, and they listened for five solid hours—three in the morning and two in the afternoon—and went home very satisfied. This winter, it looks like we could make it a full time job just putting on seminars. Whenever we can put together 30 or 40 people, we will go. Education at the farm level is one of the greatest needs in American agriculture today because they are not getting it. Education is not coming down to the farm level in the fashion it should. The farmers themselves are begging for it. Our Michigan sponsors decided that they are going to bring education to every Michigan farmer through this seminar type route

ACRES U.S.A. Well, if you're going to sell something different, something grounded in principles, such as bio-agriculture, the farmer must understand the principles of why they should use this machine, this fertilizer, whatever?

SCHRIEFER. Right. Once he understands this, he will have no problem with making that transition. There are a lot of other companies following suite and it is a very good sign. I have one in mind He has followed us around to meetings and field days and he got caught up on this thing of education. So he gets the president to come down to a field day and seminar and they decide that they want all their field service representatives to go through an education program so that they can bring more to the farmer than just product. They want their men to know tillage. They want them to understand soil chemistry so that when they go out and work with the farmer, they can help him in other areas of need rather than drop a product and run. And I think this is a good sign.

ACRES U.S.A. What's wrong with farm production technology?

SCHRIEFER. Farm production basically has become an entanglement of unrelated, single-shot inputs, and there is no systematic approach to production across mid-America. Farmers have to know the interrelationship of the sands, the silts and the clays and how they interact and affect water-holding capacity, fertility management and fertility placement, the dos and don'ts of fertilizing. He has got to know how it relates to drainage and how it relates to air and water management and tillage. The farmer has got to understand these things. He has to understand soil chemistry. He has to understand what the soil audit says. He has to be able to read it and understand every line on that thing and what it is saying. He has to understand Cation Exchange Capacity and how it relates to fertility, and so on. He has to understand fertility inter-reaction within the soil.

ACRES U.S.A. In other words, he needs to have a little more than seat of the pants knowledge?

SCHRIEFER. Sure, or he is going to make a lot of mistakes and he is going to continue to do such things as putting lime on just for the sake of liming—just to satisfy the pH level, which is someone's idea of how you should lime. He is going to be out there broadcasting certain phosphate fertilizers when he has got no business broadcasting them because the soil situation locks them up. He is going to be using nitrogen improperly. And he is going to be neglecting such things as sulfur and trace elements and everything else. So this guy has got to understand soil chemistry up through the interaction of each element period. He had got to understand soil biology and he has got to understand how it influences soil and plant nutrition period. He has also to understand the green growing plant. He has to recognize that this is a factory, that everything he does affects this factory and its output. He has to understand that nutrition comes from the soil, to the plant, to the animal and that if he has missing links in the soil, they are going to be missing in the plant and they are going to be missing in the rations of his livestock in both food and feed, and problems are going to be there as a result. He had to understand, too, that when he does all he can with the soil, there are still ways he can measure how well his factory is doing.

ACRES U.S.A. During the growing season or later?

SCHRIEFER. Both. There are such tools as tissue analysis and infrared photography and good observations out in the field. We tell the farmer, "You are now elevated to the title of plant manager." And you know, having a good title sometimes makes poor pay more acceptable. But we also say that as a plant manager you have a job, and that job is to make that plant show a profit. If you don't make it show a profit, then you may just lose your title and your pay. This shouldn't be hard to understand. Farmers have tools. Education is a tool. Now what I am saying is that farmers need a bucket full of tools to think with. Without these tools, which is basic knowledge, you can't plan, you can't project and you can't be innovative.

ACRES U.S.A. Well, can't a farmer learn these things?

SCHRIEFER. This is the way that I feel about it. Anything that I can learn, a farmer can learn. And it is that simple. I don't consider myself any smarter than a farmer. The problem is that it is going to take people to bring this knowledge to the farmer in an orderly sequential fashion. You don't start training a scientist up on the top end. You start him with the basics and you build him up from there. We have a responsibility and a big one. We have to recognize first of all that we, as advisors, don't have all the answers. We also have to let our client know that we don't have all the answers. And when he knows that, then he will understand that to get to the bottom of a problem a cooperative effort is required. A farmer has to deliver some input. You can't just tell him everything.

ACRES U.S.A. How can you tell whether you're getting through to a farmer?

SCHRIEFER. It isn't always easy. I have often said that a little knowledge is dangerous. It is very easy to overexcite people who are less than half informed. You can get them right up there on cloud nine. You can get them out there making all kinds of mistakes. So we have got to lead our people. We have lead them through progressive education. They have to understand why you are asking them to do certain things. If they understand why, then they will do it. They will make a lot better client and you are going to have less of a turnover in your clientele.

ACRES U.S.A. Let's take tillage. How do you reduce tillage to the simplest terms?

SCHRIEFER. To start, tillage has to be seen as just one aspect of a totally integrated system. The plow is still predominating as a tillage tool across a major part of the Midwest. The only thing that is going to change that is education. I can give you some good examples of where we have gone into areas with large groups of farmers across a large geographical area. We have explained to them tillage management principles—what they should be trying to do, how, why! In these areas the plow has practically disappeared, often in a couple of years. But education was required to do it. There are all kinds of systems being used. But in tillage there is no one single system that works everywhere because we have to recognize that there are lots of differences in soils — the way it was laid down by the glacier. There are differences in slope drainage, seasons, rainfall patterns, rocks, soil types, crops, and so on. No one system will work everywhere. What we can do quite simply is to outline the principles in a one-two-three-four fashion in the order of their importance—tillage principles. And you may have to give or take, but by understanding the principles, you can adapt a system more intelligently to meet your particular situation.

ACRES U.S.A. What is the prime tillage problem—or is any one problem prime?

SCHRIEFER. There is one problem—compaction. If that is the number one problem, then the next logical question may be, What causes compaction? And I have to give an equally simple answer that anyone will understand. The thing that causes compaction is simply the force of gravity. That is all there is. The force of gravity is that force that tried to pull everything to the center of the earth. It is the thing that gives tractors and farm machinery their weight. It is the thing that causes rain to fall from the sky and to have weight and inertia so that when it hits the soil, it acts like a little bomb. Gravity. Now, fortunately we have gravity, because a world without gravity would be a strange place, and particularly at the farm where you want to get some traction out of a tractor. The next question, then would be understanding compaction and what causes it, what are the things that counteract gravity? And that is a relatively simple answer.

ACRES U.S.A. Okay, how do you counteract gravity so as to avoid compaction?

SCHRIEFER. There are two things basically that counteract gravity. Number 1 is the product in the soil that we call humus. Humus, as you well know, is the end result of complete decomposition of residues. Humus coats soil particles. And if that coating is there in sufficient depth, then that humus acts as a shock absorber between soil particles. It expands with moisture and even with heat. It expands and contracts, but it acts as a cushion or shock absorber between soil particles to prevent them from locking horns and touching each other the way they do in the subsoil. The second thing that counteracts the force of gravity is the thing that makes humus, and that is biological activity that is in the soil. When bacteria go to work, they cause a coating of soil particles around the colony that is developing. That pulls soil particles together into clumps or clusters. We call them granular or crumb structure. And they have a purpose for doing this. Because by clumping these soil particles into a granular till structure, they build in a ventilation system in the soil. This protects the soil from going anaerobic or losing the oxygen supply which they totally depend on. So they protect themselves by building in a ventilation system and this also benefits crops.

ACRES U.S.A. You have, of course, named two things many companies try to complement with the addition of products now on the market.

SCHRIEFER. Yes. There are some soil conditioners available that can complement these processes if used intelligently.

ACRES U.S.A. So basically, there are two things that counteract the presence of gravity. But you haven't included the plow. Why not?

SCHRIEFER. Well, people say plowing must be good because today still we have plowing contests. You win a plowing contest if you can perfectly leave the surface bare—nice and level and bare. As a matter of principle, I quit attending plowing contests or having anything to do with them 20 years ago because plowing just doesn't fit. There was a day when it fit because we had to use low power to lift the soil. This was at 2 million pounds per acre—lift it up on an inclined plane and flop it over. We had very little power after that. The soil was different then too. It stayed looser where we could prepare a seedbed in the spring and that is what tillage is all about—getting the crop planted. That worked for a long time. But now today, we recognize that the plow is creating fantastic problems. Nature never intended for the soil to stay bare, seal over when the rains hit it, and turn anaerobic so gravity could pull it back down and settle it in and hold excess water, warm up slow in the spring, stop the biological decay, and so on. This is general fall tillage practice. It creates plow pans, air locks, and sets the stage so roots can't go down and capillary water can't come up.

ACRES U.S.A. Well, the same thing is true in spring?

SCHRIEFER. Absolutely. In spring seedbed preparation, we go out there and try to salvage with iron and try to build the seedbed with iron when it is already saying that we are trying to correct a bad situation that started last fall. That's why soil warms up slowly. It contains more water. So we dig iron down into it now, and we smear the soil colloids. We lift them up. We expose them to the sun and the rain. And we have what we call a clod. That is one reason that some cartoonists call farmers clodhoppers because clods are man made. Nature never makes clods—we make them. Anytime you have clods in your field, you had better believe you are doing something wrong. You are making some mistakes.

ACRES U.S.A. What is the real purpose of tillage? We've heard about "no-till" and we all remember Plowman's Folly?

SCHRIEFER. The purpose of tillage is to grow crops. That's why we should ask the plant what it wants from us and reverse the cards a little. Let's ask the plant. The plant, if you ask it what it wants, is going to look up or down on you, depending on the stage of growth, and since you ask, it will probably look up to you. The plant is going to say this. You as a plant manager out here managing this soil—we want you to do basically four things. Number 1, we want you to give us plenty of ventilation. We want air managed around this root system. Out of that air, I want the oxygen, and if I have plenty of oxygen, then my roots will continue to grow throughout my whole life. They will take in nutrients easier because both growth and uptake of nutrients is an oxygen demanding process that requires a lot of oxygen. I can take up water better even when water is short if you give me plenty of air. So air management is the first thing that plant would want. Secondly, the plant would say, I want you to manage the water underneath me. I don't want this basement of mine flooded because when you flood this basement, several things happen. Number 1, you drive out the air and there goes my oxygen and I am right back where we started, the same thing compaction does. Number 2, if you let this soil go anaerobic, we are going to lose fertility because anaerobes require oxygen also and they are going to get their oxygen off the nitrate molecules, off the sulfate molecules and they are going to turn these things into gas and they are going to escape or volatilize out of the soil so you are robbing me of nutrients. The third thing it would say, I want things down here to decay because it is tripping me up. If you get things to decay, we will have better soil structure. We are going to have the organic side of plant fertility. You are going to produce enzymes, you are going to produce hormones, you are going to produce antibiotics and a whole bunch of other things that you don't even know about—unknown factors—that the system requires if it is going to be healthy and produce abundantly. So get that stuff to decay.

ACRES U.S.A. And how do you manage decay?

SCHRIEFER. We are doing a lot of things to manage decay. I personally believe that within a very short period of time we are going to be able to guarantee a farmer that his stalks will decompose within a given time frame. Right now we are doing a lot of work in balancing the soil chemistry, making sure sulfur is up, calcium is up, but we are also going into those stalks with a stubble digester which breaks down lignited materials into simpler substances so bacteria can attack sooner. We are combining that with a little bit of nitrogen to reduce the carbon-nitrogen ratio so that the bacteria can speed up. We can almost guarantee that residue is going to decay.

ACRES U.S.A. Are we going to forget balanced soil fertility?

SCHRIEFER. No. That's the fourth thing. We don't overdo it and we don't under-do it. We get the fertility of this soil in the ballpark. I have seen systems where tillage was improved, air and water management was improved and decay was improved. It made a poor soil analysis look like it did not know what it was talking about. I have seen low fertility come through like a shining star when these things happen. Now I simply say this, a farmer is losing more yield potential because of mismanagement of tillage than probably anything else he is doing on the farm. When you mismanage tillage, you mismanage everything. You get compaction, you lose air, you lose water control, you lose decay, so it is very easy to see why this is the area of greatest loss. It is also the area of greatest cost reduction on the farm. I believe personally that a farmer can reduce his tillage cost by 30 to 50 percent easily if he really goes after if. And that is money in the bank because it is money you don't have to spend. If the plant says, Mr. Farmer, you give me these things and then step aside—get out of the way and I will do the rest because your worries are over. You are going to get the most out of that crop for that particular season. That is important and that is what tillage is all about.

ACRES U.S.A. In other words, it all means getting back on speaking terms with nature?

SCHRIEFER. That is all it is. Let's take the fall tillage goals. Fall tillage should relate to what you are doing in the spring. It ought to take the weather hazard out of the spring situation. That is a very important thing that plowing doesn't do.

ACRES U.S.A. And subsoils?

SCHRIEFER. Leave the subsoil down there. Subsoils are being farmed in the subsoil. Fall tillage programs should incorporate residue leaving enough on the surface so as not to interfere with seedbed preparations or warm up in the spring, but enough to increase the water properly, to use pipettes and wicks to let water and air in and to stop erosion. A fall tillage program should leave the soil in a rich position. It should position the soil over winter in ridges. It would be nice to do it in one trip or as few trips as possible.

ACRES U.S.A. What kind of tools do you recommend?

SCHRIEFER. There are some good tools on the market—Glenco, Landoll. Some farmers have created their own tools. They disc ahead, they put chisels on and put wings on them to mix and ridge. That is great. You can do that in several ways. Fall tillage must have goals and if you don't reach those goals, you are compounding your problem in, the spring. Now why don't we talk about a ridge? The ridge concept is pretty hard to kick a hole in. It may not fit everybody but it is generally pretty hard to kick a hole in, because you increase your surface area, you get faster warm up in the spring, you are less likely to have an anaerobic situation because of too high water up that ridge. When you warm up quicker, biology comes to life and the thing you will notice about the ridge is that nature builds a seedbed for you in that ridge through its biological activity, through faster warm up and better water and air management. Then all you have to do in the spring is go out and level seedbed material and most any type of barnyard furniture will fit that. It gives you more options in the spring, but when leveling these ridges don't go too deep. Don't go down to the cold, inert soil below the ridge. Leave that stuff down there. Just level the seedbed material and go to planting.

ACRES U.S.A. In other words, let Mother Nature help you build the seedbed, and fall tillage must relate to spring tillage.

SCHRIEFER. Right. If it doesn't, you haven't got a system. You see, the farmer is paying a price for plowing. I just heard the other day that we are losing two bushels of soil for every bushel of corn we harvest in the Midwest. Now it takes a long time to lose 100 tons of soil at two bushels a clip. But it is going and we cannot afford these types of losses. It has got to stop. There are some other innovative practices that can be used. We have innovative practices for the management of alfalfa. There are tricks you can pull out with alfalfa that will really put you on top of the market. In the spring, we can go out and put on 150 pounds of ammonia sulfate. Ammonia sulfate is not a bad word. It contains nitrogen and it contains sulfur. And with alfalfa, the two areas that we are lacking in are in nitrogen and sulfur to get maximum yields. The nodules will not necessarily handle themselves. By doing this, we are getting more tonnage but we are getting quality and we are getting faster comeback of the second crop and the third crop. But the sulfur is the key because when you put ammonia sulfate on you are getting the sulfate with the nitrogen in the form the plant uses and here is what it does: it builds keratin which is vitamin A activity in the animal's body. It improves photosynthesis and efficiency of the plant and it improves the biological quality of your protein because it raises the two essential amino acids that carry sulfur in the plant. So you improve the biological value, when you start shoving it into the cow, you are going to see a big increase in milk production and herd health. That is something that very few people are doing.

ACRES U.S.A. Have you tried broadcasting rye early in spring with alfalfa?

SCHRIEFER. Yes. This works. Let it grow up with the alfalfa. There is no better feed in the world. It is even better than alfalfa in some respects. The first cutting, if it catches good, you can pick up a good weight and you have a good grass legume mixed. When you cut it, that is the end of it. The roots stay in there and keep the soil open. On the second crop, you hardly see a spear of it. Next year, do the same thing. It is very effective. Don't overdue the potash. Keep that in line. Don't let it luxury feed because it reduced the feed value and fouls up magnesium in the plant and sodium-potash balances. There are a few tricks you can pull with alfalfa.

ACRES U.S.A. What do you recommend for row crop support?

SCHRIEFER. The start of the plant is important. If it starts right, it will go. We are doing some strange things down in that row. We are putting in some enzymes, some surfactants, some nutrients. That is the name of the game because seed quality is not necessarily what it ought to be. It depends on where on that ear you pick that seed rather than how it was grown. Is it physiologically mature? Is that nutrient package complete around the germ—not necessarily. Don't bet on it. We do everything possible to help that situation.

"No Magic Bullet"

An Interview with C.J. Fenzau

Originally Published: October 1981

It has been almost 50 years now since the famous corn blight episode wiped out so much of the American corn crop in 1971, the year Acres U.S.A. was founded.

At that time, C.J. Fenzau, (birth information not available), was a consultant in the Midwest, and using the Albrecht formulas with appropriate soil tests in tow, he entered a plot in the annual contest sponsored by one of the chemical entities.

Years later, speaking into an Acres U.S.A. microphone, C. J. told us how did. He fall tilled the soil and applied the required materials of fine, powdered calcium lime, Sul-Po-Mag. No other entrants in the 100-acre demonstration field had done any work in the fall.

"We were the only ones," C. J. said, "and everybody was laughing at us. The next spring we went out and planted on the 25th of April. Everybody else was trying to get the soil worked up, but it was too wet, and that delayed spring tillage, or they had to work the fields on the wet side. Then we had continued wet weather all the way through until June 10. And then it turned dry—dry as a bone, and hot. It was during the heat stress period, right at the critical stage of growth—about the 10th to 15th of June—that things were set in motion for corn blight."

We included this interview because it shows how an American farmer found a unique system based on observation and trust in nature, while convention began to accelerate our country's family farm consolidation and destructive practices. C.J. Fenzau was also instrumental in one of the earliest books on the subject, *Eco Farm: An Acres U.S.A. Primer.*

ACRES U.S.A. Do you have blight on your plot?

FENZAU. No, not at all. Strange as it may seem, we had the same variety of corn as all the other 100 acres of five-acre plots belonging to the fertilizer and chemical companies. We used Pioneer

3557—I believe that was the number—and it was the identical seed corn used on all the plots with no variables in seed. Everyone started with the same seed count. The others got blight. We didn't.

ACRES U.S.A. What did the corn industry say was the problem?
FENZAU. We were told that the variety used was grown with Texas male sterile cytoplasm, and so it was susceptible to corn blight. That was the seed industry's answer to the problem, that we'd used Texas male sterile cytoplasm.

ACRES U.S.A. But you used the same seed and didn't have blight side by side with plots that became infected. What were your experiences in the commercial arena in that Indiana area?
FENZAU. We had over 100,000 acres of corn growing in Indiana, Illinois and Iowa, all in areas where blight was severe and where corn crop was literally destroyed. None of our farms were touched with blight that year regardless of the variety we planted. So we felt variety didn't have anything to do with it. It was all a case of plant nutrition and soil nutrition that set the stage for the blight to remain dormant in our fields in spite of environmental conditions that would be favorable to it.

ACRES U.S.A. How did the fertilizer and chemical people face the people in view of their good horrible example?
FENZAU. In all of the five-acre plots in that 100 acre demonstration area—and there were about 20 companies involved—there wasn't a single sign out front of any plot, not a field man or salesman to invite the farmers to look at their great success with herbicides and highly rated fertilizers and their overly enthusiastic methods of growing corn. We had a choice location, right on the first corner to the left—right off the main walkway next to the demonstration tent where you got all the traffic. We were harvesting 142 bushels of dry corn in early October, no corn blight on any of our plants. We had hundreds of farmers stop by to look at this demonstration. They would watch us harvesting and follow with our fall tillage. We were doing it right there. We did a once-over, ripped up the stalks and bedded up the soil ready for planting next spring. All the rest of the plots had zero to five bushels per acre. The ears on their plants were sometimes the size of a thumb with maybe two or three discolored kernels on those ears. And the more nitrogen they used, the worse and blacker the plant.

ACRES U.S.A. How did your demonstration plot impress educators and scientists from the university or the seedmen?

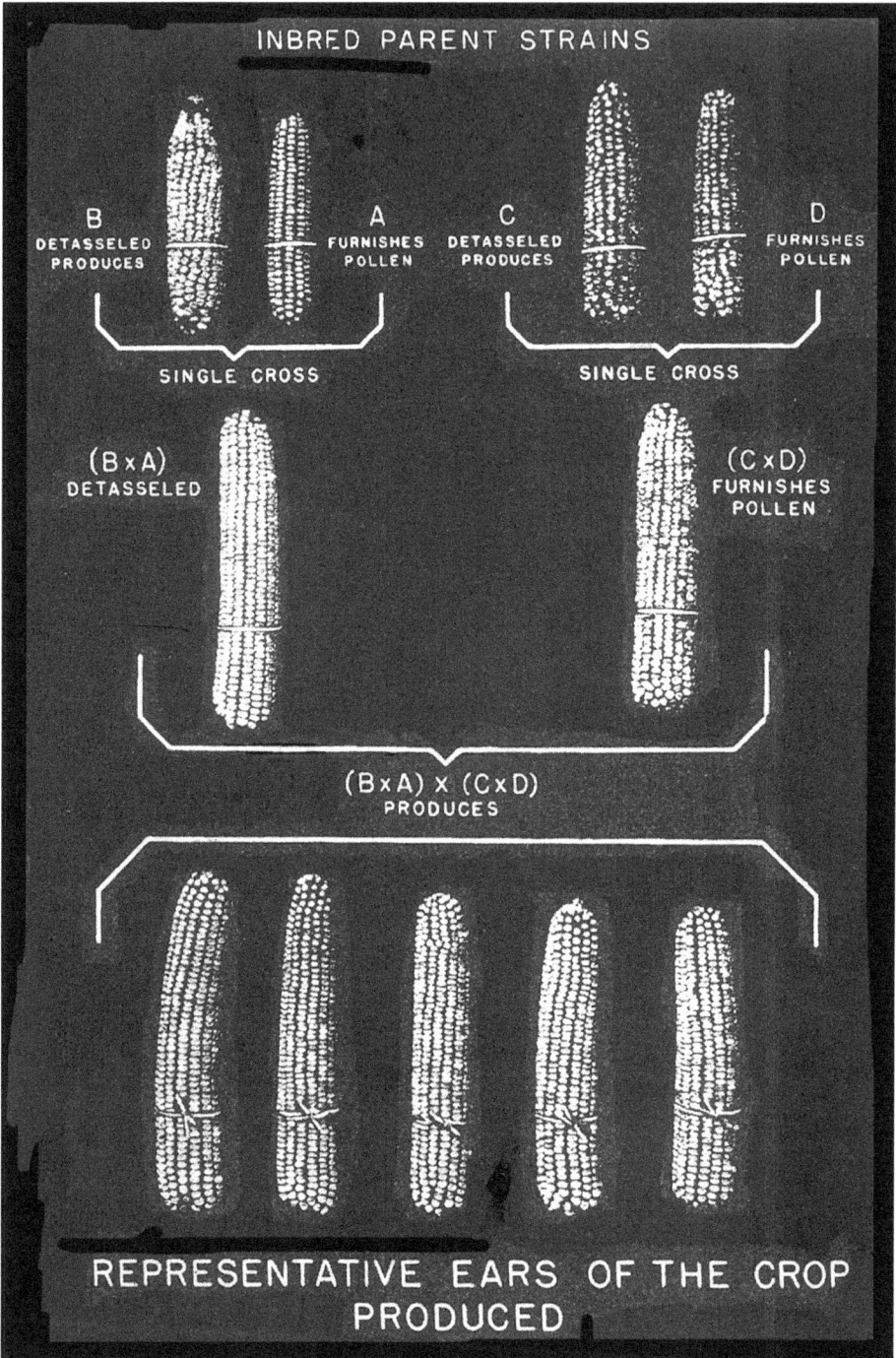

INBRED PARENT STRAINS

B
DETASSELED
PRODUCES

A
FURNISHES
POLLEN

C
DETASSELED
PRODUCES

D
FURNISHES
POLLEN

SINGLE CROSS

SINGLE CROSS

(B x A)
DETASSELED

(C x D)
FURNISHES
POLLEN

(B x A) x (C x D)
PRODUCES

REPRESENTATIVE EARS OF THE CROP
PRODUCED

From *Eco Farm: An Acres U.S.A. Primer*

FENZAU. Not a single educator or scientist or seed-man—not one institutional man was interested enough to wonder why this beautiful susceptible crop had escaped the blight when there was total failure all around.

ACRES U.S.A. Did you pursue this publicly?

FENZAU. No, that was the last time I became interested in being involved in a public display of the principles I was working with. It became apparent that display simply would not attract the attention of the institutional people because they already had their obligations and were not able to consider what we had to reveal. It also became apparent to me that they wanted to put their heads in the sand and hide and therefore they refused to admit that what he had to show even existed.

ACRES U.S.A. That was an unusual season. What happens if you have the opposite kind of weather?

FENZAU. That's an interesting question. In that same county the following year, we had exactly what you mentioned, and that year Stewart's disease was common throughout the area. This represented an entirely different problem that was the consequence of an opposite sequence in weather from the time of planting until the plant was in the death stage. This usually came on in late August or early September, and any of the weaknesses that existed in the soil at that time just made the plant susceptible to the new invasion. The second year after the blight, we had a new problem in the cornfields—aflatoxin. So you had the government coming out of its corner on that one. They were going to disqualify the farmer from selling aflatoxin corn, there would be penalties, they were going to put black lights on the corn to measure the presence of *Aspergillus flavus* as the corn came, into the elevators. So you see each season has its variables in weather, heat and light and moisture, and these all have a direct influence on which kind of invasion is going to occur.

ACRES U.S.A. The same general principles hold for insects, do they not?

FENZAU. Correct. These principles relate to insect proliferation as well. That's why putting together a package of good nutrition enables the farmer to correct for these limitations. The right management of nutrients will allow the plant to obtain the necessary minerals in the right form for maintenance of the right hormone and enzyme potential needed. The lack of certain minerals or an excess or dominance of one mineral over another—these things set the stage for insects of various types. If you invite some maladjustment of the hormone process, you invite some insect. But you can cope with some mistakes and you can correct some situations by providing the proper foliar materials, or you can deal with the problem in the seedbed that sets the stage for the plant's life, and then correct any weaknesses that come from the variables such as weather. We have been very successful in using the right prescription when it became necessary to react to a stress situation.

ACRES U.S.A. Did you have foliar materials when the blight problem came along?

FENZAU. We didn't apply the amino acids in that case, but we had done everything in the soil. We had everything working right because of our fall management and the steps we took to correct pH. By managing decay in the fall, we didn't have residue left at planting time. Everybody else still had corn stalks sticking out of the ground, and that was just the wrong system of decay under wet conditions. In short, we had the job accomplished, and this made the difference.

ACRES U.S.A. In cotton country it often gets too dry. If the soil is corrected properly, will it enable the farmer to handle the drought stress better?

FENZAU. Absolutely. The entire capillary system is influenced by how you handle your cation exchange balance, how you manage pH, and so on. If it is functioning properly, you can draw deeper on that capillary water. This in turn has a bearing on the insects you'll see.

ACRES U.S.A. Let's take boll weevil. When do you see it most?

FENZAU. Basically, it almost follows the soil that has a low level of calcium. Infestations are more severe when the structure is tighter or cakey. And on the more loamy soils with better organic matter, the humus system is tougher, so you won't have the same severity. You'll have various degrees of infestation within a field in direct proportion to the soil's limitations. Many times you may see a plant that is not only affected by poor soil conditions in terms of drawing the boll weevil, it also has fungus infection that prevents it from having a normal growth.

ACRES U.S.A. When farmers spray for the boll weevil, is it possible that the chemistry helps set the stage for the tobacco budworm?

FENZAU. That has happened in California. Spraying has actually destroyed all the friendly insects that normally regulate the tobacco budworm by refusing to let it emerge. So if you remove these friendly insects by spraying, you'll find the tobacco budworm expressing itself in that system.

ACRES U.S.A. Why does cotton get the boll worm as opposed to the boll weevil?

FENZAU. You don't get boll weevil in California. It's the pink boll worm. The boll weevil is chiefly in the south. In the south they have a lower pH because of the rain. In the west most acres have a higher pH. However, the pH in the west is of an entirely different construction. This has a direct influence on the hormone systems that will allow the egg to emerge or require it to remain dormant. And that hormone process in the soil in the living systems has a very important cycle or system if the insect or disease is going to emerge. You always come back to the ability to regulate air, water, temperature and decay. Those four factors could be different year to year, and the intensity of emergence of a specific weed or insect is going to be influenced by those conditions.

ACRES U.S.A. Let's take the cotton aphid. This is an insect that appears all over the country. It is available in the West as well as the South!

FENZAU. I don't think they're entirely different. They come to attack the sugars in poor-doing plants. They will find those photosynthesis sections of plants that are malnourished, that are not supplied with an adequate balance of nutrients for the synthesis process to function. Various restrictions in light could figure. It could be related to various restrictions in the supply of carbon dioxide. So these aphids will always reflect poor synthesis of sugars produced by the photosynthesis process, and it all comes back to plant malnutrition. The plant that does not have the ability to break those sugars properly or quickly as they are being produced by the heat energy and light energy of the sun is the plant that is in trouble. There's an accumulation of sugars in the upper portion of the plant. Usually that's where malnourishment will appear first. So the aphids will always find those regions of plants, the succulent ones: The aphids suck the sugars. On the other hand you can correct for photosynthesis weaknesses with a supply of nutrients that will allow photosynthesis to be effective. And the aphids will not attack that part of a plant. The aphids may be present in the field, but the only place they will attack is where the plant is weak. I often find aphids in July and August during the hot part of the year when the sun is working its best, and shade starts to be a factor in the growth of a plant. You may find that aphids may be present in the fields, and the only place you'll find them is on the lower side of the first one or two leaves of a corn plant. Those leaves are dying and not working. They've served their purpose and have no further function in the life of that plant. Their job was to get the plant emerged and out of the ground, and that's what they did. Now the aphids will find the only part of the plant that is malnourished, and they'll continue to harbor down there, but they don't do too much. They're kind of slow and sluggish.

ACRES U.S.A. You see them on cucumbers and squash and produce like that?

FENZAU. They'll find that part of the plant that fits their needs and they won't bother the rest of it. And then a little later in the season—if a plant just runs out of potassium supply needed to keep an enzyme system working during the maturing phases, and they're a little weak and unable to supply the potassium due to too much or not enough water—you have the insect. Often times, the shucks on the corn plant feeding that ear to mature it will run a bit short of potassium, and then you may find aphids sucking on those portions of the plant, but not causing any real damage.

ACRES U.S.A. There are insects that are somewhat different and yet somewhat the same. Take the flea hopper, the cotton flea hopper, and the leaf worm—do you have any observations on why one rather than the other becomes a problem?

FENZAU. It is always a trace mineral problem. It's always related to one or some combination of trace minerals. You may have the one and this one likes it, and the other one doesn't. It relates back to the amino acid complexing forms of trace minerals. If they are powerfully synthesized into that spiral-matrix of the amino acids particle—and sometimes just a certain single molecule of iron or zinc positioned in a certain way in that molecule disestablished that amino acid character—so it's able to function completely. Otherwise it may not continue to function like it should. Of course, this is where those trace minerals have to be metabolized into and if they don't get there, then you have a kind of a ersatz amino acid complex. Then it is not able to complete itself into forming finished, mature proteins. This is what human life or a higher balance of life has to have. These trace minerals affect the health of animals and human beings based on the complete thoroughness of the functions of the amino acid structure while proteins are being built. And any plant or vegetation system must have the capacities to mature those amino acids into finished ripe, edible, acceptable, exchangeable proteins. And if they're missing a link in there, then that's the time for them to be destroyed in the food chain. And that's the function of the insects.

ACRES U.S.A. Why do Colorado potato beetles show up in one area, and not in another?

FENZAU. I have had that happen. They would come right up to the middle of the plot and stop at the soil division line, and that was where the soil was treated differently. I can also relate that to another factor. Even if I have good soil nutrition, I could take an improperly grown seed from the year before, one grown under adverse conditions—take that poor quality seed and just at random scatter it throughout several plots. Those were the only plants that the potato beetle would find in the early stages of potato growth. I have done this and the beetles found every single plant, even though I had it in a good plot.

ACRES U.S.A. Well, what was the difference between the plots? What treatment got you the beetles, and where didn't you get them, generally?

FENZAU. I got them strictly in the commercial fertilizer area: 300 pounds of nitrogen in the course of a year, 200 pounds of phosphate, and 200 pounds of potash. In the next relative spot to

it I didn't have any phosphate or potash. I applied a ton of compost, put about 20 pounds magnesium/sulfur in the soil and that was basically all I used. Now the magnesium/sulfur corrected the pH and caused a different humus system to function. Trace minerals were being released that sustained vigorous growth. In fact, I got almost double the yield from one end of the plot compared to the other. And potato beetles would just go to a dividing line. They would not cross that line, not a single plant in a 36-foot plot—eight rows—would have the beetle. Now, I could put foliar sprays on two rows, four rows, and leave four rows without it. The intensity of beetle damage would be a little less on the fertilized foliar bed rows, and more severe on the non-foliared rows, but they were still present, even on the foliar sprays. So the foliar sprays by themselves, without an adequate nutrient system in the soils are not going to give you optimum protection. The same thing applied where we've used a seed treatment. If you don't do other things in the soil to release an adequate supply of nutrients, the seed treatment itself wouldn't have a chance to function. Neither does a single foliar spray or one single ingredient.

ACRES U.S.A. In other words, no magic bullet?

FENZAU. No magic bullet. Now, right alongside it, in another plot, I had four rows of potatoes with my program on both ends. But the variable from one end to another was compost and no compost. Everything else was the same. I had no potato beetles at either end. And I had four rows of potatoes and four rows of sugar beets. Now on the sugar beets we had the downy mildew on two rows and not the other two rows, from one end to the other. It didn't make any difference because of the soil variables. The variable was in the foliar spray. We had pH release factors. The compost didn't correct the nutritional supply in the sugar beets enough to provide a deterrent to the downy mildew. The mildew was on the two rows, from one end of the plot to the other, without foliar nutrition. The foliar nutrient sprays were used on the other two rows. Now I was always asked the question by everybody who came in to look at this thing, How come mildew, a virus that floats through the air and lands on all of the plants, why doesn't it become active on the row right next to one totally infected by it? Now again we had beautiful fields with mildew on it on a group of farms in California this past fall.

ACRES U.S.A. What happened?

FENZAU. We had sugar beets and we had common downy mildew on beets. Soil data showed low calcium and an extreme excess of magnesium and failure of potassium to be available to the plant. You know that we need deposits to build sugar in beets. Then there was a very low copper and low boron level. The soil was really bad, but they had beets. And that field has a history of very low-low sugar in beets, and the mildew came into that field. A half-mile away the mildew wasn't that bad. It was something peculiar to that field because it was not taking every field in the valley. There were parts of the field that had much worse development of the fungus than other parts of the field. That tied in directly with the amount of gravel in the soil, and the color of the soil. We went in with a recipe of foliar materials consisting of amino acid enzymes materials, a calcium chelate, an iron chelate, and some copper/sulfur, eco-powder. That stopped the fungus, but it didn't kill it. The weather during this time was perfect for the fungus. It was warm and balmy—just the kind of weather required to have mildew run through the field fast. But the mildew was only working on the older, larger, tired leaves around the perimeter of the plant, and the young leaves coming out into the world were clean and were not being attacked. We had an opportunity to repeat that foliar treatment. About two weeks later we threw the same recipe on again and that stopped the fungus cold.

ACRES U.S.A. In other words, you achieved better photosynthesis?

FENZAU. We got a deeper leaf, better photosynthesis, enlarged veins and the fungus just retired to those older leaves around the outside and everything coming up new in the middle was clean as a whistle and the fungus didn't even get a toehold.

ACRES U.S.A. What was it that defeated the fungus?

FENZAU. It was the whole nutritional package. Most growers will apply quite a bit of floured sulfur on a field of mildew. That's a lot of sulfur on the surface of the weeds. And they expect that will toxify the mildew. But the sulfur by itself does not get at the cause of the limitations, which is nutrition. It is just another insecticide or fungicide that can be used to dry out the surface and influence the vitality of the mildew. Well, the last few years, a lot of people in the beet producing areas of the West have learned that sulfur applied to the foliage of sugar beets can subdue and retard and hold back mildew. But on the other hand, they often have to do it three, four and five times during August, September and early October. It doesn't last.

ACRES U.S.A. Expensive, too?

FENZAU. Right. They'll spend $45 to 50 an acre trying to control it. With the foliar treatments we spent $13 an acre—$9 for materials plus $3 for the airplane. Of course, airplane prices are going up. A repeat treatment would double the price, of course. But you see, not only did we have sulfur in there, we had the copper as well as the calcium to correct the weakness of the supply in that whole system, which had a bearing on the metabolism of phosphates. And so now we are generating good, vigorous new leaves and good, strong growth to take that plant into winter dormancy. Next spring it's ready to come out and really have potential to finish a good yield and a higher sugar content. We supplied nutrition at the same time even though the mildew was still present on the old dying leaves at the base of the plant. There's still more nutritional support from those anyway. Those were all deficient, weak leaves. The mildew was present there, but it just couldn't get a toehold to reinfect the new growth. And that's what we're into here—you just don't have to spray it every five to 10 days and constantly keep washing it because once it gets fired up by a warm night, it just takes off and goes off in a whole field.

ACRES U.S.A. The general observation, then, is that poisons to kill the insect is primitive stuff?

FENZAU. Well, let me continue. We observed some very interesting things. We really startled some pesticide field men who were working for one farmer. There are various species of caterpillars that eat down right in the crown, right in there in the very smallest leaves of the growing point. They like to get in there and chew around. Well, we could dig into these plants, and peel back these leaves and we could see the individual worms. But following these applications the worms were sitting in there, but they weren't eating. And that was observed by three separate field men working for a different group, walking that field. Maybe we gave them sugar diabetes, or something. The insects were just no longer interested in eating that plant.

ACRES U.S.A. In the wake of your foliar flying arrow, what are the conditions that bring on the corn ear worm?

FENZAU. I have observed that it is quite rare to see the corn ear worm present in an area where the corn borer is present. You will not find both in the same section of the field. Now there are parts of the country, usually in the higher pH areas of the soil where the corn borer just isn't existent. And you will see more of the corn earworm or the corn earworm beetle. During the stage of early silking, the corn earworm beetle comes out. It attacks the silks. It starts as a moth. And, of course, in the beetle stage they'll tend to invade that emerging silk and just chew the silk off because the silk does not have the right maturing protein capacity to be a good, wholesome vehicle through which pollen can be taken out efficiently for pollination. So it reflects the mature protein structure of the silk vehicle itself.

ACRES U.S.A. Which means an improper form of nitrogen?

FENZAU. Right. And I've seen beetles emerge and fly around the fields and never cause damage, and then in one solid area they invade, it just seems, like it was candy and ice cream.

ACRES U.S.A. Why would you have it and not the borer?

FENZAU. A different hormone system is involved. And I know in particular the hormone called the DIMBOA relates to influencing the resistence to the corn borer. And when dimboa is present in certain concentrations, the corn borer will take about one suck, and that's just about all, and he's done for.

ACRES U.S.A. How do you get dimboa into a plant?

FENZAU. DIMBOA is just a natural phenomena of a properly growing plant with a good supply of trace nutrients that allow that formation of hormone systems to work and to be present in its normal equilibrium level. Good growing corn plants have the proper level of dimboa, and are never infected by corn borer. I don't care how many are in the soil or around the area. The corn borer will only invade certain areas in the field—usually by soil type and soil variables conditions—and just stay in one solid area. They will not go across the lines. The same is true with the alfalfa weevil.

"That Was the End of the Roman Empire"

An Interview with Dr. Phil Callahan

Originally Published: May 1984

Very few people in the eco-agriculture canon were farther ahead of their time than Dr. Phil Callahan, 1923-2017. Not only did his research and writing in the 1970s and 1980s provide a jaw-dropping account of the relationship between insects and plants, the research is only now being understood and applied 50 years later. Dr. Callahan served as a navigation and electronics specialist during World War II at a station in Ireland, where he played a crucial part in anti-submarine warfare. At the close of hostilities he resumed his education at the University of Arkansas and later at Kansas State University, where he received his PhD in 1956.

In this interview, Dr. Callahan had just returned from Egypt's Nile Valley, where he had toured the country with an Acres U.S.A. group, and was interviewed while aboard his flight home by Charles Walters.

For two weeks prior, Dr. Callahan had revealed to the Acres U.S.A. group in Egypt a lot about ancient cultures in an in-depth exposition of his concepts about low-level energy. Also, during the Egyptian adventure, far-out subjects such as dowsing, radionics, paramagnetism, low-level energy utilization in food production all commanded a lot of attention.

In the meantime, in process at the Acres U.S.A. home office was Callahan's latest book titled *Ancient Mysteries, Modern Visions,* and subtitled *The Magnetic Life of Agriculture*. The lectures on the trip and the manuscript of the new book stood prominently as background to the opening of the interview.

For those who are not familiar with Dr. Callahan's famous works, *Tuning Into Nature*, and *The Soul of the Ghost Moth*, it must be pointed out that Phil Callahan was an internationally famous entomologist and was responsible for breakthrough discoveries in both areas. He was also a generalist. He wrote seven important books, and accounted for at least 80 articles in professional journals.

Acres U.S.A. Dr. Callahan, you've made some interesting comments about weak energy in nature, and you have mentioned that these are really the ones that control life. Can you elaborate?

Callahan. To start, we have to point out that life is lived at very specific temperatures. A human body isn't heated up to 3,000 degrees like a light bulb. We don't have a lot of voltage and current running around our body. We have currents that are very, very low. Our skin temperature is about 96°F. This is an area in which the energy that is coming from molecules is fairly weak, but also very specific. I have been able to look at one part of the spectrum—the infrared part—which is where we exist. We live in an infrared environment because infrared radiation—the main part of it—occurs at room temperatures.

Very weak frequencies are given off by molecules at such temperatures. These frequencies have one very important attribute. They are coherent. That is, they have additivity because the waves come out together instead of separately. When you add these waves up you get enough energy so that you can tune in to it. It is a lot like a radio. An incoherent radio wave cannot be tuned in. A coherent radio—you can tune to it and carry a voice with it, you can use it.

Coherent radiation from molecules can be used to control things even though it is very weak, and it takes a very expensive Fourier spectrophotometer to see and measure their energies. Low energy electromagnetic frequencies is what we encounter when we lead with weak energy forces in nature. Now you can break electromagnetic and magnetic. If you look at what we did during the trip to Egypt, we were looking at stone structures.

Back up a moment. When you take a magnet and cut it in half, you just end up with another magnet—a dipole, a north and a south pole. But suppose in nature, under certain conditions, you had the two poles separated. They would be called monopoles. My work reveals that most plants are diamagnetic. That means not susceptible to magnetism, or like the north end of a magnet. Most stones and rocks and good soil would be the south end of a magnet. Therefore, we could say the south would be equal to minus in an electric field and north would be equal to plus and minus.

Acres U.S.A. You make it sound like the human system or the plant system is really an electrically wired system. You are not going to plug it in the 220 or anything like that. But you are plugging it into nature and the environment around you. Is that what you are saying?

Callahan. That is exactly what I am saying. It is not plugged into a battery or 110. It is not even plugged into 2.5 volts from a battery. It is really plugged into the sun. The sun gives off visible radiation. It gives off radio waves. It gives off light. It gives off infrared. The sun is heated up very, very strong. But by the time these energies reach Earth, they are what we are talking about. They are very, very weak. Although they may be intense and hot at the sun, by the time these energies reach Earth they are almost like a radio wave from a satellite. It has gone from 10 watts if it is up in a satellite, to about 10 to the -17 watts when it comes down hundreds or thousands or millions of miles away. We are utilizing these very weak energies. A human body is utilizing these very weak energies from the sun. So are the plants.

Acres U.S.A. And these energies are coherent at certain frequencies?

Callahan. They are coherent frequencies, and actually there is a moderator of these coherent frequencies from the sun. They do not necessarily start out as coherent frequencies, but when they hit the molecule, which is an oscillator, the molecule turns them into coherent frequencies. In other words, they aren't coherent coming from the sun. They are what you call a pumping energy. They put energy into a molecule. But the molecules that your body utilizes take that pumping energy and turn it into a different energy, which is coherent. In the magnetic spectrum, the same sort of thing happens. When you have big magnetic storms—that is what the sunspot is—these energies come out. Stone is the receiver or the mediator that turns these magnetic energies into a coherent form. You always have to distinguish between coherent and incoherent, between static and tunable. So, you have these magnetic forces coming from the sun, which are like static, but when they hit the right shape of a rock or a temple or the soil or whatever it is, the receiver converts the incoherent magnetic energies into a more coherent energy. Plants use this energy to grow. In this case of healing, the body can tune into this energy to heal.

What I have seen in a lot of these temples, and what the ancients did is this: they would take people and stand them in a certain spot, or they would lay them on a rock, or they would pass them under a rock. These rocks would be like antennas collecting not electric field, but the magnetic south monopoles, and focusing them into the body or into the plant.

There is nothing contradictory about this in a conventional science. All you are saying is that you have, instead of a metal antenna, a rock antenna, which is collecting radiation and turning it into a signal that controls the body.

Acres U.S.A. This radiation doesn't all have to come from the sun, does it?

Callahan. No.

Acres U.S.A. What about from the planets?

Callahan. It can come from the planets. It can come from the night sky. When the sun goes down, you have 1.3-micrometer infrared flash. You can't see it because your eye can't see infrared, but the sky suddenly is lit up by infrared in the 1.3-micrometer region. A micron is 1/1,000 of a millimeter. This very narrow light, in fact, lights up the sky. It radiates the molecules in the air around you. Also, it reflects off your skin. Your skin reflects radiation like a piece of alumi-num. Now suppose we had a mosquito and it wanted to see you, and you suddenly light up with a light the mosquito will sense. You can't sense it. It is invisible to you. But it is not invisible to the mosquito's antenna. The mosquito will go to that reflected radiation because the signal is as strong as an Indian signaling with a mirror. Even though it is dark outside to you, that light is there. It is invisible, but your skin is reflecting it, and as you walk you are like a flashing mirror. The mosquito tunes into you, you might say. It zeroes in on you. Most of these radiations are invisible. You could say the same thing about magnetic forces coming from rocks. We don't have a good way to detect them.

Acres U.S.A. But they do exist and this gives a lot of credibility to people like Dr. Rudolf Steiner, who wrote material that seemed too far out for most scientists to grasp when he wrote it. He was seemingly light years ahead of academic science because he was talking about weak energies that could be utilized, that could be focused in on plants.

Callahan. Correct. Weak energies are what run living things. When these weak energies no longer work right, or when, say, a human being gets sick or dies, it would be correct to say that the weak energies were no longer tuned in. People like Steiner didn't really know how to put these concepts into words that modern scientists could understand. Yet he instinctively knew what he was talking about because he observed nature enough to see these things working. He wrote in German. It is a hard language unless you grow up with it. You lose a lot in translation. In the second place, Steiner didn't have any of the conventional scientific background that we have today, so he couldn't put his concepts into language that scientists understood either. Even today, most people do not understand him. But that is exactly what he is talking about. He was talking about weak energies that he knew existed, even though it might have been an instinctive thing with him.

Acres U.S.A. Well, maybe it was and maybe it wasn't. He seemed to have a good handle on what ancient people had done in this regard. He spoke with a tremendous knowl-edge of anthropology and history in tow, and he called attention to many of the same things that you are calling attention to.

Callahan. That is right. Again, perhaps he had the same insight we both have. I have an instinct for ancient systems, so I go back and start studying them, and pretty soon I say, Well, what is going on? And then I try to figure out what that something is. But modern science requires that in order to believe something, you have to be able to repeat an experiment. Not many people really want to do it. They look at these ancients as being superstitious, and therefore the things we're talking about are considered superstition. Modern medicine views ancient cures as superstitious, and this is ridiculous. When you look at the ancient literature, you can see that certain cures worked. I am sure that Steiner was doing exactly the same thing. He was looking at ancient people, but you still have to have a feel for it. There is something besides just being able to go back and trace history. You have to have an instinct for cross-correlating what happened, and not be blinded by our modern overview. You have to disassociate yourself completely from our modern technology, really.

Acres U.S.A. Dr. William A. Albrecht would always start his course in soil science by first calling attention to geography. He would say, "This is what you are going to farm and you have different conditions in different parts of the country and world." One of the maps he used illustrated radio receptivity according to soil types. What would that have to do with it?

Callahan. Well, what he probably meant was the character of the Earth itself as an antenna. For instance, when astronomers wanted to pick up weak radio waves from the universe, they had to hollow out a big sphere in Puerto Rico. It was pointed like a parabolic antenna. So even today we are using a form of the Earth as a reflector for weak energies. The astronomers themselves are doing that. They made a big parabolic reflector out of the Earth itself and put a sensor above the middle of it. It is a couple miles wide, or more.

So by the shape of the Earth, the type of the rock, the type of the soil, you determine the coherence of weak energies. That is what he was talking about. Energies radiate out from different forms of earth.

Acres U.S.A. Isn't this what we are doing when we put a dish in the backyard and point it at a satellite floating over Omaha?

Callahan. That is the exact same thing. What you have is the utilization of weak energy, because by the time the energy leaves the satellite—which is 300 or 400 miles up—and gets back down to this dish, you have picked up an energy again that is probably—well, I don't have the tech orders on these things, but you are probably dealing with a millionth of a volt of energy.

Acres U.S.A. Okay, back to this plant and this animal system. If you viewed it like a house, it would have a lot of wires running through it?

Callahan. That is right. If you view all of these different plants and animals and people as being sort of tuned antennas, which can collect various energies and utilize them, you get into what I am talking about. By looking at the weak energies of the different shapes that are on an insect antenna, those shapes tell you something. You can't go to the nerve endings like the entomologists and say, "This is the way it works." You have got to look at the shape of the antenna and correlate the shape with something. The minute you do this with the spines on a plant or the spines on an insect, you are saying, It must be a collector, it must be an antenna collector of weak energies.

Acres U.S.A. Some people have been working since the early 1920s—and even before that—with the idea of focusing these weak energies. Is this a realistic approach? We refer to radionics.

Callahan. Yes, it is a realistic approach if we only apply it in terms of what we know absolutely about modern day weak energy communications. You mentioned satellite dishes. If you apply that type of science to what radionics people are doing, you can probably begin to get a toehold into what they are doing by just experimenting randomly. In other words, sometimes they have success and sometimes they don't. It will be necessary to focus energy so that there is success every time without the present dependence on the operator. Most of radionics depends on whether the operator is tuned in and how skillful he is. So it doesn't always work. But if you knew exactly what the frequencies were, then you could make maximum use of the human body as a living antenna. In Athens, I bought one of these rock statues of a god because I think it is made out of a paramagnetic stone. I am going to take this little statue home, and if it is not paramagnetic I will probably make is paramagnetic by coating it with clay or probably with one of the modern acrylic paints.

Acres U.S.A. Well, is limestone usually paramagnetic?

Callahan. That is the problem. You never know until you test it. Some limestone is paramagnetic and some is diamagnetic. Those are two different forces.

Acres U.S.A. Do you know which is which?

Callahan. Not by looking at it.

Acres U.S.A. In other words, if scientists would tackle this radionic proposition, we could get some real definitive answers?

Callahan. We could really get some definitive answers, but they don't want to do it because they look on it as being superstitious, and they look on all these ancient cultures as nothing but superstition.

Acres U.S.A. Meaning science has left this field in the hands of amateurs for development, and a lot of them are talented, and a lot of them aren't?

Callahan. An amateur can be just as smart and just as talented as a trained scientist. He often has an advantage in that he doesn't have a mental block. Most trained scientists today are specialists. If you get the talented amateur working, he doesn't have the mental block of the specialist, and he might be just as good of an experimenter. Of course, it would help him if he would learn a little of the scientific terminology. He should learn the basic terminology of science. Like you are an editor. You have basic words that stand for basic things. What is mainly wrong with people who study radionics and towers and ancient ways of healing and homeopathic cures is a lack of familiarity with concepts that would be of value to them.

Acres U.S.A. Can you be more specific?

Callahan. Well, how do you make a transistor work? You do it by putting a very small amount into a crystal. It breaks the crystal structure. That works because the molecules have more room to vibrate if they are further apart. If you cram too much into something, it is like two violin strings that are too close, you can't vibrate them. If you get them further apart, they vibrate.

So, homeopathy is a good example of solid state physics. So the practitioners ought to learn the terminology of solid state physics, and then they could run experiments and exhibit meaningful results. On the other hand, the scientists ought to quit making fun of people doing that sort of thing and start studying it themselves. You have a bias both ways, I'm afraid.

Acres U.S.A. From our point of view, it appears that we are no longer entitled to ignore things like homeopathy and even radionics.

Callahan. You have to take a look. You cannot ignore them. If you continue to ignore them, you will by default contribute to the building of more and more high energy, uneconomical, tear-apart systems for agriculture. If you go back and look at what solid state physics really is—you'll see that the transistor is a low energy system. You dope a little crystal with something and you get a good signal out of it. You have got to go back and look at these low energy systems, learn how to manipulate them, and you will grow better crops. These low energy systems cannot be ignored any longer. There are not enough resources in the world to sustain high energy systems being used.

Acres U.S.A. Therefore use of low energy becomes the linchpin of sustainable agriculture?

Callahan. It is God's energy instead of man's. The energy we are using is man's energy. It is made by burning oil. The other is God's energy. It is made by utilizing the frequencies, magnetic and electric, from the sun. That sums it up. Teilhard de Chardin said that science should get you closer to God. So the more we learn about low energy systems, the closer we will get to God because we will be farming God's way, utilizing the sun—the energy from the sun—instead of using Standard Oil's system.

Acres U.S.A. How strong is the energy you have in microwaves, in radar and things like that, compared to the energies we are talking about?

Callahan. Two things are involved. Physicists measure energy in terms of electron volts of energy from the molecule or from the atom. Electron volts is the energy that you would have coming, say, from an atom or a molecule. The shorter the frequency, the more the energy. In the X-ray region, energy is so strong it can kill people. The ultraviolet region is not as strong, but it is still strong enough to harm you. In the light region it is fairly strong. In the infrared your energy has gotten weaker and weaker, and the radio wave has gotten very weak.

Acres U.S.A. Where is radar?

Callahan. Radar is radio, so essentially it is weak, but like any weak signal, you can amplify it and make it damaging. The radio waves from the sun are very weak. It is like taking a light beam which doesn't hurt you—a flashlight for instance. I shine it in your eyes and it doesn't bother you. But, I can take that same electron volt energy that is coming from those disturbed atoms that are putting out the light—and that is what light is, disturbed metallic atoms that are vibrating in the filament—I can take that and put so much energy into that flashlight that I now have an anti-aircraft battery search light. If I shine that in your eyes it would burn your eyes out.

So, you can amplify weak systems. If you are talking about a basic level, light, infrared and radio are very low energy alongside X-ray and ultraviolet. But, if you are talking about amplifying these energy systems, you can take radio and make it so strong that you can cook with it. That is what they are doing. They are making a weak energy so strong that you can cook with it.

Acres U.S.A. Which is your microwave oven?

Callahan. Correct.

Acres U.S.A. How does it work?

Callahan. The energy is so strong from this klystron or this radio generator that as the waves go through the meat or whatever you are cooking, it sets the molecules to oscillating so fast that it heats up the food—and when it heats up, you cook it.

Acres U.S.A. So you are really cooking the cell from the inside out?

Callahan. Again, correct. Which is not what fire does. Fire just heats the outer surface and works its way in. This radiation penetrates straight through and cooks the whole thing evenly. However, in my opinion, it is a totally different type of cooking. I have timed the time it takes something cooked in the microwave to cool off as compared to something that is cooked and allowed to cool off after conventional cooking, and the microwave cools off quicker.

Acres U.S.A. What does microwave cooking do to the digestibility of food?

Callahan. Well, we don't know. Wee have all sorts of information that says it works, but nobody researches it. It might not hurt you at all. I can't say whether microwave cooking hurts or doesn't hurt you when you eat it. The units themselves are safe. They are well shielded. They are not going to hurt you if you have a new modern set because they are shielded quite well. So there is no leakage, which used to be a problem. But when you eat food cooked by radar—nobody has investigated whether it has killed all the enzymes, or if it is bad for you.

Acres U.S.A. Whether it is breaking down your immune system?

Callahan. Nobody knows what it is doing, so as a scientist I can't say it is bad or good—but as a scientist, I can say, I suspect that it is bad.

Acres U.S.A. How is the microwave oven different from the radar gun the policeman shoots at the care coming towards him?

Callahan. It is the same thing. The same system is involved.

Acres U.S.A. In other words, you have taken a weak energy and magnified it to the nth degree so that it will bounce back and give a reading rapidly.

Callahan. That is right.

Acres U.S.A. That would make exposure very high, wouldn't it?

Callahan. You are being radiated. We don't know whether it is good or bad for you. Again, my educated guess would be that it is bad for you.

Acres U.S.A. Well, a cafe operating with a radar range used to post a sign warning customers with a pacemaker not to remain on the premises. If this is bad, why is it possible for a policeman to shoot a radar gun at people on the street at random? He may have a heart patient with a pacemaker coming at him!

Callahan. It is the same thing. It is just as bad. The radar guns should be outlawed. There is no question, they should be outlawed. The policeman is pointing radar at you. During World War II, they would tell you, Don't get in the radar beam. What did we used to tell kids? Don't sit too close to a TV screen. Now they are telling you to sit as close as you want to a TV screen because it doesn't hurt you. That is ridiculous. You shield a TV set because some sets have so much power in them they are giving out all kinds of things like X-ray. You shield it so you don't get radiated. Even then, they say don't sit too close. So if somebody points a radar gun at you and actually radiates you, that is ridiculous. There ought to be standards for this sort of thing, and there should be research to reveal exactly what the police are doing.

Acres U.S.A. What is the radar gun doing to the policeman who sits by the machine all day long?

Callahan. No telling. I can tell you one thing from my experience with ground control sets in a GCA, meaning ground control approach. We used them after World War II to bring airplanes in to a zero landing. I have operated those things. I have repaired them and worked on them and even helped design antennas for one unit in Tokyo. After a couple hours working one of those things, I would come back totally fatigued. There is no question that it affects you, which is why I was somewhat sympathetic to the air controllers' strike, not the labor part of it, but what they were asking for was just. It is the most fatiguing job in the world. I know because I have done it. You are being radiated all the time. I have experienced it with GCA sets, I couldn't take it for more than a couple of hours.

Acres U.S.A. The observations you're presenting are serious. Why are scientists staying away from this issue?

Callahan. It is a matter of not listening. Scientists have to listen. It cuts both ways. Scientists have to listen to mystic types, and mystic types have to start listening to scientists. They bad mouth each other. People working with low energy are bad mouthed by the conventional scientist.

Acres U.S.A. Isn't that understandable? How can they get listened to when we see people creating phony scientific tests? All we can get out when we ask a very pertinent question is more academic freedom and intellectual honesty quotes?

Callahan. That is the cover-up.

Acres U.S.A. So we end up bad mouthing each other, which is bad both ways.

Callahan. You saw that on the Egypt trip. You have just got to keep calm. If somebody asks you a question, you tell them what you know and if it is not what they want to hear, and they get mad, you just have to quiet down and say, I am not going back at you. Just cool off a while and we will see what the problem is. If I can help you, I will. But if you don't want my help, I won't be mad at you. It doesn't hurt me whether they take my advice or not. It won't change my life any. You just have to be calm in all of these things. You have to think.

Acres U.S.A. So is it correct to say that the biological agriculture we deal in is geared to low energy?

Callahan. Biological agriculture is low energy agriculture. That is the way God created agriculture, and we are destroying it with our high energy anhydrous ammonia that is used to dope the ground. If you get enough dope in the ground, you have had it. You stifle those low energies; you kill them. Just take the tomato that you ate in Egypt that grew without all this junk on it; then take one from Florida that is grown on nothing but worn out soil that is fertilized with anhydrous ammonia and some form of P and K. The one from Florida tastes like cellulose. Your taste and smell tells you. You can't even smell them anymore. Tomatoes have a beautiful smell. The American tomato ... you can't even smell it. It tells you something is missing. Your nose is one of the best molecular detectors there is. When something doesn't smell right, you know something is wrong.

Acres U.S.A. Where do we go from here?

Callahan. I think we are to the point now where you need to get organized within the scientific community and also the farming community. The pragmatic people who are growing food utilizing low energies need to harmonize their knowledge with our whole agricultural system, and to do that you are going to have to put it in such a way that it has a scientific basis. In other words, everything has to be based on experimental evidence and on a repeatability factor. It is the same as when a scientist says, If I make an airplane and I do everything right, the airplane will fly. If I plow a field up, the minute I plow the field I am changing nature. It is no longer wild. It no longer has many plants. You have to set a table with one type of plant; how do I harmonize that one type of plant; how do I harmonize that one type in this environment utilizing low energies. The biggest problem in organizing this sort of thing is to convince the conventional scientist to quit being a specialist and to see the ramifications of the whole thing. The way that the energy is utilized by nature is the way a farmer is going to have to manipulate his farming field in order to get the most out of nature.

Acres U.S.A. This means bringing this entire field out of the underworld of science and putting it on the table with replication, and the usual scientific system can function.

Callahan. Exactly. It means bringing it out into the open. Don't be afraid to say it regardless of what other interests say won't work. Conventional agriculture has to hit things over the head with a hammer to make them work, and that is not the way nature works. But you have to get your principles into science so that everybody can be working on it—meaning everyone knows and understands science.

Acres U.S.A. If we could understand the circuitry of a brain, for instance, and the thought process, we might even learn how though transmission takes place between husband and wife or close relatives?

Callahan. Exactly. You are dealing with weak energies again.

Acres U.S.A. Weak energies, if you could develop a coherent method.

Callahan. I think so. I think that is what happens accidentally with ESP. You have, say, a mother in the U.S. and a son, say, in Vietnam, and all of a sudden the son is hurt or wounded, and she knows it instantly. This has been verified in war after war after war, and in one of the best verifications of ESP in my opinion is the case reports of things that take place during traumatic experiences in war. The son's electronic circuit brain is very much like his mother's. He has 50 percent of her circuits. Therefor, his brain puts out a lot of energy. If you can scan the earth from a satellite with 10 to the -17 watts, there is no reason why your brain isn't putting out much more than that. In fact, your brain is probably putting out, I would guess, 10 to the -12 watts and 10 to the -17 is less. Yet you can make a TV picture and turn something from a satellite into a TV picture with the 10 to the -17, and that is a trillion, trillion, trillionth of a watt. Your brain putting out 10 to the -9 or something like that is certainly a stronger signal and would go around the world 40 times. Of course, signals do not go around the world in nature. You have what you call Schuman Resonance. Schuman Resonance is when you have harmonics from lightning bolts that go around the world at about eight to 20 cycles, and who knows what they are controlling. You have thousands of lighting bolts all over the world, and the ionosphere above and the earth below act like a big hollow cavity, and they go around and around. You can tune into them. Nikola Tesla did this. He sent waves around the world. He was no doubt utilizing Schuman Resonance to do it. He was ahead of his time. Schuman Resonance wasn't even discovered until about 15 years ago, but Tesla was doing this back in the 1890s.

Acres U.S.A. Most of his work has been lost to the U.S., hasn't it?

Callahan. His work is in about 110 patents and is not lost, really. I have all 110 of his patents. His work is ignored. He was the father of the radio. He was the father of the fluorescent tube. He was the father of the AC electric motor, the father of automation. He made an automated boat and controlled it by radio before Marconi was sending static. He did work with generators and turbines. He invented one of the best turbines, and they still aren't using it. It is better than any turbine on the market right now. He has a patent on it. His work is ignored. In many ways I admire him, but he was a high energy guy, really, and the world has taken his stuff and carried it to extremes. He would probably agree with that assessment if he were alive today. But he has received little credit for any of it.

Acres U.S.A. Why does official science and the conventional wisdom have such a propensity for ignoring things, holding them back many decades?

Callahan. I think it is the same thing in every field of mankind endeavors. There are people who are ahead of time. They are laughed at. Most of the great novelists of the world, the painters—Van Gogh, people we say are great painters—were laughed at. Van Gogh's paintings were thrown around. A Van Gogh now would cost you a million dollars. His work was just thrown in a pile back then. Most people who are ahead of their time are laughed at and ridiculed and never recognized until five or six generations after they lived.

Acres U.S.A. We hope you don't draw too much laughter, but we think you are one of the few scientists who has the wisdom to see the necessity for taking a look, which is what Acres U.S.A. likes to do, even though others point a finger at us shake their heads sadly.

Callahan. When people point a finger at you, you just have to do what you think is right. Sometimes you are wrong, but if you are really doing what you think is right, you are probably right. If you are doing it from a profit motive, there is nothing wrong with that motive. You have to do that to live. But if you do everything from just that motive, then you are in trouble. That is not the only motive. There is nothing wrong with that motive in moderation, but that is not the only motive we can live by.

Acres U.S.A. One more thing, how do the theories of Louis Kervran fit into what you are talking about?

Callahan. Very well, because he is talking about transmutation of one basic element and changing it to another. If you can do it with high energy, with nuclear energy, there is no reason why nature can't do it at low energy. In fact, the Japanese have repeated some of his work. So his work is a part of what I am talking about, a solid state, low energy system for changing signals, manipulating signals, and there is no question in my mind that he is right. It is transmutation that makes biological life unique. That life can transmutate. It can change. It can control signals. That is what evolution is all about.

Acres U.S.A. Do your observations square with the ancient mysteries you talk about in your new book, *Ancient Mysteries, Modern Visions*?

Callahan. All of these ancient mysteries were things that worked, but I think the priests kept it to themselves. We lost them because they were too stingy. Many of the ancients did the same thing we do. When they found out something worked, and they wanted to make a fast buck, they kept it to themselves to impress people, or for power. So they were just as responsible for losing it as we are responsible for not trying to get it back. You can't point fingers and say the ancients were wise. They were no wiser than we were. They were darn good low energy scientists, that is what they were.

Acres U.S.A. Bust as far as the institutional arrangements for passing it on, with all our faults, we probably are superior.

Callahan. Yes, I think we pass information on far better than they did, with the exception of maybe the Egyptians who carved what they knew in rock walls. But then they set everything down in a coded system which is hard to understand. It has taken me years to even get a few little inklings of what they were talking about.

Acres U.S.A. We seem to have lost the reason for things like the Irish round towers, the pyramids, Stonehenge?

Callahan. Yes, I think they were all stone antennas, but it is difficult to decode how they worked. That is what I have been working on for the past 20 years, besides my insects, of course. What they are doing with modern agriculture is the same thing they did in the medieval ages with dungeons. You throw a man in there and you feed him bread and water, and he doesn't last very long. He slowly dies because he doesn't have everything he needs. Bread and water doesn't have everything. It has some, but not everything. Even good bread, the best bread, doesn't have everything. What you are doing is taking your land and asking it to live on bread and water, and believe me, it isn't going to work over the long haul. Whenever you take a small family farm, the guy who loves the land and takes care of it, and annihilates that entity, you've made a mega-blunder.

This happened in the Roman empire when Cato manipulated the Roman Senate to start a war with Carthage. He went to Carthage, defeated the Carthaginians because their agriculture produce was competing with the Romans. He sterilized the soil with salt and came back to Rome and decided the most economic crops were nothing but olive oil. Gradually through, political laws, he gathered in all the land of the free holders, and the free holders moved to the city and became the Roman mob. That is documented in Will Durant's series of books, *The Story of Civilization*. Durant was a marvelous historian. Rome eventually passed laws that made it uneconomical for the family farmer to farm. Thus the state gained all the land of the free holders, planted it with high intensity crops. It was the end of vegetables, wheat and everything else. That was the end of the Roman empire. It looks like we are trying to do the same thing in our country.

"I've Had to Stand Alone"

An Interview with Christopher Bird

Originally Published: August 1984

Christopher Bird, 1928-1996, is well known as the co-author of *The Secret Life of Plants,* a book that opened the eyes of many to intelligent communication between plants and insects. What is lesser known is that he is also the author of *The Divining Hand, The 50 Year Old Mystery of Dowsing*, a much more controversial work that still, to this day, is discarded by most "serious" agriculture professionals.

"To dowse," says the author of that definitive work on the divining art, "is to search with the aid of a hand-held instrument such as a forked stick or a pendular bob on the end of a string—for anything. This can include subterranean water flowing in a narrow underground fissure, a pool of oil, or a vein of mineral ore. It can be buried sewer pipe or an electrical cable, or an airplane downed in a mountain wilderness—a disabled ship helplessly adrift in a gale, a lost wallet or dog a missing person, perhaps a buried treasure."

Dowsing has a rich, controversial, dubious, hopeful, and even racy history, but it also has its most serious side. Even in 1984 when the interview was held, the nation was running low on water, as we are today. No matter what you believe about dowsing, the interview uses the tool to talk about a much more sinister topic—groundwater pollution. In the mid-1980s, America faced one of its peak of water pollutions, including the near annihilation of Phoenix groundwater via contamination with industrial solvent, trichloroethylene, and with other toxic chemicals. Dowsing, or not, is really a way to discuss the importance, and vital nature, of keeping water available to grow our food.

Acres U.S.A. Mr. Bird, we have heard from a lot of people who think that dowsing is nothing but quackery. Some think dowsing is the work of the devil and therefore shouldn't be touched. As a start, what is dowsing?

Bird. Dowsing is quite a few things at several levels. Number 1, if you take the subject as runs on a ladder, the bottom run would be what we call field dowsing. I would say it is known in every single community of the United States. If you should stop off in any one of America's communities with a population of more than a thousand, you would find a dowser in that community. This person would know quite a bit about field dowsing. He would be a practitioner or a hobbyist. This person goes out to your property—be it a homestead, or a farm, or something else—and with a device in hand—an old fork, stick or a wire rod, or some other device such as an L-Rod, locates a site where water can be found. This dowser will walk over a given section of ground to locate a good site for a water well. That is field dowsing. Now, we jump up another level on the ladder and we get to know what is called remote dowsing, dowsing at a distance. Let's say it's a cold day out there. This dowser is in a truck and doesn't want to get out in the snow and walk the property. This dowser will sit on the corner of the property and will inspect the property by eye. When this dowser's eyes get to a certain point in your field, he'll get a reaction on the dowsing device. Then he drives over to another corner of the property and he does the same thing. Two lines he's looking at cross and that will be a point he establishes without even getting out of his truck. On the third level, we get what we call map dowsing. Unbelievable as it may sound, there are many successful cases of it on record, all documented to the hilt. Here, the same man no longer uses his truck. Let's say it's gotten cold. He sits in his kitchen and say his kitchen is in Portland, Maine, where Henry Gross's kitchen was. Take Kenneth Roberts, the great novelist (*Northwest Passage*, etc.). He wrote three—not one, but three books about dowsing. Or take Charlie Walterson with a map of his farm in Missouri. In his kitchen in the winter time, with the wood-stove burning and the muffins baking, he takes his map and he uses a pendulum to map a dowsing path. Now, if you want a little better pinpointing, you may want to call in the local dowser to actually pinpoint your map, because that large-scale map pretty well says exactly where to drill the well, but a few feet can make a lot of difference.

Acres U.S.A. These people have track records and proof?

Bird. Oh, absolutely. I can introduce you to the ones with real track records. I always like to make this analogy. People ask, "How do you know a good dowser?" I say, "Well, how do you know a good piano player?" And the answer is that really good professionals are few and far between. Take athletics. To become a gold medal winner at the Olympic Games, it takes something extra. You're not going to find them under every bush. So, the ones with track records have real track records on the map dowsing. This can be documented and I can provide you the documentation.

Acres U.S.A. How do you get around the charge that dowsing is voodoo? More specifically, is there a real rationale for it or at least a good concept as to how it works?

Bird. The rationale is primitively worked out. I like to think of it in the following way: it starts as a mental act. Today, of course, we're becoming aware of the fact that the mental act, the intent, is becoming more and more powerfully recognizable in science because it relates to some of the observations in quantum mechanics, namely that the observer is part of the experiment. The observer then is the dowser in the experiment he is doing, and it starts with his intent to find something. He translates that intent into focusing on a given target. Call it the piece of ground he's walking over, or the map of that piece of ground, which is merely a paper representation of the same thing. Now you've eliminated the distance factor.

Acres U.S.A. Dr. Phil Callahan suggests that the water vein sets up a field.

Bird. Yes, well Phil Callahan can suggest that. I addressed that in my book, which is called *The Divining Hand*, which you've read. But the minute you move up the scale—and I want to stick to that ladder with the rungs—the minute you move from field dowsing, the water vein sets up a field that can be contacted by dowsing the transient across it. What happens to Phil Callahan's explanation when the dowser does the same water vein on a map 900 miles from the property?

Acres U.S.A. We have no answer for that. We're not sure that we have a question.

Bird. Neither does Dr. Zaboj V. Harvalik, who for 30 years rolled up a reputation as an expert chemist. I discuss him in Chapter 14 of *The Divining Hand*. Harvalik will go on about fields and war games and provide lots of evidence, but he simply will not, as a scientist, address the problem of how map dowsing might work, although Harvalik, by the way, is a really good map dowser.

Acres U.S.A. He knows the question, but doesn't know the answer?

Bird. You've got it!

Acres U.S.A. So far, we discern that dowsing is more art than science. You suggest intent has a lot to do with it, and in your book, you mention Arthur Middleton Young, the inventor of the Bell Helicopter, who holds that one of the most amazing examples of animal behavior is the motion of an amoeba, which can reach out by extending a pseudopod, devoid of any musculature, from any place in the body. You quoted Young as posing the question, "What causes the projection if it is not attention and intent?"

Bird. Yes. Like amoebas, dowsers project an intent to find, or a request for the location of a given object or target. What is projected? Perhaps a mental or psychic pseudopod of possibly infinite length. Whatever it is called, an answer to the request seems to be fed back via their bodies in the form of molecular movements, which because they are usually not consciously perceived are called involuntary. The muscles cause the dowsing rod or pendulum to move, thereby objectifying the muscular action that, self-generated by the requester, cannot really be termed automatic.

Acres U.S.A. But aren't there a lot of dowsers who maintain that some mysterious force moves a dowsing rod, not muscles? And if the muscle theory is correct, why is the dowsing rod needed at all?

Bird. Well, there are particularly sensitive people who dispense entirely with an instrument. They seem to feel physiological changes in their hands, feet, stomach, throat, or body organs. And as I mentioned earlier, how do you account for force being a factor when someone is map dowsing miles from the scene? As a matter of fact, there is a strong likelihood that this mystery has been solved by Dr. Jan Merta, a Czech born physiologist, psychologist, gifted psychic and professional deep sea diver now working on oil rigs in the North Sea. Dr. Merta came to a meeting of the American Society of Dowsers in Danville, Vermont, in 1970. He was very interested in the water dowser's art as related to the extrasensory—map dowsing, for instance, location of lost children and successful searches of missing objects. Merta did a lot of experimental work at McGill University in Montreal. His findings caused him to suspect that the movement of the dowsing device has to be directly connected to muscular contraction in the body—specifically in the arms or hands. He therefore reasoned that if he could build an apparatus that could simultaneously record both the movement of the dowsing device and any muscular contraction, he would be able unambiguously to determine which came first, the contraction or movement of the device.

Acres U.S.A. Did he come to any solid conclusions?

Bird. He did. He electrically wired the carpi radialis flexor in the wrist area of the forearm. The instrument translated what was happening to ink and paper. After the several tests were finished, Merta concluded that the dowsing devices react only after human beings operating them pick up a signal, which stimulates a physiological reaction. He further concluded that if the dowsing device were only an amplifier magnifying a sensation, then dowsers should be able to teach themselves to pick up such sensations without recourse to any dowsing device whatsoever.

Acres U.S.A. How does Merta explain it to laymen?

Bird. He suggests that a projected request for information in dowsing is analogous to the number selection, the bodily reaction to the workings of the vast telephone switching system, and the final muscular twitching or neural response to the ring of an appropriate telephone on the other end of the line. In other words, a successful search depends chiefly on the accurate formulation of the requests. Or as the computer people have it, "garbage in, garbage out" is the recipe for failure.

Acres U.S.A. Are we talking about wavelengths, as Dr. Callahan does, and are we talking about low-level energies, about the human being as a part of the instrument—in short, about the subjects covered in our interviews with Phil Callahan and Jerry Fridenstine— and the off-the-wall observations in last month's leading article, the one in which we told about lemmings marching to the sea, and birds finding their targets thousands of miles away.

Bird. I think so. We've all heard about dogs that find their way to their masters over incredible distances. The ruby-throated hummingbird—all of three inches long—trips from the Panama to the East Coast of the U.S. and back again. The Pacific golden plover makes seasonal nonstop flight 3,000 miles from Alaska to the Hawaiian Islands with no landmarks whatsoever. But the fact is that good science has established that birds can detect minute radiant changes in the Earth's magnetic field ,which may very well help their orientation during migration.

Acres U.S.A. We've had one dowser explain that all people have an aura to radiate energy. He said some people's go way, way out, some probably doesn't escape them by more than an angstrom. Is this meaningful at all?

Bird. Well, one of the best ways to get at an aura is to dowse it. The aura certainly is not fixed at any given length. It depends on state of mind, state of health and on the time of the day. It has been proved through measurement by dowsing that the aura can extend after certain practices have been done—such as meditation, or just quiet relaxation. It has been noticed that if a person would sit over a certain piece of ground on the earth where there were known telluric radiations, and I would at this point have you refer to 50 years of German work that is very important to medicine, if a person sits on one of these places they can definitely act to shrink his aura, in a very small period of time. So the aura can be looked on as a pulsing field around a person, which is known in metaphysical literature as the etheric body. It varies. It is never static.

Acres U.S.A. So the dowsing society would not set up its operation in Telluride, Colorado?

Bird. The dowsing society has many interests, and this is one, but the focus I think that we're trying to get from the journal now is not to forget that basically dowsing has to do with finding water. There is a national water crisis coming in this country, and that is one of the reasons I believe Acres U.S.A. finally wants to allow the topic of dowsing in the magazine.

Acres U.S.A. You are perceptive. You've mentioned map dowsing. Is there a third rung on the ladder?

Bird. Actually there are several more rungs. When we get to the next one, it's what we call information dowsing. Now with information dowsing, we access knowledge through it. We don't need a map anymore. We just want to know how to build a, let's say, given device that no one has been able to build. We may want to check what the minerals are in the soil and we don't want to waste the time to send it to some lab. So we sit down and ask a bunch of questions, each one has a yes and no answer. We bring down that chart and pendulum and we look for calcium. We look for manganese. We look for trace elements in the soil. If we're good dowsers, then we can determine the amounts, the percentages and so on just by that method. That has been done and verified. Again, you get to the problem of who is the artist and who isn't.

Acres U.S.A. Again, it is more of an art than a science?

Bird. It's both, in a sense. One time I was on the radio and the commentator wanted 10 successes out of 10 tries. The radio man sneered. He said, "Well, that's not very good." I said, "Well, just think right now we're in the World Series and the leading team at this moment is batting .331 and is being paid a million dollars a year to be right one time out of three. I just told you that good dowsers are right eight times out of 10 and you still complain."

Acres U.S.A. How far back does dowsing go?

Bird. The next issue of *The American Dowser* is supposed to have an old Egyptian drawing I have been looking for. One guy wears a kind of loin cloth made of a leopard skin. That makes him an aristocrat or a king and he's holding something in his hands that looks exactly like a dowsing rod. However, when I wrote my book, I called it the 500-year-old mystery of dowsing. Why? Because the first documentation in printed literature shows dowsing started in the mines about that long ago. In the mines on the Czech-German border, they were looking not for water, but for silver veins. Documented dowsing started around 1450. There are other traditions of dowsing that probably go back to the beginning of civilization.

Acres U.S.A. Can you teach anybody to dowse?

Bird. The answer must be stated in the form of a question. Can you teach anybody to figure skate? Can you teach anybody to play the violin? The answer is yes. Some of them never get beyond Turkey in the Straw on the violin and the skaters fall flat on their seat every time they go out there. There's going to be a Jascha Heifetz at the end of the line with the violin and a Dorothy Hamill in the other area, right? So you have a spectrum in between. You can teach anybody to dowse. It's partly due to practice just the way violin and figure skating are and partly due to inherited talent and partly due to whatever else goes into figure skating, violin playing, or into speaking a foreign language. We operate a school for dowsing in Danville just before the Conference. This year it will be the 10th and 11th of September, whatever Monday and Tuesday is for that week.

Acres U.S.A. A modern schooled dowser can be the beneficiary of a lot of knowledge earlier counterparts didn't have. Still, there was that movie, The Water Witch Drowns in His Own Dry Hole, and there is the National Water Well Association, which derides dowsing as "three giant steps backwards." How do you answer?

Bird. With the facts. Take Wayne Thompson. His dowsed wells come in 95 percent of the time. The best anyone in Marion County, California, has been able to do is 85 percent. The U.S. Marines took up dowsing in a big way in Vietnam. The art was used to located water, to be sure, but also booby traps. I related the story of Major T.F. Manley and Louis Matacia, a land surveyor, in my book. Matacia proved his skill sin a village used to train Marines on their way to Asia. Scattered throughout the village were booby traps similar to those devised by the Viet Cong—punji pits into which a man could step to have both sides of his legs from ankle to knee impaled by sharp steel spikes covered with excrement to insure instant infection. The village had trip-wired grenade by a single VC in hiding. Under the ground, there were secret tunnels, weapons caches, stores and hiding places. Matacia dowsed his way through it all with great accuracy. Later, *The New York Times* reported on dowsing techniques being used in Vietnam, using the colorful words, "coat hanger dowsing".

Acres U.S.A. Among the interesting case reports you detail in you book, you have priests and scientists, farmers and philosophers, men and women from all walks represented in dowsing's 500-year march through recorded history. We find most interesting very old theories—the Cartesian corpuscular theory to account for Jacques Aymar-Vernay's ability to track human beings.

Bird. Yes, Pierre Garnier published a report: *Philosophical Treatise in the Form of a Letter to Monsieur de Seve, Seigneur de Flecheres, in Which it is Proved that the Extraordinary Faculties by Which Jacques Aymar, with a Dowsing Rod, Followed Murderers and Robbers, Discovered Water and Buried Silver, Reestablished Landmarks, etc., Depended on a Very Ordinary Natural Cause.* That's the title. I don't know if the same guy who wrote the title also had the energy to write the report—if you can allow a joke! Garnier held that corpuscles exhaled by the sweat of the murderers at the moment they perpetrated their crime were of a different pattern than those they would normally emit. This is ancient literature, and yet we find Callahan telling us that mosquitoes hone in on the ketones of the breath. Even today, this is the character of the subject, something mysterious, something requiring a vision.

Acres U.S.A. The fact is that we live in a scientific age. You mentioned the scientist, Dr. Harvalik. And we gained the information from your studies that Harvalik's experiments in effect said that dowsers who are successful are skilled in programming and may be more sensitive to weak amounts of radioactivity than available physical instruments. Does this make these people the equivalent of human Geiger counters?

Bird. That was Harvalik's assessment. He suggested that these "Geiger Counter" type human beings might well dowse for natural mineral resources especially if radioactive elements in rocky environments are associated with them.

Acres U.S.A. You did say Harvalik became expert in map and information dowsing?

Bird. Yes. Let me illustrate the point. While in Australia, Harvalik met with a Sydney Water Board engineer who rejected the idea of dowsing. The engineer gave Harvalik the name of one of Sydney's many reservoirs. He wanted to know where it was. After dowsing for both answers, Harvalik drew an azimuth on a piece of paper and called the distance at 12.6 miles. Next, Harvalik was asked to determine how deep it was. He determined it was 58-feet deep. The engineer called him on his error—the reservoir was recorded at 75-feet deep. A few days later, the two were driving in the hills when they came to the reservoir in question. While standing at the edge, both noticed the water level had dropped. The engineer asked a worker what was the present level. The answer: 58 feet.

Acres U.S.A. Is there any connection between dowsing and the radionics boxes one sees now and then?

Bird. Let me answer it this way. Harvalik took an interest in these boxes with their especially designed electronic circuits and capacitors, resistors and inductors. With the aid of calibrated dials, these boxes are to be set like rheostats to alter values of energy held to the non-electromagnetic coursing through them. The boxes, he said, are tantamount to dowsing devises in that they combined with their operators to form bio-physical units to search for and identify radiations emanating from both animate and inanimate objects. He said that in the eyes of conventional scientists radionics machines are wired nonsensically, or in such a manner as to make reception and physical identification of radiation impossible. He concluded that the devices were merely to help operators program their minds to react to and recognize stimuli of a given and distinct pattern.

Acres U.S.A. Did this scientist ever set down his complete rationale for dowsing?

Bird. He left an affidavit. In it he noted that, "Even now, many people, including scientists, believe that field dowsing is grounded on superstition and has no scientific basis. My research establishes a mechanism for it as follows: small magnetic field gradient changes or electromagnetic radiations stimulate the sensors in the adrenal glands and in the pituitary, which form the three-dimensional perception system permitting even pattern recognition. The stimulus is transmitted to the brain, which contains a signal processor. The brain gives the arms the command to twist, which is accompanied by increased blood flow through the finger capillaries. The twist is very minute but can be made visible with the dowsing rod, which is nothing but a mechanical parametric amplifier and can be made of any material. Since the magnetic and electromagnetic field patters are contingent to their generator sources, dowsers can program themselves, by asking the brain to command the arms to twist only if a certain pattern is present in the stimulus reaching the sensors, i.e., attributable to underground water, oil pipes, etc. The foregoing shows that field dowsing can be explained scientifically and in an understandable manner." Although he has no such explanation for remote, map, or information dowsing, Harvalik in an essay, *Radionics and Programming*, does not close his mind to the idea that they may one day be explained. Since, as he holds, programming is a mental act that influences the appearance of a dowsing signal, he concludes that, after mastering the programming method, "One could attempt to program oneself for energy forms of magnetic and electromagnetic nature and discover structures that would lead to the detection of such esoteric energy forms as postulated by certain radionics practitioners."

"I Started Looking at the Soil"

An Interview with Dan Skow

Originally Published: September 1984

Dr. Dan Skow, 1942-2013, is a veterinarian who discovered quite early that not all answers could be found in a little black bag. Accordingly he helped form Animal Nutrition, Inc., a company that services veterinary accounts by testing basic forages and grains, and with computer software and vitamin and mineral supplements builds textbook-correct dairy, beef swine and equine rations. Skow soon discovered that correcting feedstuffs for balance didn't entirely solve the problem. Toxicity and poor quality couldn't be pardoned by megadoses of supplements. This led him to the soil, and the subject of this interview. Why is his voice so important? He is also the moving force behind consultation based on Dr. Carey Reams' biological theory of ionization, a foundational tool for 20th and 21st century ecofarmers.

ACRES U.S.A. Dr. Skow, as a veterinarian you seem to have departed a bit from your principal profession in dealing with soil systems and agriculture. At least that is the way most people view it. What prompted you to become a consultant and a crops and soils man?

DR. SKOW. I was doing a lot of nutritional counseling with farmers, a lot of ration work. I have my own computer system for balancing rations based on the available feedstuffs plus vitamin and mineral supplements. I was doing very well and getting increased production for about five years, starting in 1968. From then on it became almost impossible to get basic improvements in what I was doing. So I started looking for other answers and this led me to the quality of crops. I discovered the basic quality in general was decreasing at a fairly rapid rate. It always required higher vitamin and mineral levels to maintain production based on those crops. After a while adding more vitamin, mineral and other extra supplementation didn't improve production. That's when I started looking at the soil.

ACRES U.S.A. In raising a better quality crop, is that how you made your contact or connection with Dr. Carey Reams?

DR. SKOW. In a way. I actually got involved with Dr. Reams a little earlier because of some individual health problems. I also was looking at the health food movement. I read a lot of books and decided that supplements and vitamins didn't provide the complete answer to the thing I was dealing with. So I got into it a little deeper by studying the problem from the standpoint of the body chemistry. I discovered that some things are good and some are not so good, depending on the individual.

ACRES U.S.A. Dr. Reams has a slightly different approach to balancing the soil, to dealing with the calcium and the phosphate levels, and especially with maintaining the conductivity of the soil. Can you give our readers a broad overview of what the aims and procedures are, and then we will get into some specifics later on?

The terms anions and cations have a special meaning in the Skow/Reams lexicon. Their definitions are explained in *The Farmer Wants To Know* series, specifically in one title (now in process) called *The Basics Behind Soil Chemistry and Fertilization.* — *Acres U.S.A. magazine*

DR. SKOW. Dr. Reams' basic plan is probably more misunderstood than really different. What he has done is develop a system whereby you're really checking the electrical potential of the soil. According to his biological theory of ionization, plants grow from the electrical and heat energy that is created in the soil. In order to do this, plants require nitrogen, phosphate and potassium—and calcium and carbon elements—in the soil to bring about rapid and high quality plant growth. They also need trace nutrients. That aspect of it hasn't been considered until recently. So you have here a new concept and new terminology. But Dr. Reams still deals with the basics, namely N-P-K and calcium, as major plant foods. These nutrients plus humus and organic matter govern how well a plant will grow. I don't feel that there is a discrepancy in this. It is just a case of bad terminology being used to describe plant elements in terms of electrical conductivity within the soil.

ACRES U.S.A. There were some writers in the early 1920s in France and in the U.S. who used terms somewhat akin to those Dr. Reams is using. For instance we have in mind a fellow named Georges Lakhovsky. This scientist talked about the circuit, so to speak, and the energy flows going past the capacitors and resistors creating heat. He wrote a book called **The Secret of Life**. Are you familiar with that?

DR. SKOW. I have read it. I have the book.

ACRES U.S.A. Is this something akin to what Dr. Reams is addressing?

DR. SKOW. Yes. There are some similarities there. The thesis is akin to understanding where he is coming from. But he is basically talking about an electro-conductivity through the soil. This is essentially the basis of what we are teaching in seminars. There is probably quite a lot of similarity.

ACRES U.S.A. Dr. Reams has terminology he uses to measure conductivity. Not many people seem to be familiar with the grammar he uses for the subject. In some cases, this has been confusing—for instance, Dr. Reams' use of the terms cation and anion seem to be in conflict with standard agronomy.

DR. SKOW. This point is brought up many times. When Dr. Reams originally started to do this work, he had to develop a vocabulary and a set of terms for reference to explain the things he had observed. Perhaps some terms should be changed. But in order to understand and teach, he developed his method, and perhaps in modern language or current thinking, his grammar doesn't correlate with standard soil dictionaries. So you have to look at the program from the standpoint of definitions, which we define explicitly in the classes. I have just finished a booklet on that aspect of it, the objective being to define the terms so that one can understand the methods he uses. This booklet is now in draft form, and has a working title—*The Farmer Wants to Know About Anions and Cations.* It is my introduction to the terminology we use, and it is only to be used for the things that we explain, and cannot be applied to other approaches using the terms anions and cations.

ACRES U.S.A. The Mormons use the terms "saint," but they don't use the term to mean what the old Catholic Church meant by "saint". You are talking about cations and anions, Reams style. What is the first thing you try to analyze when you determine the crop or forage to be grown?

DR. SKOW. The first thing you have to determine when you are planning a fertility program is the crops. For instance, are you growing the crop for seed or are you growing it for the forage or leaf? That is the first thing that has to be determined.

ACRES U.S.A. Why is that?

DR. SKOW. When you grow a crop for seed, the fertilizer program is completely different than for leaf production.

ACRES U.S.A. Let's say we decided to grow for baled hay?

DR. SKOW. If you want to grow for baled hay, then we want to minimize the elements that contain what we call cationic energy, or positively charged elements. We want elements that have electrons on the atoms spinning counterclockwise. In that situation then we have elements with less energy, but they promote growth. The primary element defined as anionic—which I know current thinking has classified as cationic—is calcium. In a test tube it may be cationic, but in the soil it responds differently. In hay production, the primary element you want to work with in the soil—and make available so that maximum growth can be achieved—is calcium.

Dr. Skow explains how energy content of a chemical is computed in The Biological Theory of Ionization:

One great advantage to The Biological Theory of Ionization is that it provides a means for one to calculate the energy content of any chemical compound. This allows one to determine the relative value of various fertilizers, the present situation of one's soil, the amount of fertilizer needed, and the estimated yield from a given crop. One first needs to know how and where to obtain the energy values for the various elements. One starts with the *Periodic Table of the Elements*. Each element is given what is called an atomic weight. This is written, most generally, in the upper right hand corner of each element. These values were arrived at based upon carbon, which has an atomic weight of 12. In *The Biological Theory of Ionization*, the atomic weight is the key to energy calculations for it determines the number of electrons in the shell or orbit of the element. These electrons can be either anions or cations depending upon the element. Both of these last two statements are contrary to orthodox chemical definition, but one has an extremely difficult time calculating energy under the orthodox definitions. As mentioned above, anionic elements have a cationic core and an anionic electron shell; cationic elements have an anionic core and a cationic electron shell. The anionic elements are calcium, potassium, chlorine, sodium, and sometimes oxygen, hydrogen, and nitrogen. All other elements are cationic. Since anions vary in millhouse units from 1 to 499 and cations vary in millhouse units from 500 to 999, it is obvious that not all calciums are the same, or all potassiums are the same, or all water is the same, speaking from an energy basis.

The variance, in the millhouse units of energy, accounts for the theory of valence in orthodox chemical terms. Elements may lose or gain varying millhouse units of energy giving the impression of either losing or gaining an electron. If an element loses or gains an electron, then it is no longer that element. For instance, if calcium lost an electron, it would be potassium and if it lost two it would be a new element not yet discovered.

— *Acres U.S.A. magazine (published in 1984)*

ACRES U.S.A. Isn't this true with almost any crop?

DR. SKOW. Yes. You have to have a base of calcium for any crop you want to grow. Some of the indications of insufficient calcium in the soil are hollow stem crops, especially in hay crops where there would be hollow stems. If there is sufficient calcium in relation to the other elements, then there will be a solid stemmed stalk, completely so with small grains as well as hay crops. When you are growing a hay crop, it becomes necessary to deal with this from the nutrient element standpoint. The ratio of phosphorus to potassium must be kept in a 4:1 ratio. In other words, four parts of phosphate to one part of potassium are a requirement for maintenance of that ratio to get optimum growth.

ACRES U.S.A. Yet maintaining this ratio is not the same thing as dumping so-and-so much phosphate or potassium on the soil?

DR. SKOW. Absolutely not.

ACRES U.S.A. What governs whether you can achieve or maintain this ratio?

DR. SKOW. This is determined by using a soil test. We use the LaMotte soil testing system from LaMotte Chemical Company in Massachusetts. It is simple yet quite adequate for our purposes.

ACRES U.S.A. You can do this in the kitchen, can't you?

DR. SKOW. That is correct. The general consensus is that the margin for error is greater with a small kitchen test than with big fancy laboratory procedures, but at a practical level we question whether there is much difference. We talk about margin of error and this type of thing, but the science of soil testing has a long way to go. We don't really know how a plant takes nutrient from the soil. When we use acids and extraction solutions that are not normal with the soil, it is not easy to determine exactly what really is available to a plant even if it is in the soil.

ACRES U.S.A. Well, we'll have to agree that the plant doesn't go down to the drug store and buy a reagent so it can go about the business of extracting nutrients and growing.

DR. SKOW. That is right. So, at best, any method of soil testing is only as good as the man who works with it and interprets it in a local area on an ongoing basis with years of experience.

ACRES U.S.A. What is the role, then, of the microorganisms in the Dr. Reams' view?

DR. SKOW. The microorganisms are extremely important, especially soil bacteria and some fungi in particular. This is probably one of the biggest single problems we have in American agriculture today. We have done about everything we can to reduce their numbers or alter their species to the point where they are not producing the nutrients in a highly available form for plant growth.

ACRES U.S.A. What has been most responsible for this, in your opinion?

DR. SKOW. This is kind of a bold statement, but as far as I am personally concerned—and I think this is becoming more and more evident—there is one product on the American market that has probably done as much damage to crop production over a long term: nitrogen.

ACRES U.S.A. Don't they have to use nitrogen in some cases?

DR. SKOW. You can live with nitrogen a little easier if you don't put so much on at one time. But if you use products like anhydrous ammonia, you can achieve the same dismal results, although they won't be quite as bad as with the chlorine. The thing with nitrogen is this: If people would take it and make it into an aqueous ammonia and add something like molasses or a liquid humate to it and then apply it, it would work very nicely.

ACRES U.S.A. Is this because the mixture supplies carbon?

DR. SKOW. Yes. If you would attach that product to a carbon of some kind—it would work beautifully. What some of us are doing now is, this: we run anhydrous through water and make what we call aqueous ammonia, or we buy a liquid nitrogen that way and then mix it with a carbon source. Most of the time it is simplest to buy molasses. Some of the liquid humates on the market would work very nicely. Also, and sometimes just plain table sugar can be used.

ACRES U.S.A. If you could find a boxcar of contaminated or spoiled table sugar you would be home free, wouldn't you?

DR. SKOW. That is right. It would work very well. Never put on more though than about 12 pounds absolute maximum of white table sugar per acre, preferably 6 pounds.

ACRES U.S.A. Which will give you a very readily available form of carbon?

DR. SKOW. That is correct. With molasses, never apply more than 3 gallons per year per acre.

ACRES U.S.A. What kind of a calcium level are you targeting?

DR. SKOW. It again depends on the crop. In hay production one has to be careful. We like to gradually work from an existing level—using the LaMotte method of testing—up to somewhere in the area of 4,000 to 8,000 pounds of available calcium. We don't take into consideration the cationic exchange capacity commonly used by most laboratories. What we use is a standard conductivity meter. It is available from most instrument houses.

ACRES U.S.A. You want to know the conductivity of the soil. That is what you are more interested in?

DR. SKOW. Yes. We want a complete soil analysis on a regular basis. We want good records kept. We want conductivity readings frequently during the growing season to see whether or not we are getting an exchange.

ACRES U.S.A. It appears that you're proceeding as they do in solid-state physics, using the right amounts of the right materials to get the circuit back into operation.

DR. SKOW. Yes. You have to check this on an ongoing basis. Correcting it with light foliar applications if it runs too low is the best way to remedy a deficit.

ACRES U.S.A. We realize trying to go into bounds of the entire system on a tape is probably impossible. But can we summarize what you are doing? You are testing the conductivity of the soil. When it starts to falter at any point during the growing season, and until the fruiting time arrives—the part the farmer sells—you move in with a foliar application, or even with a fertilizer of some sort. Is that what you are doing?

DR. SKOW. Yes. I would like to clarify what affects the conductivity reading. This is just a rule of thumb we go by. The lower the humus content in the soil, the more unstable the soil. We define humus as the fraction of the soil that contains carbon that no longer can be identified in its original physical form. For instance, if you formerly had corn in that soil, and the cornstalks were there and you could still visibly identify that cornstalk material, then we classify the residue as an organic matter source. When it goes beyond that point, then we classify the carbon-containing material as humus. That is the fraction of the soil that contains your organic acids. It is the fraction that can release elements in the presence of water off your clay colloid. That might be defined by, or termed, a form of cationic exchange. The lower that humus content is, the more unstable the soil is to weather conditions—for example, rains, heat and wind. We define the energy in the soil by use of the conductivity meter. So it becomes extremely important to do everything humanly possible to improve aeration and tilth and humus content of that soil. We need much less foliar applications or minute applications of different fertilizers to maintain a stable soil environment if we have a good humus level. So I can summarize. As the humus of the soil and its organic matter parent are wasted away, you have a more unstable vehicle with which to operate. In fact, you are pursuing what I call crisis farming from one day to the next. As that humus and organic matter drop down, and your electrical conductivity across the surface of the land becomes unstable, you have erratic weather conditions if enough acres are involved in similar procedures.

ACRES U.S.A. Do you use seaweed extract?

DR. SKOW. You can use seaweed extract if you have a good quality available that works and is quite soluble. But we have discovered that our cheaper formula works scientifically. The reason we use this concept is that this combination supplies all of the basic plant needs in highly available form through the leaf. The reason we use fish probably over the seaweed—although seaweed does work well—is because there is some small amount of oil in the emulsion, and when you put that into the spray tank and agitate it well, it will draw all the molecules of ammonia in phosphate and surround them so that they won't evaporate as rapidly into the air when the spray leaves the nozzle.

ACRES U.S.A. When do you recommend doing this spraying?

DR. SKOW. We recommend spraying at anytime the temperature is above 85 degrees and the humidity is down to zero. The optimum time to spray is if there is a slight dew on the plant, preferably in the morning between 4 and 8 a.m. The other time is in the evening from 7 p.m. on. We have used this formulation with an ordinary field sprayer and a mist nozzle, always getting fairly good results. We do prefer a mist blower-type sprayer, but if a farmer doesn't have one, just a regular field sprayer will make a fair improvement.

ACRES U.S.A. Under the system of biological ionization, are you treating field crops? We have talked about growing alfalfa. You mentioned that fertilization would have to be different for seed crop. What do you do differently?

DR. SKOW. If you raise a seed crop, then you change your phosphorus to potassium ratio to two-parts phosphate to one-part potassium. Instead of having a 4:1 ratio for growth, we switch to a 2:1 ratio.

ACRES U.S.A. This to force more potassium into the seed?

DR. SKOW. Yes. You need potassium for seed production. I would like to clear up one thing on this business of using a lot of potassium to grow a good quality alfalfa hay. There is no denying the fact that it does help grow a crop temporarily, but it limits the height of the growth, and you don't get the volume or tonnage per acre. We have a number of farms in Indiana and in Wisconsin that are harvesting 4 tons per acre on three to four cuttings each year. This is the principle we are working on, and we are finding that it works very well. But if there has been a lot of potassium, particularly muriate of potash used in the past on the acre of land, it is difficult without a lot of good supervision to make the transition from that method over to the method that we outline.

ACRES U.S.A. Do you ever deal with fruit groves?

DR. SKOW. Yes. I haven't personally dealt with them nearly as much as has Dr. Reams. That is where he did a lot of his original work and the basic principles used still apply.

ACRES U.S.A. But you still have to maintain the calcium levels, the phosphate levels, the ratio of phosphate to potassium.

DR. SKOW. In fruit groves the ratio of phosphate to potassium is still 2:1, just as with seed crops.

ACRES U.S.A. That is quite different again from the field crop, where you want the 4:1 ratio?

DR. SKOW. The 4:1 ratio—I want strictly for leaf crops. The minute you go to seed crops like corn, soybeans, then you go to 2:1 again. And that holds for apples, peaches or oranges. One of the things I would like to clarify is that when you grow a crop, phosphate is a catalyst. It recycles. Over a long-term program, you end up adding more potassium to the soil than you do phosphate. The reason is simple. Soybeans have a high requirement for potassium. You still want to try to maintain a 2:1 ratio of phosphate to potassium, but the phosphate really recycles in the plant and in the soil, so the actual amount of phosphate removed from an acre each year isn't as great as the amount of potassium removed from an acre each year.

ACRES U.S.A. In what form do you like to put down your phosphate?

DR. SKOW. There are several forms to build the base phosphate in the soil. We prefer to use a product called soft rock phosphate.

ACRES U.S.A. You want it in the rock? You don't want it acid treated?

DR. SKOW. No, we do not want acid treated phosphate. We do use acid treated phosphate to release energy in a procedure I call topdressing. We make use of minute amounts because they do release energy. But if you put too much, they do just the opposite.

ACRES U.S.A. Would it be correct to go back to Dr. Phil Callahan's language on that. Your small amounts to repair the conductivity problem really deal with the soil in a homeopathic way?

DR. SKOW. That is correct.

ACRES U.S.A. Then in fact you are doping the soil to get that flow going, but you don't hit it with a sledgehammer?

DR. SKOW. I would agree with that.

ACRES U.S.A. Now Dr. Skow, you've grown these crops. You've followed a system that we can't detail here 100 percent because you spend week-long seminars going through all of the points involved. But you recognize that you're dealing with weak energy forces?

DR. SKOW. That is correct.

ACRES U.S.A. Dr. Skow, how do you determine the merit of fertilizers—in terms of quality production and in terms of crops capable of protecting themselves from insects, bacterial and fungal crop destroyers?

DR. SKOW. There are a couple of things I would like to mention in connection with the use of nitrogen and particularly fertilizers in general. It is very important to know the source of fertilizers, what source or form it is in. For example, nitrogen comes in two forms in the commercial market. One of them is ammonial form, and there is the nitrate form. The nitrate form—if you get an excess of that on the soil, you get nothing but plant growth, and it is conceivable that the crop will not set seed if you are looking to raise a seed crop. If you get an excess of nitrate nitrogen and a very high calcium level in the soil, often you run into conditions where there will be a reduced seed set or no seed produced at all. One of the common examples we present in the seminars is that we find a lot of people who raise tomatoes, for instance, they get nothing but growth. The blossoms may start to set and then drop off. When this happens, there is an insufficient amount of ammonial nitrogen in the soil. As a result of this, you just get excess growth and no tomatoes. This also can happen with corn and soybeans, whereby there is a reduced yield—in fact a drastically reduced yield—because of this factor. So it becomes important to know how to use nitrogen and fertilizers properly. If we have adequate humus in the soil, nature by temperature as the season warms up will allow the production of ammonia in the soil. The microorganisms will take care of this along with heat through the normal growing season. But since we have reduced the humus and basic carbons in the soil to the point that they no longer can function properly, and the microorganisms cannot do their job, we have a serious problem. It is affecting our yields and it is very difficult to maintain a stable level of ammonial nitrogen during the growing season.

ACRES U.S.A. What form do you like for your potassium?

DR. SKOW. There are several forms available. One of them is potassium nitrate. It is a potash that comes from Chile, and it is a very good source.

ACRES U.S.A. Is that still available?

DR. SKOW. Yes. It is available in a number of places again. There is what we call Arcadia nitrate of potash. It is made in America and contains chlorides. They use sodium nitrate and muriate of potash to make that, and that is one you do not want to purchase. That is why it is important to know your sources and who you are purchasing your products from. Another source of potassium is available in liquid form such as potassium hydroxide. A very good source. Another product that they use in the eastern states is called "greensand," a 6 percent potassium. Another source is potassium, sulphate. That is a 50 percent potassium in sulphate form—a very good product. Other sources that are very good are wood ashes, nut hulls, straws, compost and materials like that.

ACRES U.S.A. One thing you haven't touched on and for which Dr. Reams is somewhat famous, or at least alone in recommending, is the use of cage droppings.

DR. SKOW. He is very strong on the use of chicken manures, the actual cage manure. We are finding that this has been a real problem, and he is not quite as strong on cage droppings as he was because of the fact that soils we are into are already suffering salt build up. Since there isn't sufficient carbon there for the bacteria in the chicken manure to consume, it hasn't worked quite as well as it did for him at one time.

ACRES U.S.A. Wouldn't that work better in warmer areas anyway?

DR. SKOW. I think so. We are finding what we are doing that seems to work just as well—and this is dependent on getting our ratios right—is to use some form of compost if it is economically feasible. The thing that is working nearly as well for us is a mixture of 28 or 32 percent nitrogen, whatever you can purchase in an area. We mix that with molasses or liquid humates. That is currently our substitute.

ACRES U.S.A. In terms of systems that you use, what is the long-range effect?

DR. SKOW. The long-range effect is that we are trying to only add products to the soil that in the long run will improve the soils, forestall erosion, and so on. The big thing we are trying to do is bring back the electrical conductivity across the surface of the land, and the way you do that is to improve the availability of calcium, phosphorus and your carbons. For the future we have the use of an electronic scanner, whereby you can take soil samples and pictures of the area of the field and by scanning you can determine which fertilizer will be the best and the amount to be put on per acre.

ACRES U.S.A. Will this be something like infrared photography?

DR. SKOW. It is a form of that. It uses a different method. The principles are similar. In fact I am doing extensive research on this right now and finding there is a very nice correlation and excellent response by doing this. What we are looking at is a procedure in which we never put something on the soil that doesn't benefit it. For example, on alfalfa if we want to add a top-dressing after you cut, we take a soil sample from the field and a Polaroid picture, and then we take the different fertilizer products that are available in the area and we match them with the soil sample and picture. It is not difficult to pick out, which ones will increase the energy in that field. We have done a tremendous job on a field of oats up to now, and it is about ready for harvest.

ACRES U.S.A. Have you run across people who are trying to give that energy flow an assist with outside systems, such as towers of power, and so on?

DR. SKOW. Yes, I have a few comments on that. I think there is some benefit to them if they completely understand how to use these systems. It reminds me of when I started with the biological ionization method. It is difficult to get people to completely understand how to manage the systems. We are getting more and more people all the time, and gradually we are all learning and improving our capabilities.

ACRES U.S.A. Are you using any of the biological products in any part of your system?

DR. SKOW. Yes. The humates would be one of them I would consider. I understand their misuse as well as their use. Many products work very well and seem to have a fairly good response in soils that are toxic and high in basic salts. They are helpful in buffering. But if the soil already has a fair humus level there, they may be superfluous. Some of your fish fertilizers as a foliar application are very nice to use. They are also valuable on interim basis in loosening the hardpan, or getting permeability in the soil. In my opinion they could be very beneficial. I think the one final thing I would like to say is, I can't emphasize enough that the farmer keep records. I emphasize this tremendously in seminars. They need to know the volume taken off a field every year. They need an accurate record of what they put on each field each year. They need a consistent conductivity reading during the growing season, and they need to record this. I feel they need to take sugar readings with a refractometer during the growing season for their own information and they need to keep track when they apply something, whether there was a response or not for that individual farm. You cannot take something five or 10 miles away and say, "Okay, this guy got a tremendous response by putting this product on, so I am going to use it too!" It may not be the thing for you to use on your farm.

ACRES U.S.A. Do you always take your refractometer readings when there is juice in the plant? Is there a way of doing it on, say, corn seed that has been harvested?

DR. SKOW. Yes, there are ways to do it on dry seed by grinding the dry seed into a flour of real fine grind, and then weighing out a measured amount of seed and a measured amount of distilled water, and then taking your reading that way and convert it to yield brix of sugar.

"When I Met Bill Albrecht"

An Interview with Neal Kinsey

Originally Published: July 1989

Truly, Neal Kinsey is still an essential figure in the modern eco-agriculture movement for his role to take what we can learn in the classroom and apply it to large-scale commercial farming. When he first started, he was traveling the world trying to find a soil health expert to help him start a business in the early 1970s, and discovered Dr. William A. Albrecht around the same time as Acres U.S.A. founder Charles Walters. Albrecht's teachings were fundamental to Mr. Kinsey, who went on to carry on his legacy and become the most pre-eminent expert on managing Cation Exchange Capacity in the soil. Kinsey, the author of *Hands-On Agronomy*, has helped transition millions of acres to organic or sustainable farming methods in his lifetime.

ACRES U.S.A. Mr. Kinsey, how did your clients fare during the summer of 1988 when so many people were being burned out by the drought'?

KINSEY. In terms of my clients, 1988 was one of the best years we have ever had. It's not necessarily that they made the most yield, but in terms of the way the prices went in 1988, and in comparison to what they thought they might do, last year was a very good year.

ACRES U.S.A. What steps did you take to protect your clients? We saw many burned out. Some got 10 bushels of corn an acre, and their soybeans were literally annihilated in the field and we know there were spotty rain situations, but let's take a dry area. Your clients did well?

KINSEY. Yes. First of all, there is nothing we can do that will take the place of rain altogether, but there are some steps that can be taken that will help us take advantage of whatever moisture there is. I contract with farmers to come in and do soil analysis work. Using that soil analysis, we try to build up fertility levels, which are the very first defense against dry weather. If we don't have a soil audit to go over with the farmer, then we really don't have a starting point.

ACRES U.S.A. One of the problems that has been pointed out by the biological side of agriculture is that if you are to use salt fertilizers to any extent and not very judicious, you better be sure that you've got the moisture to go with it. This year we certainly didn't have the moisture. What is your position on rescue chemistry? Do you crutch with it once in a while, or do you have clients who don't use them at all?

KINSEY. We use chemistry and physics in terms of the structure of the soil to help the farmer. We have farmers who are strictly organic all the way. And we serve farmers who are commercial in every respect.

ACRES U.S.A. How did your organic people make out in 1988 compared to some of the others?

KINSEY. It's always a difficult thing to make a generalized statement. I had one man who called me just as his corn was tasseling and he said, "You know, Neal, I have trouble understanding what's happening. You can drive down the road and you can look at some fields of corn and they look like they are just dead. And others look like they are surviving tolerably well. In my own case, I've had 2 inches of rain since January and my corn is tasseling today and we have received another 3/4th of an inch. My corn has stood all the way through this, and I would never have believed that we could make it with two inches of rain from January until the corn tasseled." I just talked with him the other day and he said, "Well, I really didn't have a bumper crop year." This man sometimes makes 160 to 170 bushels of corn. In 1988, he made anywhere from 70 to 100 bushels, but he certainly didn't have a flop like some of the fellows in his neighborhood with 50 bushels or less.

ACRES U.S.A. How long had this farmer been practicing biological agriculture?

KINSEY. When I started with him I stressed that the very most important things were calcium and magnesium at the proper levels. He was getting what a lot of farmers were missing in terms of the right air and water space in his soil as well as the minerals and humus there. I have worked with him personally for about five years now.

ACRES U.S.A. Five years—so he had his soil system in pretty good shape?

KINSEY. He had a few fields that were not exactly where we would want them to be, but he had basically 65 to 70 percent calcium and his magnesium levels ran around 11, 12, 13 percent, which on an upland type soil is absolutely ideal.

ACRES U.S.A. Well, Neal, when you get this audit, are you following the basic Albrecht formulas of 70 percent calcium base saturation?

KINSEY. In actual effect, when I was studying under Dr. Albrecht in the early 1970s, what he was teaching at that time was 65 and 70 percent calcium and between 10 and 15 percent magnesium. This was the ideal calcium-magnesium ratio. It would depend on the laboratory that you used and his numbers may have varied depending on how they were running the analysis. For example, we have some laboratories that show the calcium at 68 percent and another analysis will show calcium at 80 percent on the very same soil. It all depends on how they are running the test.

ACRES U.S.A. These are all approved procedures so you have to literally stay with the same laboratory, don't you, to get some consistency?

KINSEY. That's right. As a consultant I have to stay with the same laboratory, and that's one that uses the methods that I was trained to understand and interpret for the farmer. If I get off on someone else's soil test, it could be that the soil test is just as reputable, just as good, run with absolutely modern equipment, but since I wouldn't understand how to interpret it, I could mislead the farmer.

ACRES U.S.A. Is it a good idea to have the laboratory interpret the soil test, or should you do this out where you are working intimately with the field itself?

KINSEY. I tell the farmer he has to have someone interpret that test, someone who can understand what the numbers are saying and who can keep in contact with what that farmer is doing. To me, logically, it's the man in the field. I also stress to the farmer that he has to work with the person who understands what those numbers mean, whether he's out in the field or not. In other words, if they don't have any representative out in the field and that lab is making some recommendations, they'll probably know more about their numbers than I would. We encourage our farmers to give us a test. Don't tell us which areas of his farm are the best, but after we go out and pull the soil test, we sit down with the farmer. If he keeps good records on his soils and he knows how they have been performing—and most farmers, especially the biological farmers can do this— we can tell him which areas of his farm are doing the best, which are average and which are poor just from that sail audit.

ACRES U.S.A. Neal, we have all heard horror stories of people who sent the same soil to several laboratories and one said, "Add so and so many tons of lime," and the other one said, "You shouldn't add any lime at all; you should do something else." Farmers almost get to the point where they wonder whether any of them really know what they are doing?

KINSEY. I've even experienced that with some of my own clients. I recall a man and his son who hired me to come and analyze their soils. After the analysis was done, I told them what they should do in terms of fertility for their next crop. The father looked at me and said, "Well, Neal, I hope you won't get upset at us, but the only reason we had you run the soil tests was that we wanted to see where we are now because we have a product that is suppose to release the phosphate and potash in the soils." Once you start using it, then you eliminate your need for fertilizers. It makes it available from what is locked up in the soil. They were getting ready to put 4 tons of high calcium limestone per acre based on the soil tests they were using. When we pulled the test and sat down with them and went over the data, on some areas they need a ton, on some areas they needed 2 tons and on some areas they absolutely didn't need any limestone at all. If they had put 4 tons on, they would have caused themselves a lot of problems, especially in a drought year. First of all, it will be three years before you see the full effect of putting on high calcium lime. But that calcium would have loosened the soil and the soil would actually have dried out much more easily with an extremely high calcium level in relation to the magnesium level.

ACRES U.S.A. The late C.J. Fenzau was always conscious of the fact that there is such a thing as overliming and there is certainly such a thing as not treating even one field exactly the same way from end to end. We used to take our soil audits according to field pattern, according to weed pattern, to identify the various shadings of treatment that any certain area would take. Is this a procedure that you like?

KINSEY. It certainly is. We were doing some samples today. This particular land was in the government set aside program in 1988. You could see every place that they had even run a tractor or a vehicle through the field because it packed down the soil and certain weeds came up thick. Weeds are an indication of what's right and what's wrong with the soil. If you compact an area to a point where the land can't breathe, then you have trapped carbon dioxide and all kinds of unhappy things happen. We emphasize to our farmers that you have to take a good representative sample. If there are weed patterns, if there are growth patterns to the crop, if there are differences even in the color of the soil—heavy red, light yellow, black or a good dark brown—these things must be recognized. I've had farmers go with me to sample their farms. One man was complaining to me at a meeting how these soil labs were terrible because you could never get anything consistent: After we pulled samples for a half day, he said, "Neal, you know I realize now that it wasn't the laboratory's fault. It was mine because I put the heavy black soils in with the light soils because they said to get an average of the field. I took some of every soil and put it all together so that I would have an average."

ACRES U.S.A. Mapping out the farm and mapping out the fields is a pretty good procedure, and identifying samples according to variations that you find is also a very good procedure?

KINSEY. Yes, and we tell our clients when it's sampled properly, the analysis will reflect each area. If you have a problem area, then as long as you keep on continuing to treat it just like the rest, it's going to remain a problem area. Problem areas have to be sorted out and specifically treated for whatever that problem is.

ACRES U.S.A. What kind of a laboratory do you like to use in getting a readout so that you know how to plan for the next year's crop?

KINSEY. We use a laboratory, first of all the man who runs the lab understands how to run the tests the same way I understand how to interpret them—and that is based on what Dr. Albrecht taught—and secondly that test needs to reflect the clay content and the humus content of the soil as indexed in the exchange capacity. We have to consider the pH. But we also like to read sulfate, phosphate, calcium, magnesium, potassium, sodium—the base saturations of the cations plus hydrogen. In addition, we measure manganese, iron, copper, zinc and boron.

ACRES U.S.A. You mentioned Dr. William Albrecht. What was so different and so unique about this man, in your opinion?

KINSEY. When I met Bill Albrecht, he was already professor emeritus of soils at the University of Missouri. The only reason I had the privilege of getting to know him was that my office was on the second floor of Mumford Hall and his was on the first floor. A friend of mine asked me if I had ever visited with him. I made it a point to go down and talk with him because he actually understood! I lived in southeast Missouri. He understood what the problems of our soils were, and he started describing those soils to me. He could describe the soils of Arkansas and Texas. He was just an excellent student of the soil and when I say student, a lot of people learn by reading books and so forth. Albrecht taught as a student. He would say, "Study nature, not books." That wasn't just a saying with him because he would actually be out in the fields working with what was happening and working not just with the plants, but also feeding what was produced to the livestock for several generations to see what the outcome of the fertility was.

ACRES U.S.A. He was a champion of the biological test. In other words, what does this crop do when you run it through the animal'? Never mind what the lab or the chemist says, what does the animal have to say? Albrecht also went to Australia, Russia and Europe and he made the scene! In other words, he knew soils, their geological construction from day one, irregardless of what part of the world it was in.

KINSEY. That's correct. During the last few years, I have been doing some work in Europe. When you go over there and mention the name of William Albrecht, the professors still know who he is. There are a number of professors who still use his works. When one of my clients over there was showing a professor of soils what we were doing with his farm, after about 15 minutes, the professor looked at him and said, "Yes, what Mr. Kinsey is telling you is exactly what we still teach our students."

ACRES U.S.A. Bill Albrecht not only visited these places, he lectured there and the scientists took in what he had to say. When we started Acres U.S.A., contact was made with him, guess how? We were reading a magazine published in South Africa, and the article was telling about this great man at the University of Missouri who no one around here knew much about. Acres U.S.A. called down to the University of Missouri. They told us he was too hard of hearing to be interviewed. So with that we swung our car down 170 and made the scene. He welcomed this editor with the usual statement, "Enter without knocking and leave the same way." Have you ever heard that?

KINSEY. Yes, I have, many times.

ACRES U.S.A. In any case, we are together on his basic premise that calcium was the king of nutrients? Albrecht's next premise was that your calcium, magnesium, sodium and potassium had to be in an equilibrium for crop production?

KINSEY. Yes, and that's why we have base saturations of each reflected on the soil test. It does matter what order those are in, in order to produce crops.

ACRES U.S.A. At one time, Albrecht helped the Brookside Laboratories as a consultant and also designing their program. Are you familiar with that?

KINSEY. Yes, I am familiar with it because after leaving the University of Missouri, I actually was able to study under Albrecht through the facilities of Brookside. I became a certified agronomist through the program that he taught for Brookside Laboratory.

ACRES U.S.A. I would tell my clients that one of the very most important things you can do is to use fall tillage if you have a compaction layer. Neal, let's take a typical farmer you are advising, and he didn't get any rain in 1988. What is the first order of business?

KINSEY. The first order of business is to try to tell that farmer what we want to build for him. We want to give him the textbook ideal of an excellent soil. That is 50 percent minerals and humus, 25 percent air space and 25 percent water. The question comes up, "Well, how does the farmer know when he has it?" The only way to tell is by measuring and correcting the amount of calcium and the amount of magnesium in that soil. All soils need to be adequately limed, but too much lime or calcium and/or magnesium from a natural situation can be harmful to the crops or the soil. I explain to the farmer that calcium loosens the soil. It puts more space into a soil and that's good up to a point. Magnesium, on the other hand, is more or less the thing that holds the soil together and makes it tighter. If you get too much of it, then it's going to tighten that soil up to the point that you'll either not have enough air or water, and in the heavy clay soils there will not be enough air. The very first rule that we have for a farmer is to favorably influence the cation levels in the soil by putting on the correct amount of the right type of limestone, whether calcium carbonate or magnesium carbonate. That holds true basically for the East, the South or for the Midwest.

ACRES U.S.A. We have a communique from the University of Missouri that advises farmers not to use fall tillage at the end of a drought year. They should wait until spring to accomplish their tillage. What do you think?

KINSEY. First of all, I would tell my clients that one of the very most important things you can do is to use fall tillage if you have a compaction layer. You must eliminate that compaction layer and that's one of the reasons we have a farmer to do it in the fall. If you go in and try to work your soil deep in the spring, it creates so many air pockets that you are not going to have it settle out enough. As a result, you are going to lose moisture even faster.

ACRES U.S.A. So this advice is something that you are going against at this time?

KINSEY. In relation to how we work a program, yes. After we work with a farmer—and let's say we have his calcium-magnesium ratios 65 to 70 percent calcium, and 10 to 25 percent magnesium—the very next thing we concentrate on is to get rid of any compaction layer not just as a means of getting the root down for fertility but also as a means of getting that root down to the available moisture. This can be accomplished via several means. One that everyone tends to think of is mechanical. If you are going to deep rip, well then we tell the farmer to do that in the fall and there is a specific set of rules that you have to use in order to make that type of a program work.

ACRES U.S.A. OK, can we go into some of those rules?

KINSEY. First of all we tell the farmer, do it in the fall and do it when your soils are dry enough. Don't go through a wet soil in order to try to eliminate a compaction problem. You are just going to make things worse. We encourage our clients to get an extra man if he has to. Don't wait until everything is harvested before you start thinking of going in to eliminate your compaction. After the crop is harvested, get in there and do it if it's dry enough. The next rule we have to look at beforehand, and we have to look at it on our man's program before we would ever tell him to start ripping—and that is his calcium level should be at least 60 percent base saturation on a medium to a heavy soil. It's a must. Hey look, we need to have the air space and the water in the correct proportions and the calcium-magnesium is the real key to having that ideal soil.

ACRES U.S.A. What's wrong with putting a little calcium down real deep for root return and for flocculation?

KINSEY. If you are talking about getting a little bit down there, we want that calcium to penetrate as deeply as possible because we want a good mix all the way through. That's one of the reasons for advocating that we rip, and there are a lot of people who think that calcium won't be carried down through the soil. But if you get it open, the limestone that we put on top will actually be carried through and will saturate that soil pretty deep with the amount of calcium that is needed. On the other hand, in terms of research that I've seen, you concentrate on correcting the soil as deep as the fencepost rots. This is something that I have seen in Acres U.S.A. It's something that I have seen in publications that come out of Germany and it's something that Dr. Albrecht taught. It is just a common sense principle because that's where the microbial activity is. If you can get that top six to seven inches of soil into a fertile balance, then even if you have a toxicity problem underneath, those tap roots will penetrate that area and it won't be a problem.

ACRES U.S.A. Many farmers have the idea that they can deal with a compaction problem with iron and machinery alone. But you seem to say that if you want to have a soil with good tilth, you must get back to where your microbial livestock in the soil does its work?

KINSEY. Yes, you must have the mineral balance there to feed the microbial activity, and that's another reason why we stress to the farmer, Hey look, we need to have the air space and the water in the correct proportions and the calcium-magnesium is the real key to having that ideal soil. Once you have that ideal soil there, it isn't just a matter that we have the ideal mineral balance, it also has the ideal physical structure and it also provides the ideal environment for the microbiology to really take hold and prosper and function.

ACRES U.S.A. So you are going to have to manage some water, some air in the soil, and going down to the elevator and getting 17-17-17 or whatever they are selling this week is probably the last consideration. Is that right?

KINSEY. Absolutely. We have actually sampled farms and found that the farmer was using too much fertilizer. This is hard for most farmers to really grasp. But they are putting on so much of the regular fertilizers they are tying up some of the micronutrients. Instead of helping their yields, they have, in fact, lowered their yields from overuse of fertilizer It isn't necessarily just one or the another. For example, one thing that has been said for years is that if you overuse phosphate it will tie up zinc. If you overuse potassium, it will tie up boron and manganese. If you overuse nitrogen, it will tie up copper. In fact, if you overuse phosphorus, it can tie up copper or zinc. If you overuse calcium, it can tie up boron, magnesium, potassium, iron, manganese, copper or zinc. We can overdo any one of these things and cause a problem. On the other hand, some farmers haven't gotten to the basics. They have extreme deficiencies. No matter what they try to do in terms of trace elements, it is never enough because if they don't correct their limestone, trace elements aren't really going to be the answer.

ACRES U.S.A. So operating by the seat of your pants is out. It's a one-way ticket to bankruptcy?

KINSEY. You can't farm under the conditions we have today by the seat of your pants even if you have somewhat ideal weather.

ACRES U.S.A. Back to tillage, water management, auditing the soil and finding out what the situation is, getting the right materials and using them judiciously, there is one school of thought that does not brook any use of commercial N-P-K fertilizers what-soever. But you seem to say we have to use these materials judiciously to fit into the pattern that the soil audit is telling us exists. Is that a fair statement?

KINSEY. That's correct.

ACRES U.S.A. To pursue this a little further, the farmer should have an idea as to resources available, how things come together. How far do you dig into that? Do you help the farmer with his management problems or are you simply a soil man and let the farmer worry about finances and that sort of thing?

KINSEY. I don't go into a farming operation and say, "Now you have to tell me what your budget is, and so forth." But we can put his priority needs into a first, second and third order. This way he can know what to do next, and when his fertilizer budget is used up, that's where he stops. We don't want the farmer to overspend. We want him to use the fertilizer budget he has set aside most judiciously. When that is gone, it's time to stop. Our program is not to go in and tell the farmer that he needs to allocate more money for fertilizer, but to say that taking the money that he has allocated for fertilizer, this is how he needs to go about using it

ACRES U.S.A. Neal, we've been discussing the anatomy of soil audits, what they tell you, how you answer them, how you deal with them in the soil, how it works with water, how it works with managing air in the soil and so on. But you know what the average person says, "Yes, well this is nice theoretical stuff, but how does it work out there in the field?" You told us about one client, but why don't you walk us through a couple of other clients and tell us a little about each operation, how they survived the 1988 drought, and what kind of crops they got.

KINSEY. First of all, we try to work with the farmer and with whomever he deals with for getting fertilizers. We're not really in the business of selling fertilizer. I have one fertilizer dealer whom I've been working with. He recently made the comment, "Neal, your services for my customers would be five times greater if there was no irrigation in this area." He said, "There are so many fellows who cover up a lot of their mistakes with irrigation." In southeast Missouri, it was quite dry in 1988. On 360 acres of soybeans, this farmer averaged 14 bushels per acre where he followed just the normal fertility program. But in trying out the program that we were working with on 80 acres, the first year you are not going to see all the benefits. So here's 80 acres of soybeans as compared with actually another 280. And the fellow was getting close to bloom in his soybeans. He said, "What can I do? My soybeans are really hurting." Remember, on the rest of the farm he averaged 14 bushels of soybeans per acre. One thing I stressed to him was that he should come in just as the beans started to bloom and apply a foliar application of nutrients right on those soybeans. Now to the row, this just happened to work out. When he came into that 80 acres, the yield jumped double, and he was making 28 bushel to the acre. That is what he averaged on that 80 acres.

ACRES U.S.A. You mentioned foliar fertilization. Of course this has been one of our pet antipathies. For 20 years on the West Coast, chiefly in the flower trade, they had developed a perfectly valid system for foliar fertilization of plants without a single university in the United States even becoming aware of it. And it has only been recently, in the wake of the development of biological agriculture that a great many people have found out about foliar fertilization. What kind of foliar fertilization are you using?

KINSEY. I learned about foliar fertilization on the east coast, working with Dr. Jim Childs, now retired. He was senior pathologist for USDA in Florida. I worked with him on citrus, specifically with his experiments on diseases of lemons as related to fertility. I got involved with some citrus people down there and found that they were using foliar application for all types of deficiencies. In fact, that's where I really got my fast-hand experience at working with a lot of the materials that are quite available, whether we're talking about foliar or whether we're talking about soil applied micronutrients. There are a number of different approaches. The one that I take after all the basic nutrients have been applied has several traces put together with a little bit of nitrogen and potassium. This is the one I use on most soybeans to get that bean to take the foliar right in through the leaf. The key there is putting it on at the right time and with enough water to make sure you get enough spread on the leaf.

ACRES U.S.A. Of course directions for use of a product are pretty much the responsibility of the manufacturer because he has to spend the years and the money it takes to work out exactly how to use this particular product?

KINSEY. Yes, I would certainly agree.

ACRES U.S.A. Have you used any of the seaweed foliars?

KINSEY. Some of my clients have used them. I have clients who have used the seaweed foliars and who have been very pleased with them in conjunction with biologicals and without them. I have a fellow I have worked with in central Illinois for quite some time. He has run quite a few experiments with seaweed type products in drought years. In 1983 he had soybeans that made as much as eight or more bushels to the acre.

ACRES U.S.A. So this is basically a refinement technology that's come along. It has developed a long way since Dr. Hanway got those national headlines?

KINSEY. Some people are scared to death of this technology because they thought they had an answer. Some got terrific results one year and then came back the next year with few results. Maybe they happened to hit on a missing nutrient for one particular year. We want to let the soil supply the basic fertility and then use the foliar to bring us through the stress period—a drought period, perhaps. I do not recommend that a foliar be applied if the leaves of a crop are limp. What we have to do is wait until late in the evening and then if the crop will recover enough so that you don't have a limp leaf there, come in the evening particularly if there's a nice dew forecast. This is when we get our best results. You require a certain level of humidity in the air for best foliar application results—in other words, for these to work and pull you through and give you that extra push in a drought situation. The one thing that I try to do is get a good root structure and get those roots on down. Now we have clients that use all different types of methods, but one that seems to be particularly effective—that I've been pleased with in Iowa, Illinois, Missouri— has been to use a soil conditioner. If you get the soil well aerated in the top six or seven inches, then use that biological, it seems to be able to take hold and go ahead. But if you use a biological and the soil is extra tight, then that's where we run into some problems.

ACRES U.S.A. So if you use a biological on the A&P parking lot, you're probably not going to have good results?

KINSEY. Not likely.

ACRES U.S.A. We haven't gotten into weeds very much, and we certainly haven't gotten into insects. But if you have the type of balance that Albrecht was talking about, weeds and insects aren't really that much of a problem, are they?

KINSEY. They're certainly much less of a problem. One of the very first men I ever worked with in Illinois was an organic farmer. He started farming organically in 1963. I didn't begin working with him until 1973. He had a tremendous foxtail problem. He had giant foxtail that was taller than his corn. And he said, "Neal, I would appreciate it if you would help me out with this." He encouraged me in fact to hurry up and finish Albrecht's course so I could go to work with him. And we eliminated his foxtail problem. This man simply would not use herbicides. He would let the weeds come and take him over rather than that, because he was convinced that herbicides were not beneficial to his land. He'd rather have the weeds. In one year's time, his fox tail problem was virtually eliminated using the principles we've discussed. And look at the results. One can conclude that we have been most successful. But perhaps we've gone too far. Of course I'm not advocating that we reverse these good figures. Rather I'm saying that humans have almost become addicted to using drugs and antibiotics as much as farmers have chemicals on crops and low grade drugs in livestock feed.

Another subject that organic farming can handle quite nicely is erosion. Popular wisdom is that the closer to no-till farmers get, the less erosion we will have. Since I live in the no-till capital of the world—Knox Company, Ohio— and am myself surrounded by no-till farmers, I believe that no-till has a definite place, even in the rotations of organic farming. I have heard of southern farmers with the longer growing seasons than I, who no-till beans without chemicals into the corn stubble and then no-till l again without chemicals a hay mix into the bean stubble as a part of a rotation that also includes deep chiseling. But continuous no-till farming—with no chemicals—is sheer biting the hand that feeds us. A good organically farmed soil has this layer of humus and tilth that prevents a lot of erosion simply because it acts as a giant sponge. So this method of farming actually helps the soil breathe better in periods of extensive rain, and helps retain the moisture in the soil by acting as a vapor barrier in the event of a dry period. Organic farming thus becomes an insurance policy against extremes in the weather. Add to this the good techniques of conservation tillage and the deep roots of a legume-sod crop in a rotation, and I suggest that organic farming methods will save more soil than all the soil and water conservation techniques from here on out. It is my understanding that by adding just one year of a deep-rooted sod crop to the typical corn and bean rotation, that one can reduce erosion by a factor of four or five.

Erosion problems have also been compounded greatly because so much of our soil is in row crops. Knox Company, Ohio used to be the county with the largest number of sheep of any county east of the Mississippi. Now many of those sheep pastures on our steeper slopes have been in continuous corn for ten years. It doesn't take a PhD in hydrology to figure out which will hold the most soil on these hills. Add to this the price of lamb vs. the price of corn, and it looks like we ought to go back to sheep for reasons other saving soil. Another thing that organic farming does is to make us consider all our resources. For example, by aerating my manure pile, I get a semi-composted product that is half way to humus when I spread it, and I only have to haul half as much material. I have been able to use tillage in place of chemicals to control weeds in my rotation. It has taken almost ten years, but I have created from scratch, modified, and cannibalized my machinery to make the most of my crop and animal residues to replace the plant nutrients for my future crops. Oh sure, I still have weeds, lots of them. The way I look at weeds is so much different today then when I started farming. Instead of trying to kill foxtail I bale it and find it makes good hay. Sure, I've got to control my weeds enough so they will not reduce the yield of the crop I'm intentionally growing, but elimination is not practical.

"Compost is the Staple"

An Interview with Malcolm Beck

Originally Published: October 1991

Throughout our history, it has been a rare thing for Acres U.S.A. to publish a lead story that isn't answered by readers asking for "more." And often, we find ourselves covering the same issues decades later. Take our cover story in October 1991, which addressed one of the biggest problems facing the U.S. today—how to handle the raw materials called "waste". Sound any different to today's conversations? Space in the landfills is vanishing, and the mountains of organic materials have not.

The report on Malcolm Beck, 1929-2014, in this issue asked more questions than it answered, and many of those questions are still unanswered today. Acres U.S.A. traveled to San Antonio and to the Rio Grande Valley in 1991 to interview him and flesh out this very significant story. The scientific basis for compost was best stated by Ehrenfried Pfeiffer when he told fertilizer producers, in effect, "You need my compost to make the salt fertilizers work," and then he added—under his breath—"but if you have compost, you won't need the salt fertilizers."

In the world of compost and inputs, Beck was a true romantic. His ability to deftly explain what we know, and what we do not, still resonates today. His real impact from his work can be measured by how much more the world knows about composting today than when this interview occurred in 1991. Because of that immensely important progression, we all owe a debt of gratitude to Malcolm Beck.

ACRES U.S.A. Malcolm Beck—your yard out there. What used to be a lush garden now looks like a batch of huge cones of various types of materials, waste materials that would normally go to the landfill. What kind of materials do you have?

BECK. The biggest pile is sawdust from a door manufacturing plant and from a bunch of cabinet shops. It is dusty and dry and needs moisture, so we spread it out and get loads of paunch manure and vegetable wastes, such as onions and jalapeno peppers and brewery wastes. The barley and the sawdust soak up all of the juice in that paunch manure. Then we pick it up, say an equal amount of sawdust with our other three products and put them all in a big pile. Then we add some stable bedding to it, chiefly horse manure, and let it ferment. We push it up in the big monstrous piles you see. We found out that we could keep our carbon nitrogen ratio spread out. Instead of 25- or 30-to-1, we will go 35- or maybe 40-parts carbon and 1-part nitrogen. That kind of slows the activity. With the activity slowed down, you never get a hydrogen sulfate smell, that rotten egg smell. If you have too high of a nitrogen-to-carbon ratio, you would run a compost pile too fast. If you have too much moisture, the compost can't get oxygen. In our Austin-San Antonio area climate, we only turn the com-post pile about once every six weeks to two months and we only turn it four times total before we have finished compost.

ACRES U.S.A. Basically, you have done away with the extra heavy capitalization, and you just use a high loader?

BECK. Yes, just one piece of equipment—that big front-end loader there. That is all it takes to make compost in this area. That and time and space. Of course, before we sell the finished product, we screen it, but that is largely for cosmetic reasons. If we deliver a load of compost and a beer can or a horseshoe rolls out, it looks trashy. That is not what people expect.

ACRES U.S.A. You sell it bagged and in bulk?

BECK. We sell it any way the customer wants it. We deliver it. They pick it up. We sell it in bags. We'll even let them bag it themselves if they want to cut costs.

ACRES U.S.A. As we walked through the yard, we noted one pile that was made up pretty largely of tree chippings. What sort of a procedure does it take to compost these materials?

BECK. First, we let the tree trimmers dump their chippings and shredded materials in the yard, and we also let the landscapers dump their brush. Then we grind it through a chipper. We use a big hay grinder and grind it into a fine mulch. For a hundred years, San Antonio has been importing mulch. Most of this has been hardwood bark or cypress bark or pine bark. At the same time, the city was throwing away millions of yards of these tree trimmings, not knowing what to do with them. They burned it and hunted holes to dump it in. I decided that it should make a good mulch. So we started grinding and cleaning it up and we found out that it is superior mulch. It has the buds in it, the leaves, it has all the carbohydrates and the proteins in it. This is what feeds the plant. If you just use bark, you have a high carbon product and it robs from the plant. Another thing is that these tree trimmings—when they go through that hammer mill— kind of fray up at the ends and becomes kind of fibrous. When you put this material down on a slope or along a bank, it sticks in place. Water doesn't wash it away the way it does pine bark nuggets. Shredded tree trimmings stay in place. All of those fine materials from the cambium layers and the buds, well, they settle to the bottom. Being high protein and full of all the nutrients, they convert into nitrogen and actually feed the plant instead of robbing the plant. So it is a superior product.

ACRES U.S.A. It is superior to maybe getting cedar bark out of the northwest and putting it around your plants and maybe creating a nutritional problem for your plants?

BECK. It accounts for a temporary nutrient tie up because the microorganism have all this carbon to feed on, but they also need nitrogen and other elements, so they rob it from the soil.

ACRES U.S.A. You do some of your own chipping?

BECK. We have two machines. We have one machine, a big chipper shredder that we run the limbs through. This one cuts the material into a sliver up to 12 inches long. We have a modified hay grinder that comes from Colorado. It grinds wood into a beautiful mulch. It is a hammer mill type machine without a screen in it—very efficient. It will grind up to 200 cubic yards per hour.

ACRES U.S.A. Is there any limitation on size?

BECK. Well, we can put more hammers in it and grind to a smaller size, but as a practical matter we keep it one size. The small machine will take up to 12-inch logs. This hay-grinding machine will only take up to two inch diameter sticks. When you get bigger than that, it either doesn't go through it or it just throws it through. It doesn't damage the machine, it just doesn't do anything with it.

ACRES U.S.A. The one that can take the 12 inch logs or better—does it have blades?

BECK. It has blades, sharp knives instead of hammers. It has power rollers to push the wood it into the blade, but if it pushes too fast to the blade and the blade starts to slow down there is a sensing device that stops the rollers and it stops pushing until the RPMs come back up.

ACRES U.S.A. You also had big piles of turkey litter?

BECK. Yes, we get in a lot of turkey litter. It has a base of peanut hulls or rice hulls or wood shavings that they run turkeys on. We use that to make a landscape soil. We mix the turkey litter with topsoil and we let it all compost together. The composting heap pasteurizes the topsoil and knocks out all the seeds and nematodes and things like that. It knocks out everything harmful and allows the beneficial microbes to remain alive. It digests the weed seeds. Then we take that product and we blend it with sand or bark or perlite and make mixes: We have special mixes for azaleas, roses, lawn dressing, and trees and shrubs. We have also developed a mix for gardening and flower beds.

ACRES U.S.A. You did say roses?

BECK. Yes, we have a special mix for roses. That is a very difficult plant to grow, especially if you are going to be in competition at the garden club. So it helps if you start out with a good soil. That is half of it.

ACRES U.S.A. Moving around the yard, there were other piles of material. Will you tell our readers about them?

BECK. Some of them are sand that comes from the little town of Poteet. They have an acid sand down there. That is the strawberry capital of Texas. We use this sand as a loosening agent. The orange sand is also a loosening agent because it has very little clay in it. Nevertheless, it has enough clay in it to hold certain nutrients such as iron and sulfur. The sand has a pH of about 6 or slightly lower. That other really red pile of sand has a high clay content, but the pH is much lower. It runs as low as 4.35 and of course all of our soil around here is highly alkaline, up around pH 8 and 8.5. Also that sand has a lot of iron in it and we are lacking in iron. It has a lot of phosphorous and we are low in phosphorous. So if we take that sand and use it with our soil and turkey litter based compost, it compliments our soil and makes a good mix.

ACRES U.S.A. Have you looked into what Phil Callahan talks about when he cites paramagnetic qualities of soils and parent soil materials?

BECK. Yes, I have read his books and I am playing around with that some. That sand we use is not paramagnetic. It is not out of the bowels of the earth, you might say.

ACRES U.S.A. But the iron in it, if there is any, probably would be paramagnetic?

BECK. It probably would be. They tell me lava rock would be a better paramagnetic rock. We use granite sand and I have a load of lava sand coming in. I have to get a fix on exactly what each material does for the final product.

ACRES U.S.A. How do you get granite dust?

BECK. The granite dust-is from just north of San Antonio, about 60 or 70 miles. It is solid granite and they have a bunch of mines up there, and they mine the granite. Years ago I would go out there and scoop up the dust. It was cheap. Then they found that there was value to that granite sand, and they put a big price on it, and that is fine. There is value to it, and I don't mind paying for it. We used to import our granite sand from Lithonia, Georgia. Then I found out we had granite sand in Texas. I got to looking through some old geology books. The University of Texas once did several analyses on granite deposits north of San Antonio. I compared these data with data from Georgia. It turned out that we had a higher quality sand here in Texas than I was getting out of Georgia.

ACRES U.S.A. Georgia granite, according to Callahan's readings, is highly paramagnetic!

BECK. I'll have to trust him. I don't have any instruments to measure it. In any case, anything that is igneous or that came out of the bowels of the earth would be highly paramagnetic.

ACRES U.S.A. You had other materials in your compost yard, and they don't often fit into the compost scenario. First of all, weren't there some rice hulls?

BECK. There were rice hulls. We compost them with shrimp heads. We call it Cajun compost. This compost is used by home gardeners and landscapers more so than by the farmers around this part of the country. The landscapers didn't take long to ask us to handle other products. They said, "If you handled railroad ties and fertilizers and edging, we will buy them from you." So I become a one-stop shopping center. Compost is the staple. Landscapers also asked me to handle lava rock. Then I found a clay pipe company south of San Antonio that had this big mountain of broken clay pipe and tiles. I got to looking at this stuff and thinking that this would make a beautiful decorative rock.

ACRES U.S.A. This tile, after you crush it up and chip it up small pieces, it is not going to float away when you have a gully washer?

BECK. It stays in place a whole lot better, and if you run over it with an automobile or walk over it, it doesn't crush or flour up. It is a superior product right here at home. Previously it was a landfill product. My philosophy is that we need to recycle these materials. These have economic value. It is really stupid to fill canyons with them. If you study nature, nature recycles everything. There is nothing new coming in on this earth and there is nothing leaving it. Everything has been recycled for millions of years. If we would just change our philosophy and not call these things waste products, rather call them natural resources that need to be used, we would sooner come into tune with nature.

ACRES U.S.A. The only time we do that, of course, is in recycling scrap iron. We have always recycled scrap iron. We treat it economically as a new raw material just as though we had just mined it out of the ground!

BECK. Yes. All products should be treated that way, every piece of paper, plastic, all organic material, broken pipe and everything. It came from the Earth and it can be used again. Why should you have to mine it and concentrate on separating it from all the other debris and minerals when here you have the pure product. If we would just separate it at the curb site.

ACRES U.S.A. Perhaps we ought not to have disposable pop bottles?

BECK. We should have reusable bottles. When I was a kid, I would walk to school or walk to work or something, and I always had a sack with me to pick up soda water- bottles. I think we got a penny a piece for them, and we always had spending money. It made the trip a lot more fun.

ACRES U.S.A. In the process of converting over to providing these gardening materials for golf clubs, municipal parks, lawns and so on, you have had to adjust your thinking and go out of the truck farming business?

BECK. Well, the dollars did that for me. When I started making compost I was getting all the manure from the polo stable and putting it on a big pile. When time and weather permitted, I would put it out in the field. People would see my big pile of manure out there. A few of them were organic gardeners. They said they would like to buy some of it. I said I needed it for my fields to grow vegetables. Finally one day a boy talked me out of about four yards of it. He was an organic gardener and he knew the value of compost. He gave me $40 for four yards. I got to thinking, if I was to take that four yards out into the field, work it in and then prepare a seed bed and plant seed and then cultivate and hoe and irrigate-and harvest and pack, it would take a lot of time to turn that four yards of compost into $40, if indeed it even reached $40. I got to thinking that maybe here was a better way to make money. The compost grew into a big business. I make as much money a week as I did all year farming, and I could do it all from the seat of a tractor. Most all raw materials for this compost operation come from San Antonio. Some of the turkey manure comes from within a 40- to 60-mile range, and some of the stable bedding comes from a larger range, but that is done on a back haul, so I save some transportation there. We compost about 300 cubic yards of material daily. All of these materials would have gone to a landfill or a dump somewhere.

ACRES U.S.A. The tree trimmers, the tree doctors who take limbs out of trees and re-move unwanted trees, they can bring these to your place and not have a dumping fee. Is that the idea?

BECK. That is right. Well, I charge them a little bit for dumping the brush because I have the expense of cleaning their trash out of it and regrinding it. It is still a big savings for them because I am much closer than the dump is and I charge them one third the tipping fee that the dump charges.

ACRES U.S.A. Are gardeners of consequence or truck farms using these products that you are manufacturing?

BECK. I would say that half of the gardeners in San Antonio are now using compost. The gardening scene in San Antonio is quite different from gardening in Kansas City where you have the toxic talk shows. We have Howard Garrett in the Dallas, Fort Worth area and we have John Dromgoole in Austin. He has been in it all his life. He has two shows, two hours each, and Howard Garrett has his shows. I advertise on four radio stations, but my advertisement is really not advertisement. I give gardening tips on the radio. The old county agent that works for me, he too, has turned completely organic and we have him a little radio show in New Braunfels.

ACRES U.S.A. But so far your compost hasn't really reached farmers?

BECK. No, but the farmers are thinking. When I first went into business, when a gardener would come in, I would asked him if he was trying to be organic, and 95 percent said, No, and only about 5 percent said they were. That was 10, 12 or 15 years ago. Now, when they walk in and I ask them if they are trying to stay organic, and they say Yes, if they can I think 100 percent of the gardeners would prefer to be organic.

ACRES U.S.A. Well, after all, they are the people who buy groceries in the store. It will be just a matter of time before they make their wishes felt on this primary producer out here?

BECK. I think that is correct. All the young farmers we are getting are trying to go organic. A lot of these farmers are just weekend farmers. They have moved into the country and have anywhere from two to 20 acres. Usually they are growing a few hundred or a thousand tomato plants. These people, because of the certification program, get- a little bit more money for their produce. They are trying to go organic, and some of them are succeeding.

ACRES U.S.A. Mr. Beck, in addition to compost, you have invented and developed a lot of products that sort of walk hand in hand with the garden and the landscaping trade. We see statuary, edgers and so on. Why don't you tell our readers about some of the things that have a revenue generating potential?

BECK. Since we got into business, we have had at least a dozen new products or innovations that didn't exist in the horticulture industry before. For example, we were some of the first people to actually make pasteurized potting soil or pasteurized landscape and garden soil by composting the soil along with the compost. As far as I know, we are the first people who have done that. We are the first people who came up with a specific rose mix and azalea mix and tree planting mix, all these in bulk. Besides that, I looked at an edging that was on the market and I saw some flaws in it so I upgraded it by putting some metal on it to stop grass from growing through it and everybody said, Beck, you ought to patent that. For fun, I asked a young girl who was scheduled to become a patent lawyer to look into the matter. She got me a patent on it. Since then, we have come up with collapsible tomato cages, real economical compost cages, and we make customized stepping stones with any kind of topping on it the customers wants. Another idea we had was the crushed clay pipe for a decorative stone, which I mentioned earlier. When we started selling the compost and the compost sand mix to put on lawns, people kept asking how to spread the stuff so I designed a tool to spread it with and make the job easier. It looks like a steel squeegee. It is very tough to spread compost with a rake of shovel but with this squeegee you can just throw down compost and spread it out until it disappears in the grass. Our small compost bin is simplicity itself. It is made half-inch by one-inch welded wire. We take a 10- or 12-foot piece of that and make a circle and then on the ends of it, we pop rivet some pieces of metal that are cut real smooth and gives it some stability, and in those pieces of metal we put some eye bolts and then you just drop a galvanized rod, down through these eye bolts and you have a circle. When you get ready to move your compost pile, you just pull that galvanized rod out and the circle comes right apart. If you want to turn your compost, you can pull the rod out, take the circle out from around the pile and set it to the side; then take a fork and put the compost back in it and it makes a real easy way to turn it.

ACRES U.S.A. We see bullet shaped "sheds," if that term can be permitted. What have we here?

BECK. They are taking out a lot of fiberglass fuel tanks these days. EPA does not like them underground because of leaks or flaws, or not meeting the standards. These are eight feet in diameter, 32-foot long fiberglass tanks and they have ribs. I found that we could take these tanks and cut them in half and stand them up, put a door in them and they make an excellent little house to store fertilizer or feed or tools. They are watertight and ratproof. There is a big company over towards Seguin, Texas, called Xerxses. They make them over there and when they take new tanks out they take the old ones as trade ins. I buy them from them. I also found I could cut the rings off. These are 16-inch rings, meaning you have an eight foot ring, 16-inches tall. Such a ring makes an excellent raised bed. Around our part of the country, we have little really good soil. Most of it is rock and most of the soil we have is shallow. That lip on the 16-inch ring sticks out. A lot of crawling insects, such as cutworms, can't negotiate that lip. This helps with insect control. When we get the tanks, they are fairly clean. Sometimes they have a little bit of residue so we just steam clean it.

ACRES U.S.A. What is the impediment to a community, let's say, a fairly large community of half a million people, setting up an operation somewhat like yours so that instead of leaves, twigs and grass clippings going to the landfill, they could be composted?

BECK. The real impediment is a lack of will. But these are all getting interested in composting, believe it or not. I have numerous people from small cities who come here and look at my operation and spend a lot of time with me. They take so much of my time I now have to charge a consulting fee, but I never charge a municipality or a city or a school. I spend the time with them that they need to launch a project. I have had no less than a dozen cities come to me recently that want to do something with their wastes of all types.

ACRES U.S.A. You haven't gotten into metal or aluminum cans or glass recycling?

BECK. There are people doing that. Let them specialize in that area and I will specialize in recycling the organic end of it.

ACRES U.S.A. Fletcher Sims of Canyon, Texas, because he is using strictly steer manure, has to go to the wind row method. What you are doing really wouldn't work for him because he doesn't have enough carbon. His is tilted the other way?

BECK. Yes. He doesn't have a bulking agent. He doesn't have enough carbon so he has to keep on beating air into the compost material and fluffing it up because it is dense product and the air doesn't circulate through it. He does have to irrigate it. He does lose moisture.

ACRES U.S.A. What would you do if you were trying to compost say, tree trimmings, leafs and grass clippings?

BECK. If you had green grass clippings, you could get enough nitrogen in there, but if the grass is dried out you would be a little bit short on nitrogen and it would take some time to compost. Your composting action would be much slower, which would be fine but it would take longer and you would need more storage area.

ACRES U.S.A. If you had some poultry droppings?

BECK. Go in with 15 percent or 20 percent poultry droppings or 25 percent or 30 percent cow or horse manure. You need to find some extra carbon. It doesn't take a lot of nitrogen.

ACRES U.S.A. If you can get sludge or manure with it, then you can move right along. Municipalities have a head start because they have about the correct ratios of everything. It is a matter of going out and scouting up the raw materials that someone has a problem getting rid of and marrying them correctly out there in those piles. Do that and you are halfway home. So, are raw materials the problem?

BECK. Not really. When I first went into business, I told the manager, "Well, in four or five years we will be bagging all this because I just don't see any more raw materials." Boy, did I miss the boat. There is so much stuff to compost, I can't describe it. All the disposal companies have already come out to make deals with me. I told them that at the moment I couldn't handle it. I told them that we needed to set it up at their place. They know that the landfill idea is coming to an end. People are going to have to learn how to select out the garbage. Garbage collectors are going to have to make the clients break down the grass, glass and the cans. The plastic, the glass, the metal and the organic materials are going to have to be separated. It is hard to teach the grown up to do that. If you start in kindergarten and start teaching reading, writing, arithmetic and recycling, it would be automatic when the kid grew up and became a household owner-operator. A kid doesn't have to unlearn anything or break any habits. You tell him this is our environment and this is what we have to do to survive. He accepts that and asks no questions. The things you learn when you are a child, you remember forever. The things you learn when you are our age, your mind starts segregating. It is not that your memory is shorter, your mind starts segregating. You remember the things that are necessary for your business. I remember the things that are necessary for my compost operation. People walk in the store all day long and I don't remember them. My wind is segregating. It has got to be started from little on. That is where it has to start. Then they have it.

ACRES U.S.A. In your area officials have made the connection between landfills and polluted ground water?

BECK. That is the concern here—the recharge zone. All they are talking about is that our Edwards aquifer is going to get polluted. I invited the Edwards Underground Water Authority out. The Cibolo Creek that we crossed on my property here, that is a recharge zone. When that creek fills up with water, it floods for a while and then the water goes right down. The water won't run. I have seen the water run across my driveway four foot deep, swift currents. Go to Selma, three miles down the road, and the creek is dry. It never reaches there. It goes straight into the Edwards. Here you have waters a hundred feet wide and four- and five-feet deep, and they never go three miles. I got worried about this. So a lady came out from the Edwards Underground Water Authority and I walked her all around. I don't do it in a car. I walk them around so that we can go to the piles and you can smell the air and everything. When the tour was over, she looked at me and said, "What are you doing for fly control? There aren't any flies back here." I said, "There are a few flies but no worse than it is anywhere else in the city." In composting, right away we mix it with a carbon product to get the moisture out of it and attract the fly parasites. She wasn't even aware of this. She said, "What are fly parasites?" I said, "Well, it is a little wasp-like insect that destroys the fly pupae." She looked at me and said, "Well, as long as you don't use chemicals for fly control, nothing here will pollute the aquifer." That is all she had to say. I was reading something else the other day about why she said that. You know, if the fuel tanks leak and we get hydrocarbons, diesel or gasoline or cleaning fluid something down in our aquifer, do you know how they could clean it up? With my compost leechates. We jump start the microorganisms down there along with some nutrients so that they can take these hydrocarbons and digest them. That is how they would clean it up. So she wasn't worried the least bit about my leechates running off.

ACRES U.S.A. When is the public policy going to be changed to tell these farmers?

BECK. Well, you still have got the big chemical industries with their propaganda out there. They have still got control of the farmers.

"Philosophy Doesn't Pay the Bills"

An Interview with Arden Andersen

Originally Published: November 1992

Arden Andersen, who was raised on a dairy farm in Michigan, had actually hoped to become an aerospace engineer when he grew up. When that didn't work out because of a back injury, he naturally fell on his upbringing and ended up becoming an agriculture teacher, but his ambition remained intact. He received his degree from the University of Arizona in Agriculture Education. Since that time, he has become a veritable question box. He became involved with others who were interviewed in this book, including Dan Skow and Dr. Carey Reams, through his work with Acres U.S.A. conferences and publications. He sought out Phil Callahan and other industry people whom he considered pioneers. Having a teaching background and a desire to teach, he summarized what all of these different people in eco-agriculture were doing, realizing that all of them have a contribution to make. He has been particularly interested in energy field medicine applications in agriculture. He published *Science in Agriculture, The Professional's Edge*, before this interview in 1992.

ACRES U.S.A. Mr. Andersen, we've considered the publication of *Science in Agriculture*, your recent book, as a red-letter event at Acres U.S.A. and we're certain you share our enthusiasm. So now we would like to have the readership of Acres U.S.A., plus growers we influence who do not regularly read Acres U.S.A., to know exactly what they can expect from your work. In short, what have you tried to do with your book?

ANDERSEN. Essentially—in my travels and seminars around the country—I've noticed two things either missing or not really very strong in our eco-agricultural system, and that is essentially a collection for farmers and consumers of data and information necessary to settle a nervousness about the alternative agricultural situation, and also something to rebut the propaganda of the petrochemical industry operating through the land-grant universities. And one other thing—we need a textbook. Certainly, textbooks are not all encompassing, but at least we need a basic text to give us a blueprint of summarizing the principles taught by certain leaders in agriculture, namely Albrecht, Reams, and so on. Regardless of whether one goes to a seminar by Dan Skow or Dave Larson, whoever they may go to a seminar with, they will have a textbook. A textbook is essential, one they can look through for basic principles of what needs to be done. Yet it also has to have documentations from the universities that verifies exactly what we're talking about.

ACRES U.S.A. From the moment we saw the manuscript, we considered it a worthy companion volume to our own *An Acres U.S.A. Primer,* which has been selling briskly now ever since it was written some 12 years ago. Do you accept that companionship?

ANDERSEN. Oh, absolutely. I think *The Primer* certainly laid the foundation for people to do further research work and reading, and certainly I would consider it a companion book to my own. Farmers should start reading things as "collections". No textbook or any book really is an exhaustive work. Certainly, when you start pulling things together, the pieces of the puzzle fall in place.

ACRES U.S.A. We noticed that you dive back into the literature quite deeply, and you reveal that many of the principles that we have had caused to surface over the last 20 years go back to the turn of the century, certainly as far back as the era of World War I. We have in mind the great work by Krasilnikov. One of your dynamite chapters is really an abstraction of everything he produced. How did it come about that these valid lessons came to be plowed under and buried so they can surface only now as the century comes to its end?

ANDERSEN. I think—as you have printed in Acres U.S.A. and in several of your books—it is very common, unfortunately, that we have more political science than natural science. It seems that the vast majority of our scientists are little more than automatons espousing essentially exactly what their predecessor professors espoused to them. There seems to be very little new and real research in academic circles. Most of the work is simply based upon what the petro-chemical industry will support. There seems to be very little new natural science coming out of these investigations. Essentially, it seems to be that because this political majority espouses a given protocol for agriculture or medicine or whatever, the rest just follow like true believers who accept no reference to reality. As a result, anything in the literature prior to our era, namely since World War II, when petro-chemical firms really got a foothold in our educational system, is erased. Up to that post-war era, scientists were involved in really great scientific research. They were doing it because they were interested in doing it. They were doing it because they had that special curiosity, much like Phil Callahan.

ACRES U.S.A. And of course we find a culmination of a great deal of this knowledge in the works of the late William A. Albrecht. Of course, Albrecht consulted with the late Carey Reams as well.

ANDERSEN. That's right. And you know what happened to William A. Albrecht because of the work he did.

ACRES U.S.A. Yes, he was retired forcibly well ahead of his time, and the soils department was handed over to an administrator who was capable of bringing in grant money.

ANDERSEN. Right and that seems to be the bottom line today—and as Phil Callahan has talked about, as have a few other former USDA people I've talked to. The USDA made a deliberate written policy decision to go from being a true, independent organization to being an industrial puppet for the industry rather than progress through science.

ACRES U.S.A. Well, your book does a great deal to dispel some of the mythology that has come out of the "chemical" camps. I am putting "chemical" in quotes.

ANDERSEN. That was my intent. In my travels and via the seminars that I have attended, I have noted that many farmers are pretty nervous about a lot of these things. Certainly their extension agents are telling them that agriculture is improving and there's only one way to go, and that's the chemical way. Yet the valid scientific literature points argue that certainly hard chemistry is not the way to go. History proved that the chemical system guarantees demise, not only of agriculture and our food chain, but the demise of our whole culture. So I wanted to be sure that we got things into the literature and into a book that reflected a consensus of work that already has been done. I am not reinventing the wheel. I don't need to reinvent the wheel. What I needed to do was to learn what has already been studied, what has already been proved.

ACRES U.S.A. In doing that, you branched out into physics, biology, chemistry, and you made each side trip not only entertaining but understandable. Some of your chapters are a bit difficult, but we find that if readers simply bypass them until they finish the book, then they can go back and reread the troublesome parts and find them quite comprehensible.

ANDERSEN. Well, I certainly hope so. One thing that a lot of farmers have a problem with comes to the fore. The old adage has been, you go to college or you get additional education to escape the farm. But the fact is that farmers need to be the best-educated people in society. Many farmers have relied upon other people to tell them how to do things rather than learning some of the basic sciences and being able to make their own decisions. So I wanted to show farmers that what we are talking about in alternative agriculture is based upon solid science. It's not based upon a nice political theory or on some religious credo. It's based upon hard core science—basic physics, basic chemistry, and those types of things. I want the farmer to be able to see that there is substance behind eco-agriculture. You don't have to believe in some wonderful theory if you just look and study science.

ACRES U.S.A. Just study nature? That's where all of science is housed and held in escrow in any case, isn't it?

ANDERSEN. Absolutely, and as Phil Callahan has pointed out many, many times, nature is a very wonderful science classroom.

ACRES U.S.A. Now taking the classroom approach you took in your book, what is it you really set out to do lesson by lesson? What is it that you're revealing for the first time, and then what is it that you merely repeating?

ANDERSEN. What I wanted to show was how to balance the system out in the field relative to the energy aspect. Carey Reams taught a lot of things about energy but some of it was ambiguous. Most of his terminology was contrary to what people were used to. Also, his system represented a conceptual departure that many, many people failed to understand—and so what I wanted to introduce was not only a common denominator, Reams, but also Albrecht and Phil Callahan as they talk about energy. Energy in agriculture seems to be a very foreign thing to most people. Hopefully I succeeded in squaring those concepts and with the practice of balancing the soil, balancing the crops and putting everything into an understandable context when talking about energy so that when a farmer looked at using or choosing various fertilizers he could understand them in terms of an energy pattern required to produce a given crop.

ACRES U.S.A. Produce and attract energy? Because, after all we're still looking to the sun as the fountainhead of all energy, aren't we?

ANDERSEN. Absolutely. That's very fundamental. Without sunshine, without light energy—not just the visual spectrum—we really wouldn't survive very long at all. Most people actually take that very much for granted. They simply assume that all we have to do is throw a seed out in the soil and dump enough water and salt fertilizer on, and everything else will take care of itself. But actually, that doesn't happen. And when that idea prevails, nature calls in the garbage crew. We have been told many, many times in several different books that if we don't balance the soil we will have diseases and insects and weeds. That concept has been brushed over a lot. So I wanted to put that in a context people would think about. I came up with the statement that weeds, diseases and insects are not there because of a deficiency of pesticide. When we think of it in that context, the light comes on. It's like, Oh yeah, of course it's not!

ACRES U.S.A. So we have to fall back on Albrecht's earlier dictum that in effect said the anatomy of weed and insect control is seated in fertility management, and not in buying a more powerful goodie from Dow Chemical or Monsanto.

ANDERSEN. Absolutely. And that proof is in the literature. There are many, many people who are coming up with very good replicated research proving that if they balance the nutrition in the soil, the problems largely resolve themselves. Dave Larson is a very good example of a field researcher who has come up with valid numbers. I saw some of his research reports from the past couple of years. He has it very well documented that if he balances the nutrition in the field, he simply does not have all of these problems with weeds, diseases, insects, poor grain quality, dry down problems, conditioning problems and so on. This thesis has been proven again and again. It's not just anecdotal evidence many school people would have us believe—you know, always the conditions were different there so what applied 50 years ago doesn't apply now. Well, nature doesn't change, as you know.

ACRES U.S.A. And these farmers need top know this isn't just classroom lectures. These are practical measures that actually work, as you write.

ANDERSEN. Absolutely. That was my intent. Philosophy is wonderful, but philosophy doesn't pay the bills unless you're a professor. The bottom line with farmers today is that their bankers want to see hard cash. They want to see success and philosophy doesn't go anywhere as far as doing that for farmers. So what I put in the book is essentially literature that is well documented. That was an important point to me, to make sure that if I said something could be done, that I backed it up with hard core data or literature that had been done by someone else with credentials and standing. As readers will discover when they look in the back of my book in the appendixes, there are many pieces of literature that have been cited.

ACRES U.S.A. But as a practical matter, our farmers haven't got the time it takes to be bibliophiles. They do not have time or the wherewithal to chase down ancient documents that probably can't be found except in the biggest libraries, so they are relying on what your codification of these things have to say?

ANDERSEN. That's correct.

ACRES U.S.A. And your codification took it from the beginning to the end and showed the actual results. Would you summarize just what it is you did in the hands on parts of the book?

ANDERSEN. I was at the prospect of the reader having no background, such as attending some seminars from Dan Skow, Dave Larson or any of the other people around the country you've discussed before in Acres U.S.A. If they haven't taken a seminar from any of these people, but they know they want to make a change, then I wanted to give them a step by step procedure. First of all, they need to start collecting data on their farms, and I give some examples of actual to that needs to be taken. Once they have data, what kind of decisions can they make? The important thing that I hope I got across in the book is that someone who decides to start farming biologically—I don't mean organically or chemically but biologically— is that biological's the middle, taking the best of both worlds.

ACRES U.S.A. Ecologically sound?

ANDERSEN. Sure. We have to be careful with terms. In Europe you say biological, and it's the same as organic in this country. So we have to be careful with our terms. Anyway, what I wanted to get across was that sound agronomy is no piece of cake. If you are going to farm in any alternative style, you better have your ducks in a row and you better make sure that the things that you do that you are well documented, and that you are documenting your own data and field observations. So many farmers like to simply plant their crop. They take a soil test and they send it off to the extension agent, and it comes back, "Oh, you just apply this," and essentially they leave the field until harvest.

ACRES U.S.A. So we've got too many farmers who think in terms of corn, beans and Florida.

ANDERSEN. Absolutely. That simply does not work. I wanted to get across this step-by-step procedure that is necessary if you are going to do alternative things. The most important thing that I get across is that they have to take down information from the field. In other words, collect data and then sit down and make decisions based on penciling everything out. It's real saddening how many farmers have very little data collection capability. If you ask them how much does a hundred weight of milk cost to produce, they can't answer that question.

ACRES U.S.A. Unfortunately these are the farmers who are the most vulnerable in the United States under present economic conditions, the seat-of-the-pants farmers! They still want to do things by the seat of their pants, and that simply won't fly nowadays, will it?

ANDERSEN. No, I don't think seat of the pants farming works very well anymore. Maybe 50 or 100 years ago when we had very little environmental pollution, when we still had a considerable amount of our topsoil left, that approach got you by. We certainly had much different conditions then from that point of view. Nature would take care of us better. But as we have eroded a lot of our topsoil, in fact, over 50 percent of our topsoil in this country, and we have changed the environmental conditions for weather. We have a lot more extremes. Most farmers can relate to for the last five years—droughts, floods, etc. We have polluted the soil with toxic chemicals. We have eroded away our biological systems. Also fertilizer is not cheap anymore. You can't just dump chemicals out there and expect everything to happen. We have to be looking at cause and effect. We have to be making definite decisions and realize that what worked last year will probably not work this year.

ACRES U.S.A. Let's talk about weeds. Why you get them, and why you don't get them and so on. We exchanged some of this material earlier. Some parts appeared in *Weeds, Control Without Poisons,* which was published by Acres U.S.A., but you set up a little different equation. You took the weed and then you told what caused it, what inhibited it, what the remedy was, and so on. You did this in kind of a ledger-type presentation. Would you recap some of that and give us a couple of good examples?

ANDERSEN. Sure. What I looked at is that weeds, essentially, are the caretakers of the soil. Many people think that if the soil is fertile, it is going to grow anything well. If you look at crops, common sense will tell you that this certainly isn't correct. Most farmers will recognize that if their soil is acid, certain weeds grow and other weeds won't grow. In certain parts of the country a given variety overwhelms and in other parts of the country, it doesn't. So I looked at that as a herbalist as well as from a medicine point of view. You have to pick out what you would consider a weed because of the medicinal qualities, different nutrient content. There is a tremendous amount of work that needs to be done.

ACRES U.S.A. Well, certainly, we haven't even started to examine DNA and RNA in weeds, have we?

ANDERSEN. Absolutely not. So, I just did a generalization and I acknowledge very much that this is not exhaustive and the other thing is that with all the different chemical systems in use now, we're getting sub-species alterations. We're getting herbicide resistant weeds and these are going to be a little bit different. For example, the giant foxtail is a major problem in a lot of areas. This weed means that we have a problem in the field with calcium, phosphorous, B12 deficiencies. This means that the weed is there because there is an imbalance in the soil in terms of relative nutrients. For example if we have a calcium problem in the field—now, quantitatively, we may have enough, qualitatively, not so—we are going to have a hard soil, a hardpan. Foxtail is nature's idea of helping to loosen up that soil. Ultimately, we get more oxygen going. Well, if we have a more oxygenated system, we're going to have more calcium and more phosphorus availability. Copper is going to start coming in to availability as do other things and then obviously some of our vitamins become activated. Now, it may take nature 100, 500 or 1,000 years to do that on her own. We obviously can't wait that long, so we need to do things to balance the system. In dealing with weeds, I cover enhancing materials. What does that mean? Well, that means things that will help that weed grow more. In other words, something that will cause greater calcium, phosphorus, copper and vitamin B12 deficiencies, things such as potassium chloride, muriate of potash, excess chemical nitrogen, salts, for example, and too much manure. Raw manure is going to add a tremendous amount of salt to the soil. Well, if you think about it, what do those things do? They knock out your organics and they solidify the soil over time. When you solidify the soil, you reduce the oxygen level, you reduce water percolation, you reduce the microbial system and so if you just think about it a little while, obviously the nutrients are going to be tied up more.

ACRES U.S.A. This is one of the things we proposed when we published *Weeds, Control Without Poisons*. Up to then, starting at about the turn of the century, the only thing anyone had to say about weeds was, this is the weed, this is how you recognize it, and this is the poison you put on it.

ANDERSEN. Right.

ACRES U.S.A. And now we think you've joined us, or we've joined you, in asking some of the appropriate questions that should have been asked at the time when they brought herbicides into play in the first place. Would you agree with that statement?

ANDERSEN. Absolutely. The comment you make there, about how we've joined each other, I think it's inevitable that if people start looking at nature, they have to come together. Nature doesn't change the lessons that she teaches us. I think it's very important to understand that if we start looking at cause and effect, we all are going to end up with the same conclusion.

ACRES U.S.A. You seem to have touched base with some favorite topics that have appeared in Acres U.S.A., well, things like the brix concept of Carey Reams, the use of hydrogen peroxide, and a number of other procedures that were not even slightly talked about before we started publishing. Yet you've brought them all together between the covers of your book and made them working tools. Would you give us a rundown on what you've covered?

ANDERSEN. My feeling is that there are no magic bullets. We have to look at any and all things that we possibly can do. It was a very early lesson for me that when we start looking at food production to sustain society, to sustain our animal agriculture, we need to look at inputs into those systems that are non toxic and things that actually help us. I think Dan Skow puts it very well—dealing with animals—that there are three basic things you've got to have to deal with raising animals Much the same is true for people. You've got to have fresh water, you've got to have adequate oxygen, and you've got to have a comfortable place to live. If we can understand those basic concepts, then we can start picking things off the shelf that are going to help produce healthy livestock. For example, hydrogen peroxide. Well, what is it going to do? It's going to help provide oxygen, but we have to be careful that we don't overuse it. There are no magic bullets.

ACRES U.S.A. Vitamin use for soils is a new concept. Before Dan Skow came along, we'd never heard of anybody using veterinary vitamins in their soil preparation mixes.

ANDERSEN. I think that's right. There really haven't been people doing that. You know, for years agriculture's been a dumping ground for toxic chemistry. People take it for granted that if we have soil out there and we put a seed in the soil, all of our vitamins and everything else is going to happen. That simply is not the case. The literature is full of various reports—and I've documented them—that vitamins do enhance microbial growth. We have found that vitamins do enhance plant growth and if we shut down the soil relative to an active, biological system, you are not going to have the vitamins produced, therefore, supplementing them to the plants is going to be beneficial. But farmers will notice, we are not using very large quantities. It's not that we are going to provide the plant or the soil with all of the vitamins—B12 or B complex vitamins or vitamin C—that crops or soils need. All we're looking to do is to jump start them into producing their own vitamins. That seems to be the most important thing.

ACRES U.S.A. You're using these materials to jump start the plant, to correct some obvious problems, but in doing so you are certainly not further damaging the soil as has been the case with the use of salt fertilizers in an imbalanced way, and in invoking the use of toxic rescue chemistry?

ANDERSEN. Correct. I think that's the basic premise in not only this book, but in several other books. The first thing a farmer needs to do in making a change is to simply stop doing what's causing his problem, and if you do nothing else, cut back on the muriate of potash or eliminate it.

ACRES U.S.A. Or find another source for potash? Certainly, the world didn't get created in terms of chlorine being dumped into the soil system, did it?

ANDERSEN. Of course not. So the first step anybody needs to take is to stop doing what's causing problems.

ACRES U.S.A. And some of these materials that are being sold, and touted by academia, and by the departments of agriculture, really don't have a place, do they? Anhydrous, for instance?

ANDERSEN. Right. As you know, those are simply products coming out of the World War II war machine. They were put together for use as a war machine. Those industries were very strong at the end of the war and rather than having to dismantle them, agriculture became the dumping ground for their wastes and products.

ACRES U.S.A. Perhaps we are getting to a point where we can summarize. Overall, what are we really saying? We know you have already made some important statements, but if we had to put all of these into a nutshell, what would you tell the people who are thinking about reading this book and using it not just as something to read but as a working manual?

ANDERSEN. My intent was simply to create a summary of the philosophies and teachings that I've seen coming out of Acres U.S.A., also out of the working agricultural industry. I wanted this book essentially to be a basic textbook to provide farmers with not only a science background—obviously it's not exhaustive—and I wanted to create an actual field manual. The most important thing is to get out in the field and observe.

ACRES U.S.A. Why were the Russians so important in unveiling many of these principles, and then having unveiled them, why did they cancel them out and not even use them? Do you have any ideas on that?

ANDERSEN. I think it's very much like what's happened in this country. The Russian war machine, essentially in the past half century, has been one of the most highly financed machines in the world. And just like our war machine in this country, there was no limit to the amount of money spent on weaponry research. A lot of these things having to do with biological weapons were financed covertly by the Russian military machine, as you can see by my documentation.

ACRES U.S.A. So basically they went on a war economy and plowed down agriculture in Russia, and now we are doing the same thing in the United States.

ANDERSEN. Absolutely. Phil Callahan is probably the best example we have in this country. Most of Phil's research work from the 70s on until he retired, even though he was with the USDA, was all funded by the Department of Defense and then it was classified so that the agriculture industry didn't have access to it. The same thing happened in Russia. The military machine came first. They really didn't care too much about the general population or the food chain, but they wanted that work because it had applications to weaponry.

ACRES U.S.A. You seem to be one of these individuals who attracts or tracks knowledge the way a magnet attracts steel filings. You've done such an exhaustive job in agriculture that we wonder if this is what prompted you to go ahead and seek a medical degree, which training you are now undergoing?

ANDERSEN. Well, I think so, and I have an interesting background from my family. My grandfather was a very interesting gentleman. He certainly encouraged me to get all of the understanding that I possibly could get, and Carey Reams was the next person who really influenced me in that. He said if you want to understand agriculture, study human health and medicine, and if you want to understand human health and medicine, study agriculture. There really is no separation between the two, and the people I've seen who've really made major contributions to both medicine and agriculture have been doing both.

ACRES U.S.A. So we're not going to lose you to medicine just because you are going to medical school at this present time?

ANDERSEN. No, absolutely not. I think that a medical background will give me the understanding to be able to combine the two, and really we can't separate the food chain from human health.

"To be Economical, Agriculture Must be Ecological"

An Interview with Charles Walters

Originally Published: July 1995

It would take a substantial amount of time to determine with certainty how many reports and stories Charles Walters, 1926-2009, wrote during his lifetime. His editorial pencil handled every word in Acres U.S.A. between its conception in 1971 and his pre-retirement in July 1994. Except for one farce, *Old Airmen Never Fly*, all his book titles have been serious studies. These are the titles he either authored or co-authored.

In 1996, at the 25-year anniversary, he passed the torch to his son, Fred, who edited and published the magazine and book line up until 2016. So naturally, they asked Charles to sit down for an interview, while concurrently asking for an exception from Dad Walters' iron-hard editorial policy.

Early on, Chuck Walters, often called CW, decreed that his picture would not be used in Acres U.S.A., there being more important fare for valuable space. This policy was violated just once or twice when *A Life in the Day of an Editor* was advertised, CW's picture being on the book cover.

The issue this interview was published, then 25 years after the first issue was printed, became the first deliberate use of Charles Walters' picture. Not that Charles Walters was finished—quite the contrary. He continued to write for years and kick off the annual Eco-Ag Conference with a resounding talk. Over the years, Chuck Walters has written many of the classics that now grace the home bookshelves of eco-farmers. Some are discussed in this interview.

Chuck Walters, who passed in 2009, was a veteran of World War II, the Korean War, a graduate of Creighton University, and held a graduate degree in economics from Denver University. His philosophy, we believe, is well stated in the timeless responses that follow.

BOOKS BY CHARLES WALTERS

- Ethical Foundations for Economic Theory (1953)
- The Greatest Farm Story of the Decade (1966)
- Confidential Alert (1968)
- Holding Action (1968)
- A Farmer's Guide to Homestead Rights (1968)
- Angry Testament (1969)
- Unforgiven (1971)
- The Case for EcoAgriculture (1975)
- The Albrecht Papers, editor, four volumes (1975, 1975, 1989, 1992)
- Parity, the Key to Prosperity Unlimited (1978)
- An Acres U.S.A. Primer (1979)
- A Life in the Day of an Editor (1986)
- The Carbon Connection (1990)
- Raw Materials Economics (1991)
- Weeds, Control Without Poisons (1991)
- Mainline Farming for Century 21 (1991)
- The Economics of Convulsion (1992)
- Fletcher Sims' Compost (1993)
- Neal Kinsey's HandsOn Agronomy (1993)
- The Carbon Cycle (1994)
- Socrates'The Lost Dialogues (1994)
- Reflections on Economic Theory, serialized in Acres U.S.A. (1995)

He also contributed substantially to several other books published by Acres U.S.A.

ACRES U.S.A. It seems hard to refer to you as Mr. Walters when everyone knows you as Mr. Acres U.S.A., founder, publisher, editor, and, chief contributor for 25 years. How did you come up with the name Acres U.S.A.?

WALTERS. It was a product of a brainstorming session. A long list of possible names was developed, most of which I've forgotten. Then Skeeter Leard suggested Independent Acres. I thought that was too long, so I modified the name and the result is what you see. I think the idea was spawned by the realization that the family farm ... was becoming an endangered species status. It represented the most economic system for producing food, regardless of what they say about corporate farm efficiency. More important, the family farm was the most successful education institution this country has ever seen. Even more important, I had in mind a special message—that the organic system was not limited to the garden, that the acre beckoned. I added the subhead, "To be economical, agriculture must be ecological."

ACRES U.S.A. Looking over 25 volumes, some 13,000 pages, articles too numerous to recite, we see a personal philosophy emerging. You seem to have allowed into the paper reports several hundred other farm journals ignore because they haven't achieved conventional wisdom status. How did you come to range so far and wide?

WALTERS. I think my big step forward came when I took up the editorial pencil for NFO Reporter. I had no education then—only two college degrees. I met up with Carl Wilken, who really blew my mind. Wilken was an engineer. He'd been involved with Charles Ray, John Lee Coulter, J. Carson Atkerson, and the National Association of Commissioners, directors and secretaries of agriculture. These people had backgrounded the Stabilization Act of World War II and its Steagall Amendment. Wilken, Ray and Coulter had developed the rationales that govern economics, mainly being stability and structural balance for the economy. Their system for maintaining a balance between the several sectors of the economy enabled the U.S. to fight World War II and generate the income to do this. In fact, a 15 percent income tax—a very low tax—would have seen the war debt liquidated. Anyway, this education dovetailed with my graduate work in economics. It really turned me on.

ACRES U.S.A. Wasn't it something of a leap to go from economics to organic agriculture? Certainly NFO never embraced eco-agriculture in any meaningful way?

WALTERS. No, they didn't. That idea was born after long talks with Arnold Paulson, after reading Rachel Carson's *Silent Spring*, then after writing *Unforgiven*. *Unforgiven* was really a book that spelled out the Wilken doctrine. Somewhere in the last few chapters I made some calculations on inputs and production. It seemed that fertilizers and pesticides were being increased exponentially, whereas production per unit of ground was coming up only arithmetically—a sort of Malthusian equation. I said to myself, "Holy cow, Rachel Carson is right," and then it struck me like a bolt of lightning. My little sister, Helen, had wasted away and died as a result of Hodgkin's disease. She had worked in one of those factories that endangered workers with toxic materials, loading cans and what have you. That's when I really got into the business of learning the biology lessons they didn't get around to teaching me in school.

ACRES U.S.A. Then you studied chemistry, soil systems and agriculture on your own?

WALTERS. Not entirely. I had formal training in chemistry, and I did quite well in the subject.

Also, I had handled an editorial desk at the *Veterinary Medicine* magazine, so I knew what Rachel Carson was talking about when she mentioned chlorinated hydrocarbons and organophosphates. They established poison control centers as toxic technology swept the scene, and Veterinary Medicine was enlisted to help publicize this "safety net" for the obscenity taking place. But you're correct, I had to learn a great deal about soils and crop lore as the first of the Acres U.S.A. papers snailed their way through the mails.

ACRES U.S.A. How did you proceed?

WALTERS. I read something about a Dr. William A Albrecht in a South African paper. He was featured as the former head of the Soils Department, University of Missouri, Columbia. I called over there. They told me Albrecht couldn't be interviewed because he was hard of hearing. I figured the way to deal with this was to go over there, which I did. We became good friends. He directed me in a course of study. Before he died he gave me some 800 papers and he passed on to me many of the volumes in his vast library. The interview really became my modus operandi for learning. Over the last 25 years I've taped about 300 interviews. Much of what was recorded and transcribed would have been lost to the mind and memory of man except for Acres U.S.A. preservation.

ACRES U.S.A. Was Albrecht your first interview?

WALTERS. Actually, Bill Graves of Hybrid Sales, Council Bluffs, Iowa, was first. Albrecht and C.J. Fenzau followed.

ACRES U.S.A. We realize that William A. Albrecht has had a commanding presence in Acres U.S.A. over the years, but how would you summarize his contribution to sound agriculture?

WALTERS. I'm glad you used the word sound. Albrecht was a first-rate scientist. Dozens of his papers spend 30 or 40 pages nailing down a single point. He destroyed a lot of mythology, including the pH myth as taught in schools these days. Long before Sir Albert Howard accused the republics of learning of preaching two false doctrines—namely partial and imbalanced fertilization and toxic rescue chemistry—Albrecht had entered his findings into the literature. I can't summarize all this here; it would take more than an entire issue. Suffice it to say that Albrecht believed bins and bushels didn't measure the true value of a crop. He worked out nutrient loads that answered a soil's cation-exchange capacity, and he communicated to the world formulas that stand up today for every crop, every latitude, every clime. I devoted a great deal of space to the Albrecht connection in *A Life in the Day of an Editor*, and of course, we published four volumes of Albrecht's papers (Editor's note: Now, eight).

ACRES U.S.A. You also wrote *EcoFarm: An Acres U.S.A. Primer* with C.J. Fenzau?

WALTERS. Yes. *An Acres U.S.A. Primer* was developed on the basis of a veritable plethora of information that came my way within a few years after Acres U.S.A. was founded. Of course, *The Albrecht Papers* figured, as did the findings of Phil Callahan, Rudolf Ozolins, C.J. Fenzau, Don Schriefer, and many others. C.J. Fenzau was good enough to read the pilot manuscript, and he made corrections. I thought it only fair to give him a byline for this service. I think the real genesis of *The Primer* was *The Land*, a journal that had passed from the scene by the time Acres U.S.A. came along. It was put out by Friends of the Land under the guidance of Louis Bromfield and Bill Albrecht. Friends of the Land was built around Bromfield's Malabar Farm in Ohio. That group actually preceded *Silent Spring* by 20 years. At one time the big soup companies journeyed to Malabar Farm to get a first hand lesson on how scientific farming was to be accomplished.

ACRES U.S.A. Why this decline and fall?

WALTERS. At the end of World War II, the fossil fuel companies descended on the university system like the vandals that sacked Rome. They gave grants with cute nondisclosure clauses attached, and in many cases they forced into retirement professors who wouldn't go along. Albrecht was one of them. A little later The Fertilizer Institute larded out grants according to the number of tillable acres in each state. And that's how toxic technology displaced not only sound farming, but the farmers themselves.

ACRES U.S.A. Displaced the farmers themselves? Can you explain that a little more?

WALTERS. In 1950, it took about 15 family farmers to plant and harvest, say, 500 acres of cotton. By switching to toxic technology and imbalanced salt fertilizers, the family farmers could be reduced to one. Jamie Whitten was head of the congressional committee that controlled farm appropriations. USDA was stuffed into toxic technology. Whitten came from the 1st Congressional District in Mississippi. As black voting rights came on, his future, rated somewhere beneath absolute zero. But with toxic technology, these black farmers could be moved out—to Chicago, Watts, Newark, Memphis, wherever. This is merely one development. Before 1960, Committee for Economic Development had decreed the countryside should be emptied. For reasons some historian may explain some day, America decided to commit suicide.

ACRES U.S.A. Exactly what was the situation in 1971 when Acres U.S.A. first rolled off the press?

WALTERS. Before 1971, Rodale's *Organic Fanning and Gardening* magazine had become established. The name was changed to *Organic Gardening and Farming*, and finally *Farming* was dropped. There were few if any organic clubs in America. Except for Natural Food Associates, there was not a single national organization. There were no courses of study built around organics, and no universities that offered degree work in organiculture. It looked to most people that the chemical victory, was complete. Far from being sunset technology, hard chemistry appeared to be the wave of the future.

ACRES U.S.A. We wonder, what was there in your personal makeup that prompted you to, well, take on the establishment?

WALTERS. I guess I was always a journalist at heart. I'd worked on rinky-dink papers, even the refereed journals, then with the *NFO Reporter*, but I wanted the freedom that went with making my own decisions without the blessings of higher approved authority. I knew that the metro papers didn't dare print anything the land grant colleges didn't approve of. Covering human medicine, they had to be governed by what scientists with credentials and standing ratified. This was part of the journalist's code of ethics. It sounded terribly self-serving to me. I didn't have the money to buy a paper, so I started one. The earnings from *Unforgiven*—which ultimately sold 50,000 copies—enabled me to do this. My idea was to deal with mundane matters—soils, animal and human health, etc., by laying the cards on the table. I didn't figure it was my job to do this—as bankers say, for your protection. I guess I was a just the facts, ma'am, type of journalist.

ACRES U.S.A. Just the facts? You do a swift turn on editorials, wouldn't you say?

WALTERS. Well, of course. The editorial abandons the role of objective journalism, I know that. Still the editorial is standard fare in newspapers, and irresistible to me. I have this genetic weakness, a longing to overuse certain lines that I find enchanting. For instance, I've been reminded that I too frequently refer to verbal killers as "slaying more people with their jawbone than Samson slew with the jawbone of that other historic ass." I think Acres U.S.A. has uncorked many good concepts editorially.

ACRES U.S.A. For instance?

WALTERS. I've always taken the position that the users of toxic technology should be required to label their production with a statement of toxic materials used. Further there should be a stiff sin tax on the use of chemicals of organic synthesis. I'm not a big believer in organic enclaves because I believe these to be impossible when damn fools are using Roundup and other such goodies from the devil's pantry. It would take a book or two to detail positions taken on things like genetic engineering, radar ranges, bST milk, NAFTA, GATT, farm subsidies, fluoridation of the water supply, and so on.

ACRES U.S.A. What is the bottom line on the use of farm chemicals? Many Acres U.S.A.

readers seem to use them much as a surgeon might use a knife, that is carefully. And, of course, the universities have a great body of knowledge on "use as directed" being safe and tolerable?

WALTERS. Personally, I consider much of that university science a mountain-sized lie. Hundreds of scientists around the world say that chemicals of organic synthesis have no safe level and no tolerance level. Further, they have no place in, on and around the food supply. Using these materials in agriculture represents bad botany, and I think the universities would issue the same conclusion except for the fact that they have become infected with grant money.

ACRES U.S.A. Are you basing what you say on a gut feeling, or is there more?

WALTERS. The gut feeling is there, no doubt. But so is intellectual support. As a matter of fact, I started Acres U.S.A. with several intellectual godfathers in tow, so to speak. I've mentioned Carl Wilken and associates for economics. I've also mentioned William A Albrecht and C.J. Fenzau and Rachel Carson. More important, I believe, was counsel harvested from the work of Linus Pauling. Pauling is the only person on this planet to have won two unshared Nobel Prizes—one for his vitamin C work, and one for taking issue with the assist of the world scientific community—with nuclear proliferation. You have to add to the Linus Pauling influence in America — Mosca, the chemistry prize winner at the Brussels World Fair. I don't think their influence has run its course by any means. It was Mosca who proved that the damage from alpha, beta and gamma radiation was the same as that of ionized farm chemicals. Frankly, I think Albrecht, Pauling and Mosca closed the circuit for me. As Rachel Carson put it, "We are in no better circumstances than Borgia's guests." Early on, I helped C.J. Fenzau make a statement on organics, fertilization and rescue chemistry that stands up as well today as when it was first written. I think it appeared in the first volume.

ACRES U.S.A. We know you wrote much of the material that has appeared in Acres U.S.A., yet you've signed precious few articles. Why?

WALTERS. I guess I didn't want Acres U.S.A. to come off like a high school newspaper laced with unwarranted bylines. Some I signed. Some, I allowed to stand with a single byline even though I could have shared the byline, as with the Fenzau paper. The objective was to promulgate the message. I really wrote those Mosca articles, unsigned. Even now, with come-lately reports on crop circles, I have to go back to those early clays of Acres U.S.A. You know, hard on the heels of World War II, our scientists tested atomic devices in the desert areas of the southwest. Harold Wills, now a staff writer for Acres U.S.A., covered that story for national media at the time. He now recalls phenomena very similar to the crop circle phenomena extant in England, and now Canada and the United States. Remember, this was before the country went crazy converting farming to oil chemical alchemy. Life's experiences have led me, Leonard Ridzon and Tom Mahoney to consider energy spins as the author of crop pictograms and circles. If you read Acres U.S.A. carefully, you'll see the source of degenerative metabolic disease as a consequence of what we're doing on the land.

ACRES U.S.A. The new management of Acres U.S.A. believes you're correct, of course, and will take the Acres U.S.A. premises forward for another 25 years. In the meantime we would like your assessment of what went wrong. We're not talking about specific steps which came first?

WALTERS. Exhausted soil always comes first. Read Walter Lowdermilk's report on vanished civilizations (*USDA Agricultural Information Bulletin No. 99*, "Conquest of the Land Through 7,000 Years"). I do not know what happened to Woodrow Wilson's health. I do know a stroke cut him down in the prime of life. Before that failing mental acuity caused him to undo *The Constitution* of a free people. He recognized this before he died, but by then he had permitted shysters to install the Federal Reserve System, write an income tax into law, protected the rich with a Foundation's law, and erased a fair slice of the genius of the Founding Fathers by changing the way in which senators represent the states. Scientific farming was on the way, and it would have supported 5 to 6 million farmers on the land, even up to our time. But too many forces—moving quietly in the background, to use Thorstein Veblen's term—permitted debauchery of agriculture first, then debauchery of the food supply. As Oliver Wendell Holmes put it, "The education of a child begins 100 years before it is born." Perhaps you can permit me to say the decline of a nation's mental acuity declines many years before it becomes apparent.

ACRES U.S.A. Has total mental acuity declined that much?

WALTERS. I think so. We have entire blocks of people who project their thinking no further than tomorrow night. I don't think many politicians think much beyond the next election. Certainly the bureaucrats think basically of legal and technical detail that lead to retirement, the Holy Grail for such a person's existence. The corporate captain is not that different from the government bureaucrat. This leaves most decisions for the society to people who make decisions without some idea of the long-term consequences involved. By way of contrast, I view the farmer who wants to leave the land in better shape when he leaves it to the next generation as the role model for civilization. Distorted education and packaged nutrition have affected even those who have resisted. As a consequence, farming has plugged in the obscene presence of toxic genetic chemistry, and lawmakers have bartered away the American dream in order to please PAC fund managers, leaving Americans to harvest the whirlwind.

ACRES U.S.A. Can you be a little more specific?

WALTERS. I think so. Look, you can get an abortion in the United States, but you can't get a fresh glass of whole milk. Where I grew up in Kansas, poverty was widespread, but it was not as bad as it is now. I could shoot a rabbit and it would appear on the dinner table within hours. My mother baked fresh bread. Packaged—meaning dead, dead, dead foods—were too costly, so they were bypassed. Our poultry ranged freely, and it had 30 percent less fat than the birds housed 20,000 and 30,000 to a building. The beef wasn't full of hormones and medicines. And milk was rich with cream, not watered down and inoculated with tropical fats. During the 1930s, Raymond Wheeler of the University of Kansas discerned that something like 70 percent of the people in Who's Who came from the calcium-rich high plains We lost that font of mental acuity the day toxic technology arrived. And we lost the ability to ask the right questions scientifically. The horror stories are too numerous to be recited in detail.

ACRES U.S.A. For example? Let's discuss at least one to illustrate the point.

WALTERS. Over the years, I have devoted a great deal of space to atomic energy, including the concept of irradiation. The scientists found a new plaything in atomic energy the minute a mushroom cloud appeared over Hiroshima. The buzzword was peaceful use and the first head of the Atomic Energy Commission opined that the atom would make electricity too cheap to bother metering. Remember, the scientific community didn't know enough about atomic energy and its effect on cells and protoplasm to matter. In fact the first big grants were released so scientists could learn a few answers. As you know, Watson and Crick got a shared Nobel prize for discovering the structure of RNA and DNA. Marie Curie's work with radiation was background stuff by now, but no one had much of an inkling what radiation with a half-life running into thousands of years would do to plant, animal and human life. This inventory of information did not slow the Brave New World people. Now we see that atomic energy generated electricity costs twice as much as electricity generated by any other means, and the big costs—decommissioning some 400 plants worldwide—have still to be met. We all remember Chernobyl, but we do not remember that some 7,000 cleanup workers died within a little more than half a decade after exposure. The genetic effects of radiation will run at least 20 generations.

ACRES U.S.A. In addition to filling the pages of the paper, you published a number of books on agronomy. What was your best title?

WALTERS. That would have to be *An Acres U.S.A. Primer.* You just about have to read *The Primer* first if you want to get the full value out of the paper itself. I enjoyed doing *Mainline Farming for Century 21* with Dan Skow. My approach was to do the word wrangling using the material of others. One of the best manuals produced in our shop was Neal Kinsey's *Hands-On Agronomy.* As for pure enjoyment in writing, I have to cite *Weeds, Control Without Poisons*. It was an outgrowth of *The Primer* section that dealt with weeds. Up to my first tryst with the subject, about the only outline being used anywhere was: This is the weed, this is what it looks like, this is the poison you use. My outline was: This is the weed, this is what it looks like, this is what the weed is telling you, these are the fertility measures you take to get rid of it. Working with Jerry Fridenstein, radionic scanner readouts were presented. Jay McCaman has used this approach to classify some 800 weeds I think *Weeds* asked some of the right questions for the first time

ACRES U.S.A. Academia scoffs at radionics, dowsing, towers of power, and so on?

WALTERS. And I scoff at academia and its dishonesty and distorted science. The farmer doesn't care about what these schoolmen think—that is, the farmers who have accustomed themselves to thinking for themselves. Many farmers are good naturalists and they make their own observations. Reductionist science is too weak for objective analysis, especially when there are many variables. Needless to say, billions of cells impose billions of variables. Besides, too much science now suffers from the defect Bill Albrecht saw in economics—so much he could hardly stand the tribe—namely cheating. The fact is radionics works. Dowsing works. As Hamlet put it, "There are more things in heaven and hell than are dreamt of in your philosophy, Horatio."

ACRES U.S.A. You didn't mention *Fletcher Sims' Compost?*

WALTERS. No, I did not. But I now remedy that oversight. Fletcher Sims is a little giant and a national treasure, much like Gene Poirot, the Golden City, Missouri, farmer I often quoted. Sims took composting out of the backyard and installed it near feedlots. He probably contributed more to topsoil restoration than all the colleges in America. The wonders he performed deserved a monument, not just a book. With *Fletcher Sims' Compost*, I sought to honor him, and also to pass a measure of his insight to others around the world.

ACRES U.S.A. You liken compost making to baking bread. Would you explain that notion?

WALTERS. Well, Sims and others tried making compost without the introduction of suitable microbial workers. He turned to Pfeiffer bacteria, then to Wensel Petrik's preparations. If you tried to bake bread with wild yeasts, the end product would be marginal, if edible. Retch told of a cookoff in which the inoculation vs. wild yeasts idea was settled for most applications.

ACRES U.S.A. Can compost restore the millions of acres that rain forest destruction and food farming have accounted for?

WALTERS. Probably not. But biodynamics can. It is the most neglected form of biocorrect farming in the United States. Alex Podolinsky, in Australia, has about 1.5 million acres under his tutelage. I've seen stuff as hard as adobe brick turned to mellow loam using biodynamic preparation 500. Hugh Lovel's *A Biodynamic Farm*, covers the various preparations. He did the editorial work on that book because I could no longer help I think the ruined acres of the world need to turn to Podolinsky's experience and to biodynamics His two books called *Biodynamic Agriculture, Introductory Lectures, volumes 1 and 2*, ought to be weighed out on jewelers' scales—they are that precious.

ACRES U.S.A. And there are a few other titles you didn't mention?

WALTERS. I didn't mention a dozen other books I've written, starting with *Holding Action and Angry Testament*, and ending with *Socrates, The Lost Dialogues*. This last was strictly a fun book. I've performed some of the material on stage, and I understand one Austin, Texas, group has turned certain chapters into entertainment. You have to remember, a lot of the material in the books got a pilot run in the pages of Acres U.S.A.

ACRES U.S.A. What were your most important stories in Acres U.S.A.?

WALTERS. Over 25 years I suppose there were at least 100 blockbuster interviews and reports. I'll recap 10 of these for the issue in which this interview appears. For now it is enough to cite Albrecht's pH/CEC story, Callahan's insect communication findings, Kervran's biological transmutation contravention of accepted science, Milt Jacobson's Bible economics, the Acres U.S.A. weed outline, Rudy Ozolin's papers, Don Schriefer's *From the Ground Up* book, C.J. Fenzau's pragmatic adaption of scientific farming, Carey Reams, the scanner, on and on ...

ACRES U.S.A. After a long 25 years, how much have you accomplished?

WALTERS. That's not for me to say. I do know that by the year 2000 approximately 25 percent of the vegetable demand will be for organic production. Of course, the inevitable seldom happens, the unexpected does. The fact is the human population is being compromised by toxic technology. The male sperm count is down tremendously. The health profile of the nation is going south in a handbasket. Even a Congress with its wits gone or a scientific community on the take will come to its senses one day. Certainly we've laid out the details. It will be up to society to correctly assess the facts and come to the appropriate conclusion. In the meantime the editorial job at Acres U.S.A. has been a rewarding one. The torch has been passed to a new generation. I only hope a lifetime of associates won't forget me.

.

"We're the Model Makers"

An Interview with Bill Mollison

Originally Published: April 1996

By the time you read this, the permaculture movement will have circled the globe, with thousands upon thousands of students teaching
other students the science and art of designing a permanent agriculture, a permanent culture. This popularity wasn't always the case.
Enter Bill Mollison, 1928-2016, a salty natural philosopher with a wry
sense of humor, belovedly known to some as the Tasmanian Devil. He
was a leader in introducing permaculture to Acres U.S.A., and was
the executive director of the Permaculture Institute. He traveled the
world lecturing to packed rooms on the techniques he devised to
save our world and ourselves.

His vitae when he lived reflected the diversity of skills he brought
to his design projects: forester, mill worker, trapper, snarer, driver,
shark fisherman, cattleman, bouncer at dances, and farmer. He was a
biogeography researcher at the University of Tasmania for 10 years,
leaving in 1978 to devote his full energies to spreading the ideas and
technologies of permaculture internationally. In 1981, he received the
Right Livelihood Award, sometimes called the "alternative Nobel
prize" for his work in environmental design.

Acres U.S.A. caught up with Mollison after a 1996 trip to the United States to speak at the Eco-Farm Conference in January. One word
of caution to the reader: To edit Mollison's language would be to
portray him inaccurately. Once a sailor, he told it like it is, occasionally
using, let's say, "colorful" language. As we wrote in 1996, "We apologize for our role in injuring any phalanges."

ACRES U.S.A. To start with, the name "Permaculture" itself stands for "permanent
agriculture."

MOLLISON. More than that, it stands for "permanent culture." Agriculture is a large element
of culture but it's not all of culture. So we have our own sort of financial systems and education
and so on. Don't leave it at agriculture. If you leave it at agriculture you are more or less in the
victim category.

ACRES U.S.A. You have written and spoken of the function of design. You maintain that systems can be designed for success and sustainability.

MOLLISON. The main problem in agriculture is it never was designed. There is no book in agriculture on the designed system. I once asked Anthony Rodale, "What is it you guys keep publishing?" He said, "Garden tips." So basically all books on agriculture are "How to do."

ACRES U.S.A. We have vivisected the whole into a thousand pieces. Your approach has been to stand way back and look at the whole, at the process of creating.

MOLLISON. Well, in Australia we are a dry continent. My classes teach that there are three important things in agriculture and the first three priorities are: water, water, water. The first part of our design is to design a whole water system for the whole farm. We had a marvelous old engineer called Percival Yeamans, and in the Second World War he was an army engineer. He was the guy who put in most of the airfields in the Pacific for America and Australia. Later he turned his attention to water and landscape and evolved a system called P-Line System, which cuts the water off at the first key point which is really the point where slopes turn from concave to convex. You can see it within a meter on a hillside. And he joined all those key points up with a large canal and puts dams at each key point and it's the highest point in the landscape where you can actually collect runoff. He made whole plans in which the water would come to the top of the farm and come down over the hill. After that it's sent to very low slopes, one per thousand, walking speed. It's got to go way to one end of the farm and then it can drop. And then it comes way to the other end and way back again. That was the first thing I had ever seen where people actually were designing landscape for production. It was in the early '50s he did that.

ACRES U.S.A. And the U.S. answer for dryland agriculture has, in general, been to harvest it through animals and grazing.

MOLLISON. Yes, we had the same outlook with a sad result, I think. I think 3 percent of our beef was produced over 75 percent of the continent, the drylands. So, you know, you're lucky to rear an animal for every three square miles. But all that's happened is that the drylands collapsed under the grazing.

ACRES U.S.A. You maintain these drylands could support plant growth, crops.

MOLLISON. Yeah, it can. What we found is that the county of Essex in England produced as much as the whole of our business—as much beef—and this is ridiculous. Why waste a whole continent to produce 3 percent of our beef? We're moving out of it. The other thing we built up in our drylands were huge herds of exotic animals—rabbits, donkeys, camels, wild horses—I've seen 10,000 or 20,000 wild horses down past the Musgrave Ranges. So between the exotic animals going wild and the beef, the desert is pretty well collapsed actually. It won't stand up to that sort of unregulated fixed stocking.

ACRES U.S.A. What do you see are the limiting factors for growth in agriculture, for plant growth and yield?

MOLLISON. Water is the limiting factor. We know that if you've got a good enough soil and water we can grow most things most places. But as soon as you've compacted the place out of cultivation by animals, then your runoff increases and all your deep arroyos date from about 1870, 1880, 1890—you know the westward spread of the Longhorn. Until that time the rivers ran. The Sonora was so named because of the sonorous sound of so many rivers. And if you get photos from the 1880s in Tucson, the rivers are running in a huge boskier of cottonwoods, big cottonwoods. Well, that's a joke when you look at the modern situation. The rivers are obliterated.

ACRES U.S.A. Well, the ancient writings said you could ride across all of North Africa under the shade of trees.

MOLLISON. Yes, and the main food of The Pima indians was fish. If you go back to the old middens, it was all fish bones. But that is like some dream today. Your rivers are filled with no water. You can probably well put that down with the Longhorns. Because as soon as they trodden out the rivers and gullies dropped from 70 to 20 feet, the whole water table went down. Then you really only get condensation water and with us that is only six percent of all the water in the rivers.

ACRES U.S.A. Your views on the utilization of the water nature gives us are based more on observation than on the human desire to control and engineer nature.

MOLLISON. That's right. There are two cultural differences between America and the rest of the world. The first is you guys really developed the oil industry and how you did it is drilling. And while you are looking for oil everywhere you strike water of different quality, and sometimes you strike gases. So when you think of looking for water you drill. And few Americans realize that you are the only country in the world that does that. You seek a well for water, nobody else does that. We don't do that. We have alkaline clay which is 88 percent or more water and it is getting into the streams and evaporating, that's how we lose 88 percent of our water and you'd the same in the drylands. It leaves you 12 percent for your crop and rainfall. When your rainfall is getting low, that's not enough. So, we have up to 25 units, up to 25 acres of runoff to grow an acre of crop, about five acres to grow an acre of crop sometimes.

ACRES U.S.A. And how does that compare to planting all six acres, yield-wise?

MOLLISON. I think it's better to use the 5-to-1. We get 16 times the grain on the one than we would have got harvesting the five. So every acre will produce 16 acres' equivalent.

ACRES U.S.A. So, not only can you afford to leave five acres idle …

MOLLISON. You need to leave five acres just for harvesting water. Not only can you do that, but if you leave it and you don't have the energy of planting it and you still get four times the grain by using a quarter of the space—it's a good tradeoff. And water is your main harvest. There are some good figures in one of your old rangeland books, I can't give you the author, and the value of the water yield from your high deserts. I'm going to go on my memory here. If we harvested the water, the land is worth $75 an acre—this is in the 40s say—whereas if we ran sheep or cattle on it it's worth $5 an acre. And they said it makes no economic sense to destroy the water yield by running sheep and cattle on the desert when we could get 75 bucks an acre annually just harvesting the water for crops.

ACRES U.S.A. We have the cattle mindset in America.

MOLLISON. Well, you've got cars and God up there somewhere. Cars are right next to God, cattle are right next to cars, and down you come.

ACRES U.S.A. You often define weeds in terms of a cow …

MOLLISON. A weed is something a cow won't eat or doesn't like eating. You look at all your weeds, they're not cattle food. But seriously, I have a friend who is the world's largest cattle producer. This is pretty weird because he lives in Tasmania. He has about 100,000 hectares. He turns out 2,000 fat cattle every week. Nobody in Argentina or America does that. He's traveled the world and talked to other cattlemen, nobody comes near it. So, he says, "I'm the world's biggest cattle producer." He lives in the little island in Tasmania—on a dry corner of it. He says it seems strange to him that he can produce more cattle than anybody else in the world. His cattle have been organic for 30 years now and permaculture for 20 of them.

ACRES U.S.A. What is different about his operation?

MOLLISON. He wouldn't have a whip or a horse or a dog on his property. So, no cowboys. Out with the cowboys, out with the rodeo and cattle rail courses. The cattle are all tremendously quiet. He can go to a gate and call and a few thousand head will follow him to the field where he's walking. He won't have his cattle rushed about. He'd shoot a cowboy because of the loss of beef he'd suffer rustling cattle about and branding them. He won't have a dog either. As he won't have any machinery; he does all of his cultivation with worms. He has a little machine—with an odd wheeled little vertical plate around the perimeter about every 12 inches—and it sort of makes little cuts across the pastures and then it scoops up. He has a scoop that follows that, so it scoops of bricks of turf where the worms are. They lay it up onto his truck and his daughter drives it and he stacks them. Then they go to a place where there is no worms.

ACRES U.S.A. So he's transplanting worms?

MOLLISON. Yes, then they lay it grass down, grass to grass, every 10 meters, or 15 yards. Each time the turf comes down it triggers a little mechanism which lets a bit of dolomite to fall on top so that they breed almost straightaway with the dolomite. He's been going over a 100,000 hectares with worms. And he's pushed production up 35 percent and he says he'll go to 75 percent in the next few years. They got rid of machinery and fuels and tractors. There is a real difference in culture here. I think you could persuade an American cow-man to, um, I don't know what you could persuade him to do, but you couldn't persuade him that worms were a good thing.

ACRES U.S.A. Well, we sometimes think that one of the problems with American agriculture is the fact that livestock has been removed from the farm so you don't get that return of organic matter. Do worms make up for lack of cattle?

MOLLISON. Oh they put out 40 tons an acre per annum of worm casts. It's a very valuable fertilizer and nobody could afford to cart in 40 tons per acre of anything. If we start cropping there, we loose 40 tons per acre of soils. So the change from losing soil from overgrazing and cropping to gaining soil by a return of organic matter is an enormous change. It's a huge factor. Another thing he's done—I helped him with this—was to lay 150 miles of water catchment drain, run-off catchment, and right along with it put in, I think, 700 million gallons of storage. So he turns a tap on at any one of his storages and it runs. One man irrigates 800 acres a day. One man walking backwards. So, that's a lot better than drilling down 2,000 feet and putting in a diesel pump and running one of these big center-pivot irrigation systems. I think Americans are real mechanically minded. He won't have that sort of stuff on his farm; he won't have pumps and tractors.

ACRES U.S.A. So part of your design process in permaculture is get rid of machines. But you're not a Luddite that suggests that we go back to horse-drawn power?

MOLLISON. No, I hate horses. I worked horses all my life. I logged with horses and I plowed with horses. I hate the bastards. I'd like to see them all turned into Big Macs or buried at the foot of trees as fertilizer. I just hate horses. And my first tractor was an old Farmall, which is an American tractor. I loved that thing. It didn't want to kill me. It didn't need three acres of hay every year. You could put it away, and you wouldn't have to worry about it for a week.

ACRES U.S.A. You're not against technology but you would rather use the forces of gravity and nature.

MOLLISON. My small tractor is good for 150 acres and it's been going probably 30 years and it'll go probably another 40. These old tractors, every farmer knows—you know the old Farmalls, Fordsons—they've usually got them pumping something, but they won't sell them. I see them. You know, I ride out through America, I see the old Farmalls chugging away on a cup and a half diesel a day. They might have a modern, new tractor but they won't get rid of the Farmall.

ACRES U.S.A. As you've done with water, you've looked at nature for clues about other design issues in agriculture. A lot of your writings keep mentioning the forest. Was that a source of education for you and why?

MOLLISON. I think one of the main problems is, America and Australia both, we saw the forest as an enemy. I took up a hundred acres when I was 20 at a pound an acre, a few dollars an acre. It was good forest and I cleared it and I burned it and I turned it into ashes and put on 30 milking cows. That was what we thought. Oh, I put a crop of cabbages in first. I grew so many cabbages there was no market for them. Then I put the cows on the grass coming through the cabbages. I'm a real pioneer, you know.

ACRES U.S.A. You've lived both sides.

MOLLISON. Yeah. I logged for some years. I don't suppose there is anything wrong with what we did. We thought we were doing the right thing. We were clearing the forest to grow food for people. That's what we were doing, and we were doing our best. All of us thought we were doing our best. Well, it's appalling now to destroy acres of forest. I would make more per year than I ever could have made out of cows. I have seen grown men break down and cry. I used to spot for a logging company and I go and offer them a million dollars for what they had left and they cried. They'd say, "I'd paid people 10 cents a day to ring back these trees and if I hadn't paid those bastards I could have retired now." Recently, in northern Australia, one of the local councils employed consultants and the forestry commission to give an opinion on what they could do with the many spare lands they had. They analyzed it and said, you gotta plant rainforest trees on them. And that will bring in $40,000 per acre per year and you can all get out of here and retire on the income. So they're doing it. Forest was never seen as a crop to agriculturists and you had to think of the forestry commission, I suppose you've got one too, it's job too is to get rid of forests. Mainly to the Japanese, mainly as junk wood.

ACRES U.S.A. Chop it down and plant it back with one species.

MOLLISON. Sometimes you can't plant it back. A lot of your forests grew in a milder time and you can't put a logged Northwest forest back because it hard frosts and it won't grow in some areas. So they say we'll never replant a lot of this.

ACRES U.S.A. What is it about a forest that lets it live for thousands of years until we chop it down?

MOLLISON. I'm glad you asked that question. What is it about a forest that lets it improve for a thousand years and maintain itself forever and no farm can do that?

ACRES U.S.A. The undisturbed forest is the definition of "sustainable."

MOLLISON. Yeah, that's right. The forest actually provides for its own growth and maintenance. It is a complex system; it's full of birds and other animals. My farm now is too. I started with an old dairy farm, a run-down dairy farm, 150 acres. I'm not a big area farmer because I'm only home for two months a year. I kicked the cows off and I mowed it for a couple of years and it converted, almost instantly, from a compacted yellowish clay to a really free loam again. And I didn't sow any worms and the worms got right through it. I have a very rich farm now and I'm putting trees in because that's my best crop. I don't have to hurry because I've got a small publishing company that supports me while I'm doing all of this. The trouble with the farmer is he's got to hurry. Once he's in the crop he's on the treadmill. One year missed is a tragedy. He's got so much back payment on big machines and he's got to plow and buy fertilizer and he has got to get that crop. If he misses it, he's down a quarter or a half a million dollars.

ACRES U.S.A. And if he's leveraged 50 percent, all the worse …

MOLLISON. He's had it. The bank has got his farm. The banks are now the big farmers in Australia.

ACRES U.S.A. So, you're building diversity on your farm?

MOLLISON. I was looking at my garden out front the other day. I started with three or four acres of trial crops, and I had nine-acre equivalents per acre. I had nine crops mixed, each of which I get the same yield as if they were on their own.

ACRES U.S.A. So your 150 acres is like a thousand acres.

MOLLISON. Right. And I haven't started, it's in the sixth year. I get the same amount of coffee as if I had only planted coffee. I get the same amount of jackfruit as if I just planted jackfruit. Same amount of mangoes, same amount of sweet potatoes, but they are all together.

ACRES U.S.A. So an American farmer does not need to "get big or get out" and farm a thousand acres.

MOLLISON. He needs to get cunning and complex.

ACRES U.S.A. The complexity is something you saw in the forest that you are recreating?

MOLLISON. I've recreated. I had what I suppose was a blasphemous thought in November 1959. I had been working for 28 years in wildlife and forestry in the field. I looked at a simple forest in Tasmania; it only had three dominant tree species and only 26 wooden species in total, understory and all. That's a real simple forest. It has only three main marsupial browsers—two wallabies and a possum. I thought, "I can build something better than this." That's blasphemous. I wrote it down in my diary. I thought this isn't sufficiently complicated. Although we took half a million firs a year out of some of those valleys—I guess that would be worth 3 or 4 million dollars—it didn't effect the forest at all. We could take that every year; it was better than any agricultural crop we could put in the valley and we still had all the trees. I think we took more value in honey annually than we could take out of the whole of the timber if we sawed it down and sold it as wood chips.

ACRES U.S.A. And the soil, was it being mined and depleted?

MOLLISON. No, the soil was totally stable under those sort of yields. A logger could make $7,000 most afternoons collecting tree seeds for Russia. And I did. I'd take 30,000 to 40,000 skins a month and the beekeepers would take about that in honey. We cut it down and sold it to Japan. And of course as soon as you cut down the forest and sell it to Japan, in come the cows and soon, in come the tractors. Really, you get nothing out of it in the end and the rivers fill up and the soil runs away. What you have done is you have reduced the public wealth very sharply and you will never get the money back that you destroyed.

ACRES U.S.A. What is it about the forest system that makes the soil stable, and can we recreate that in agriculture.

MOLLISON. My God, I've planted everything on a devastated farm. I did fertilize it a bit. I bought a bag of superphosphate and two bags of dolomite. A bag of cement because the silica was gone and I gave every plant a handful. So, on five acres that was my fertilizer. And everything grew like crazy and nothing showed deficiency symptoms—and I put a few key plants in that would show those symptoms. I had to add a little bit of borax, a bit of copper and a bit of zinc, but ounces. I haven't touched it since because nothing looks like it needs any fertilizer. The understory is one big sheet of sweet potato. I was looking at it and I noticed that every leaf had a bird dropping without exception. So I kind of figure with that much phosphate flying in there now, that it was top dressing heavier than I could ever afford to do. The only thing I've done now is to post in the field the bird perch at the foot of every tree or so. If I got a stand up in grassland I'd just drive tall pegs in and leave it to the bird to fertilize the tree. And they do. I mean, they'll put down a handful of superphosphate every day.

ACRES U.S.A. Free and accurate delivery system?

MOLLISON. Yeah, very, right at the roots of the tree. And when a tree eclipsed the post that I drive, then the birds can perch in the tree. I kind of figure all these things can't happen by themselves. And now you add the worms so you are mining the nutrients up and getting the minerals up to get on the surface. This is what I've been doing. And also about every 35 paces I put a big umbrella-shaped legume tree and it drops all its leaves once a year and boom, it's there, hundred weights of leaves come down. Then it waits three weeks and on they go again. So it's an oxygen-fixing tree, dropping an enormous amount of fertilizer.

ACRES U.S.A. How many types of oxygen-fixing trees are there?

MOLLISON. I'd have to guess. It would be about 2,000. Many of them also have large sugar-laden pods and they are cattle fodder. So not only do they increase the production of grasslands underneath, but they add that essential late-summer sugar for the utilization of dry grasses.

ACRES U.S.A. You see a role of trees in our modern agriculture.

MOLLISON. I think without trees it's moot.

ACRES U.S.A. How would a Midwestern corn farmer utilize trees in his production?

MOLLISON. Well, we would go back to old USDA manuals like the one on trees. And he would ask the USDA to give him the research on corn and trees in North Africa because they could even do that with their data banks. And he would find there are trees, we call them "farmer's trees," which if he plants through his corn 20 per acre, 30 per hectare, they will double his crop and that will eliminate his need for nitrogen fertilizers.

ACRES U.S.A. And how does he plant these, as wind breaks or random placement?

MOLLISON. No, but in line so he can still use his tractor. Not his horse, his tractor—shoot the horses.

ACRES U.S.A. Well, of course, his neighbors would talk if he started planting trees in his corn fields.

MOLLISON. Yeah, well, that's the trouble, the neighbors would say, "What are you doing planting trees in your corn field?" He'd say, "Because I'm not a bloody fool and I don't want to spend all of my money on fertilizer and lose all my soil and get half the corn I would otherwise get." I remember one of my neighbors stuck his head over the fence, he was a cattleman, and he said, "I don't like neighbors." And I said, "Neither do I." He said, "Well, we agree on that." He said, "Why haven't you got any blady grass, Imperata grass on your patch?" I say, "I don't light it every year. Every time you run a fire over your grasslands all your nitrogen and sulfur is gone for the next 30 years." And I added, "I'm not a bloody fool; I cut mine, I turn mine into worms and get all my nitrogen and sulfur back." "Oh," he says, the next year he stopped burning and started slashing it and he got good grassland out of it. But he didn't do anything about his water system. He didn't do anything about his trees. He didn't have any late summer pod forages for his cattle. I felt sorry for the cattle so I placed, all along my fence line, my boundary line, big Inga trees and the cattle spent most of their time along my fence line. I got all their manure and I threw it over the fence. The cattle know best.

ACRES U.S.A. Can you tell us more about farmer's trees?

MOLLISON. It's just a tree that you put in the crop that will increase the crop and will provide at least some of the components of the fertilizer of the crop. I well remember early so-called agroforestry attempts in Australia, that foresters and agriculturalists said well, let's join forces, because a forester knew how to plant pine trees and the agriculturalists knew how to run cattle. They had lines of pine trees with cattle in between them; it was kind of a mutual disaster. The only place for pine trees with cattle is right on the highest mound, they love them. You plant high ridges with pine trees and Casuarina trees and you get a big mattress of pine leaves and Casuarina leaves. They are like straw and the cattle camp there all summer because the wind blows under the trees and takes the flies away and the cattle love it. They are all crapping and pissing on top of the hill and you can see the clovers coming from the rundown out of that system. You see the hill starting to go green from the top down. So that's the farmer's tree on top of the hill. It's not a farmer's tree in pasture and it is certainly not a cropman's tree.

ACRES U.S.A. So what you've done is observed and thought about it. But more than these random bits of knowledge that flow from your tongue, you have formalized this into a system.

MOLLISON. All of these knolls we'll put pines on and we might as well make them nut pines because we'll make a lot of money off of nuts anyhow. So we'll make them one or another of the 50 or so nut pines, which are high value pines, close to the value of gold, curiously. There are some crops that are more per ounce than gold; not many people realize that.

ACRES U.S.A. Soybeans, I guess, are not one of them.

MOLLISON. They certainly are not. Soybeans are what I call an ecological disaster crop. In 1985 in America, I think it was 1985, we cleared more lowland forest in Mississippi for soybeans than had been cleared for all other reasons, wood chips, anything else. And the soybeans really just love rich soils and plenty of water. Well, it doesn't make soil so it rapidly exhausts them. So we took away good forest and planted that horrible little bean, that sort of poisonous little bastard. It makes it impossible for you to digest your proteins so you've got to cook it for hours. So for every pound of soybeans you've got pots to cook. It's a nasty little plant and the only good thing about it is it produces an oil which is the basis for car paint. And that's why so much is grown. Then you've got a waste product, which is the meat.

ACRES U.S.A. The more you understand, the more you can put nature to work for you the less you need.

MOLLISON. That's right. You should be thinking of getting out of work in six years time. You should be moving to where you don't have to move, you're just running around on the farm and admiring the ways the trees are working for you. This mate of mine who produces all these cattle, he's got a great big sign up on his farm and it says: "Employed on this farm are one stockman, one irrigation man, and 50 billion worms." That's all.

ACRES U.S.A. That should be every farmer's dream?

MOLLISON. Yeah, he spends a lot of time on his farm just looking, watching his cattle. It was he who pointed out to me that the only tree that liked the land are the needle leaf trees that put the mattress down. I knew that but I never noticed it. You know, many people say, "Oh, I know that I've seen them laying under the pine trees." You knew it, but you didn't know it. And this mixture of pine needles, cattle dung and urine, is an incredibly good base for clovers and as the rains fall and that stuff starts to move slowly down the slopes, in come the clovers and the protein pasture plants. And he won't plant grass. He'll plant chickory and he'll plant dandelion, he'll plant anything but grass. He says, "What's the point in planting grass, grass is going to come there anyhow. We've got to get the iron into the cattle and the zinc into the cattle. You want things that fruit down there." So he has no illness in the cattle, he has no worms because every cow on his farm can get a mouthful of worm oil when he wants it. So he only sets out herbal plants, he doesn't plant grass.

ACRES U.S.A. Let the cows visit the pharmacy when they know they need it.

MOLLISON. Yeah, and they do. To me, of course he's a cattleman who knows cattle better than I do. I've been a dairyman, but you've got to pump milk through them all of the time. He knows cattle really well and he says, "You watch them for a while." So we'll go sit by a cow and it's actually selecting, very selectively eating what it needs. The other thing that he is very certain about and so am I, he said, "No matter what size your farm is if you run cattle you've got to have 18 moves at least, so it's going to take you close to 20 months to get around to farming cattle. So you've got 18 huge fields and he moves them. So there is no fixed stocking, no patch grazing and no unlovely pastures because you put all your cattle on 1/18th of your farm and they practically eat that flat. Then he says, "Come on, cow," and walks them to the next patch. That is actually what he says, "Come on, cow." And they all look at him and follow him to the next patch. He doesn't need any cowboys. He says that the cowboys just destroy the cattle's peace of mind.

ACRES U.S.A. He divides the year into 18 parts?

MOLLISON. He moves them over 20 months. So he gives them a month and a half, maybe a bit more, on 1/18th of the land.

ACRES U.S.A. The land then rests for 17 months.

MOLLISON. Rests, and it is beautiful when it comes back. And, of course, he's putting these worms through it all the time. He's got beautiful, soft soil. He spreads the worms all of the time. And he plants throughout the year, 30,000 trees a year he plants. So, he doesn't get any moisture loss from wind. His cattle are well sheltered and he puts his mattresses down for them—he loves his cattle. And around his top yard he plants lots of walnuts to keep the flies and mosquitoes away from them. Where he's got a turn or in a corner he puts box thorns so that there are no through gates. He's a real thinking man and, of course, he is highly successful. He's a multimillionaire. I mean you don't have to spend money on fertilizers, fuels, tractors and cowboys.

ACRES U.S.A. The farmer doesn't have to figure all this out on his own?

MOLLISON. No, the cow does it for him. That's the truth.

ACRES U.S.A. But knowing which tree and the spacing and all this has been formalized into this model, this discipline called "permaculture"?

MOLLISON. That's right. What I think is eerie, very eerie, is there has never been a book of this kind in agriculture until I wrote my book. I never proposed to do a book, but I put out a lot of manuals with just hundreds of examples of how to think your way into no-work farming that builds soils. Of course, thousands of people have taken that work—or hundreds of thousands— so we've got examples of it working.

ACRES U.S.A. Not just in the United States and Australia, but around the world?

MOLLISON. Oh yes, we have institutes in over 120 countries. I don't know of any countries I haven't been to. Someone asked me the other day to come down to Trinidad and I thought, I've never been there and I don't think I'll go—I like to think there is a country I've never been to. Otherwise I wouldn't have known where to go that I've never been. I've worked in hard frozen ground in Russia and equatorial areas, all through it.

ACRES U.S.A. Do you find that some developing countries embrace these concepts more readily? Where has it seen the most success?

MOLLISON. There is no doubt about that. Any country where they are competing against global-national firms adopts it like a shot. Any country where you can walk into the nearest super-market and buy a lot of food from all over the world—and America is the net protein import-er—is not interested in it.

ACRES U.S.A. One economic newsletter we read says, "Americans are fat, dumb and happy."

MOLLISON. And that's the idea, keep them fat, dumb and happy and you can do what you like. America deliberately maintains a false illusion of abundance and it isn't in America—that abundance.

ACRES U.S.A. The fact is we are a net food importer.

MOLLISON. No, protein. You are a protein importer.

ACRES U.S.A. Could you tell us about your experience with Vietnam?

MOLLISON. Yes. Well, it was one of my students. The only thing I do that other people don't do and Americans won't do is I train students to go and teach.

ACRES U.S.A. Permaculture is not being spread worldwide because of a university system or U.N. or government action?

MOLLISON. No, I give a two-week training course. A solid training. It's solid, it's not a brain-washing. I tell them to go out there and teach now because I don't have time to go around the whole world. I met a young African last year who said, "You don't know me. I'm seven gen-erations of teachers away from you." Africans are real good genealogists and he counted all these teachers back until he got to the one I taught, so he's my great-great-great-great-grand student. Some of my teachers are in there seventh generation over 20 years. So, it's been three years per generation of teachers. And I haven't patented anything. Permaculture is in common copyright for students and journalists. You let it go. If you don't let it go, well, it doesn't go anywhere. But Americans, and to some extent Europeans, try to patent little bits of things all of the time and so it goes nowhere.

ACRES U.S.A. Intellectual property?

MOLLISON. Yeah. If there's a royalty you can get it out of a patent. This is harsh really, but thou-sands of generations of women who are now dead made those seeds and you patent them. They are not yours, they're our grandmothers'. So I ignore all that because it's obviously just another pile of crap, patenting seeds.

ACRES U.S.A. What have some of your students and grand-students done?

MOLLISON. One woman, not a pleasant woman, went into Vietnam. Learned Vietnamese in Hanoi and put out a couple of courses. She is two generations away from me, I didn't personally know her and I didn't train her. And at that time the army's concern became domestic food security because there wasn't enough food and there wasn't enough protein. The main diet for years had been salt and white rice. So people were as flat as cardboard and had all sorts of deficiencies and problems. So the Army, as soon as it stopped fighting, turned its attention to nutrition of the community. The generals were there and the platoon leaders and the troops were the farmers.

ACRES U.S.A. Did you learn something from them?

MOLLISON. They taught you a thing or two in the war. And they have immense good will toward their own country, which you can't say for most countries, Australia included. You could get rid of the whole House of Lords just about. I mean how can a cotton grower have good will toward his country? He can't. He is no patriot, or he's a scoundrel who pretends to be a patriot. Same with some of these other guys you've got in parliament. While they're lining their own pockets with one hand they are waving the flag with the other hand. And this happens in the communist world as well. There are more millionaires in Moscow after Brezhnev than there are in New Zealand and Australia combined. All they've been doing is ripping off the country while waving the red flag in the other hand. America is precisely the same, I mean, the parallels between America and Russia are incredible. They are really the same sort of system, just in one communism is patriotic and in one capitalism is patriotic. And none of them are democratic, and both of them have the top people lining their pockets.

ACRES U.S.A. What did the military in Vietnam do?

MOLLISON. They turned their attention to feeding families.

ACRES U.S.A. And how did your grand-student work with them.

MOLLISON. She was there when they did this. So they set up basically a military organization called VACVINA and VAC is an acronym for livestock, crop and water and VINA is just Vietnam. So it's the livestock, crop and water group for Vietnam—VACVINA. She said, "Would you like a permaculture person?" She gave several courses and all the generals and people came. VACVINA is a parallel to the government. It's not the agriculture department because it only works with farmers, it's out in the fields with the farmers—and the soldiers were the farmers. So, this general took the course, a nice old man, and he says, "This looks like what we are after." They declared permaculture their official policy. And it has been because they translated the book and they asked me what I wanted for it and I said nothing since they were handling the translation. They printed 150,000 and gave one to each farmer and they said, "This is what we are going to do." Well, within two years they did it.

ACRES U.S.A. Fundamentally, in Vietnam, in a very short period of time they created abundance. And abundance is a word that keeps surfacing.

MOLLISON. Yes. And community too. And incredibly built a nutritional base.

ACRES U.S.A. Abundance is sort of the backbone of the whole permaculture system. What stages or phases are there to it?

MOLLISON. It's about a five-year transition. Vietnam did it faster because it was still in the army structure and the general says, "Alright, we're going to do Permaculture. Start digging fish ponds in the rice fields." And they dug them, hand dug them. I couldn't believe what they did, like they had been digging them for years, it was marvelous. They are quite friendly with each other, it's all very cheerful. It's not like, "Go dig the fish pond or else." There is no "or else" involved. They hand dug all of their fish ponds and, of course, it doesn't take long for fish to make money and they had fish for their rice. It was magnificent. They grow an awful lot of protein. If they catch a duck and they breed it, it will lay about 60 eggs in six months. And if they've got a little dog and pups, they have pups once a year. So you see little cages of dogs and all the table scraps will go into the cages. They've got their chickens and their eggs.

ACRES U.S.A. What typically happens in this cycle, this five years?

MOLLISON. It depends on if you really do dig that pond, because the protein production out of the pond is going to be 20 or 30 times higher than it is out of the cows. Imagine taking a 30th of the protein in some land stock when you're going to get 30 times out of the fish and it's better for you. So we must grow what is best for us and what will give us the highest return because we don't have a lot of spare land. We've just got to get the highest returns. Then they built model storage systems, starting harvesting the jackfruits and coconuts, coming down through coffee and down to the rice. So every canal and fish area also has a big trellis over it made out of bamboo, which is also grown, and that has vine crops—you know, squash, cucumbers, grapes and so on. So your fish are living under a canopy of vines.

ACRES U.S.A. Beyond merely cramming as much plant and growth on the amount of land that you have, you can utilize the height and all of this diversity, this abundance of species, actually protects and feeds each other.

MOLLISON. Yes. All that fertilizer is turning over and being produced at the same time. The chickens are crapping along the barns and that washes into the fish. Everybody's eating extremely well. And there, I almost feel like I'm some mythical god. They say, with my little belly and my bald head, that I exactly resemble their god of longevity. They come up and rub my tummy with their hands because that's what you do with the god of longevity—you've got to rub his belly. I just look great, I look like the god of good food and longevity. They knew my name.

ACRES U.S.A. There is probably no greater honor.

MOLLISON. That's right. You've got a keen young man going, "Oh what guy can help the Vietnamese." So I rang him up and said, "Would you like a bit of help?" He said, "No, no, no," he said, "We're fine, we're researching and we're doing fine, we don't want any help from anyone." He said, "You've done your bit." I said, "Oh great, would you care to send any teachers here." They said, "We can come there." So they have turned into teachers. Now, if every country was like that I could go home and stay home and look after the farm instead of spending eight months a year traveling. I could stay home in a nice place and mind the farm. I have this 150 acres but everything I need I grow on an acre.

ACRES U.S.A. So you feed yourself on one acre, a diverse acre. How many types of food on that acre.

MOLLISON. Oh, 450-odd sorts of food. It's funny, a lot of the fruits can grow large; they can grow to 40-pound fruits. A decent papaya will weigh 10 pounds. When it's 40, it's big and it whistles as it comes down. Four years ago I put in trial jackfruit, which are 40-pound fruit and they'll go a hundred pounds. So I don't want to get hit by one of those because they'll actually squash a car if they fall from 100 feet—they are big trees. I put in 160 jackfruit this year. Sometimes at night the fruit comes unstuck and boom. My aim is to have a constant sort of thunking sound that's just fruit falling on the ground. It's very comforting to hear this constant rain of large luscious food. I have provided for myself. Then I took a lease out on my own farm to grow 600 mango trees because I would like to be a millionaire every year. I think that would be a nice thing to be. So I work for 600 mangoes thinking a million dollars will yield $3 million. But I don't trust my figures so I figure if 200 will do it I'll put 600 in, just in case I made a mistake in the figures. So I'm all provided for, I don't tie up the farm or its capital in the next two years. So the farm owes me nothing. So I said to my driver and electrician and all those people, "If you can see a place on the farm for it to make $30,000, I'll license you to do that. I won't lease you because we have to survive and we have to accommodate. I'll license you to grow chickens on the farm and either I could give you a license to produce eggs or produce meat chickens, whichever you wish. I licensed people to grow fish. And the fish licensee said to me, "I think you better bring a few more families on, I could make my $30,000 out of one pond now. If you're going to have 16 ponds you're going to need a lot more people." So, that's how we got it, I've issued 22 licenses.

ACRES U.S.A. But this is not a situation unique to your corner of the world—it can be transplanted.

MOLLISON. Oh, anywhere. There is one in Germany. It's got its own retail store on the farm. It actually delivers baskets throughout the community. It did close at 130 people because it is only a small farm.

ACRES U.S.A. A lot of readers may hear or read of these things and think "Oh, but that's Australia, that's New Zealand. My little farm in Iowa, in Minnesota, in Georgia ..."

MOLLISON. Bullshit. That's bullshit. That's the answer of a person who doesn't want to do anything. "Now, I got a thousand acres and I want to use all of it." Well, they can't use any of it really, that's what it comes down to. They don't want to use any of it.

ACRES U.S.A. What would be your first advice to tell a farmer how to get off the treadmill?

MOLLISON. The English farm. It has 150 cows and it's quite a nice dairy. It is only 100 acres and he grows grass and silage for the best nutrition. Some of the silage was dug into pits—in digging they found good clay. So the guy who set it up said, "That looks like good clay." He got an old brickmaker and made some molds and fired them and got beautiful bricks. They got a hold of the clay and made a big mound out of it and covered it with plastic and he said, "Oh, that'll do it for the next 15 years, that'll make a thousand bricks. Then he let it be known in the district that he would repair all the old brick fences and barns and anything built out of bricks. He got 36 people out there repairing the whole district and building new fences and new barns. These five brickmakers, who only work four days a month, are supporting 36 bricklayers. This 100-acre farm he's got, he's got about 150 people whose employment is on the farm. He wasn't a farmer, so he says to the agriculture department, "What do I do with these cows?" They say, "You milk them. You've got a milk shed so milk them." He milks them, sells the milk, and makes half a million dollars. Then he says, "Now, what do I do with all the shit that comes through?" "I suppose you could make methane out of it," though they couldn't tell him how to do that. So, he had to fly to South Africa to find a good methane system. We improved that, brought it home, put a couple of years of research into it, 13,000 pounds engineering money in that. He put in his methane system. And his cows now run the whole farm's electrical system and they also produce methane to drive the cars. But then, the bloody shit just went straight on through the methane digester and he hasn't reduced his quantity one ounce. It hasn't reduced any manure because methane doesn't use any essentially. It actually adds some nutrients. "What do I do with all of this shit left over," he says. I said, "Well you could produce worms with it." So he laid out a big straw bed and pumped the liquid manure on it for his worms, and that brings in a million-and-half pounds. "My God," he says, "I should raise those cows just for the worms." He said, "That's the real profit out of cows is the worms." So he sells worm castings and worms. I said, "When you got worms you could be growing salmon?" He said, "My God, I hadn't thought of that." You've got to up the value of your worms by growing salmon. How many people can you employ on a hundred acres? Well, he doesn't employ them either, each one of them owns their business and they pay him 10% of their net. So he pays his farm off in six years.

ACRES U.S.A. That sort of enterprise in abundance can work.

MOLLISON. I've got all the money I paid for the farm, and all the money I put into the farm, and all the money I've paid for this and that so I'm out of it. You keep the house and a couple of things you're interested in and lease it. I'll do the same. I'll be out of it in five or six years. I don't want to employ anybody. I believe that employer and employee, landlords and lessors are the same types. It's like being employed or unemployed, you know, somebody is paying you to spend the day the way they are paying you for. I don't believe that's a free life. I believe in a freedom. All men are created equal. I believe that. So they are all self-employed.

ACRES U.S.A. Geometry the combination of elements of nature play a very large role in your design process. All of the factors, the mathematics seem to have come from your study of nature.

MOLLISON. Yes, I've studied forms and patterns. I've made a few really good hits and I invented something which I think helps people a lot. It is as though we are brought up in an education system that never puts anything together to see if it works better.

ACRES U.S.A. And in nature everything is together.

MOLLISON. Everything is together in nature. Nobody stops the chicken from going into the forest. So there were lots of systems that used to happen that are not allowed to happen in modern agriculture. Just look from the air, the farms are all square, all square fields and all square buildings. It's the product of a disordered mind.

ACRES U.S.A. So laser leveling is not in your plan?

MOLLISON. No, no, no. But my water can get through the back fence alright, or can cross the entire property. It's got to water gardens and trees and cattle. Engineers have a beautiful term: it's called putting water to its duty. All my water has to produce electricity, grasses, irrigate trees and crops, and it's got to do it on its own. I'm not going to be there. When it comes through the back fence, it's got to be sent to its duty. So, I send it.

ACRES U.S.A. As you lay back and listen to the sound of falling food.

MOLLISON. Really.

ACRES U.S.A. Can you think of any parting words that you'd like to tell American agriculture?

MOLLISON. Well, American agriculture should go hang, or retire, agricultural scientists for the good of America and the world. American farmers should start getting together, like our farmers do, putting in systems, showing other farmers how to do it, and become teachers. And they should get rid of their agriculture department and their agriculture colleges. And that money can be put into the development of sustainable systems, instead of spent on a useless bureaucracy which wouldn't know a sustainable farm from a hole in the ground. I went to a big conference on sustainable agriculture hosted by one of your state departments of agriculture. They came from all over—universities and institutions—and they said we could observe if we'd like to and they said, "Would you like to talk for 10 minutes."

I said, "Sure. Really," I said, "not talk so much as ask you a question. Would you define sustainable for me." After five minutes, I said, "Well, that's enough. Obviously you don't know what sustainable is." I said, "Now I have one more question for you. Do you have an example of sustainable?"

Mutter, mutter, mutter, mutter, mutter.

I said, "OK, that's enough, you don't have an example of it. I don't think you have anything to learn here." They were meeting for something they haven't defined, can't define. They don't have any examples of it. It's amazing. And between them they are spending millions and millions of dollars. One can imagine how far it's going to get. There was a farmer in the course and he turned at the door and he said, "You guys are always the same. You're always thinking of a one-step solution to a two-step dance."

ACRES U.S.A. So you don't see any reason that a family living on the land should be owing the bank and on the treadmill?

MOLLISON. No. They shouldn't be dealing with a bank. If there is a bank it should be a farmer's revolving loan fund.

ACRES U.S.A. But they shouldn't be hand to mouth, unable to clothe their children and put them through college?

MOLLISON. No, no. They should be beautifully fed, they should have a local market they can take very mixed product to and sell locally. And they should have their own banking system and America would bloom. Farmers should link funds to farms.

ACRES U.S.A. And they can do it?

MOLLISON. They can do it. There is one little town in which we put our own revolving funds, a poor little town, a rural town with about 150 houses. One of my students went there and started up a revolving fund ten years ago. Last year they invited me to their anniversary and they gave me a lot of pottery. I said, "What are you doing giving this to me? I put a bit of money into this but I haven't actually done anything." I said, "So how has it been? We've invested $33 million in the past year and that town is sparkling, every business is smart, everything is up-to-date, every house is painted, every farm is owned by every farmer."

ACRES U.S.A. What we see in America is the countryside being decimated. Small towns dying, schools closing, people being forced to home school their kids ...

MOLLISON. They're running in circles like they do at the universities. My little town CSA, with an average investment to start with of $300 per subscriber, is now investing per subscriber at $1,000 and there is a huge surplus of money and the people run it. And they are just the nicest people and they supply the elderly people with fuel and do a lot of things which are public service. Because they are making so much money that of course they can look after the aged and put up facilities for people. And that is the community looking after itself. Government can't do that. The public servants won't do that. And banks, unthinkable, they wouldn't lend to it; the lands are too small for a bank to handle and they wouldn't lend for someone to grow organic food. No, no, no, they would say.

ACRES U.S.A. They had a lot of the same problems we have here but they have created solutions.

MOLLISON. Yes. We're the model makers. We think of something good to do, we make a model, we work it—and when all of the bugs are out of it we teach it. And we teach it all around the world.

ACRES U.S.A. Because of that you're making a difference?

MOLLISON. Yeah, we're making a difference. And enjoying ourselves. There are such a nice lot of people in permaculture. They are happy and helpful and cooperative and smile a lot, make a lot of jokes. We don't take any government money or any industrial money. And that makes us very dangerous because no one can stop us, no one can jerk our farms out from under us. It's all our money. We don't cost anything. The World Bank can't destroy us. We set up strong posts of resistance all around the world. But I figure we started a couple of decades too late. If we started 40 years ago, we would have gotten rid of the World Bank and FAA and United Nations and all of these useless, expensive agencies. We probably could have gotten rid of the departments of agriculture and agricultural colleges and possibly we could have got rid of all political parties.

ACRES U.S.A. So the race is on?

MOLLISON. We'll win. But they have done so much destruction—the political parties, the banks, and the ag agencies—that it sometimes seems like they can destroy faster than we can build.

"Get the Soil-Food-Web Back into the System"

An Interview with Dr. Elaine Ingham

Originally Published: January 1997

In 1997, the term "organism" was used somewhat loosely. Edward Wilson, the Harvard biologist, was saying that organisms—including plants, animals and microorganisms—number perhaps 1.4 million. This figure was wrong. Exponentially wrong, but that was just evidence of the times, so poorly defined were some species and so poorly organized was the literature on diversity in general.

It's gotten a little more clear now that we better understand the seemingly infinite diversity of microorganisms. Many biologists now figure the estimate of biological diversity—from microbes to elephants—is in the billions and trillions. One might say we know enough now to say an exact audit is impossible.

Dr. Ingham is an associate professor at Oregon State University. After studies at St. Olaf College and Texas A&M, she earned her Ph.D. at Colorado State University in microbiology. She is widely published and respected for her research. Her most recent work involves the analysis of tens of thousands of soil samples for microorganism biomass and activity with this information correlated to soil chemistry and plant growth. The resulting body of knowledge is nothing short of amazing and is setting academia, sustainable agriculture, and conventional agriculture on end. She humbly relates her unique understanding of the "soil-food-web" in this very special interview.

ACRES U.S.A. Well, let's go at it this way. We're told by people like Edward Wilson, the author of books like, *The Diversity of Nature*, that even in one handful of good soil you have more units of life than you have people on the face of the Earth. This becomes such an awesome concept for the average farmer. He doesn't know how to cope with it in his thinking.

INGHAM. I kind of liken it to looking at your own blood stream. When you look at the result of how your bloodstream operates, it looks pretty simple. You're alive or you're dead. But when you start getting down to the subtleties of really understanding your bloodstream, there's a lot of bits and pieces and components and complex interactions that you have to understand if you

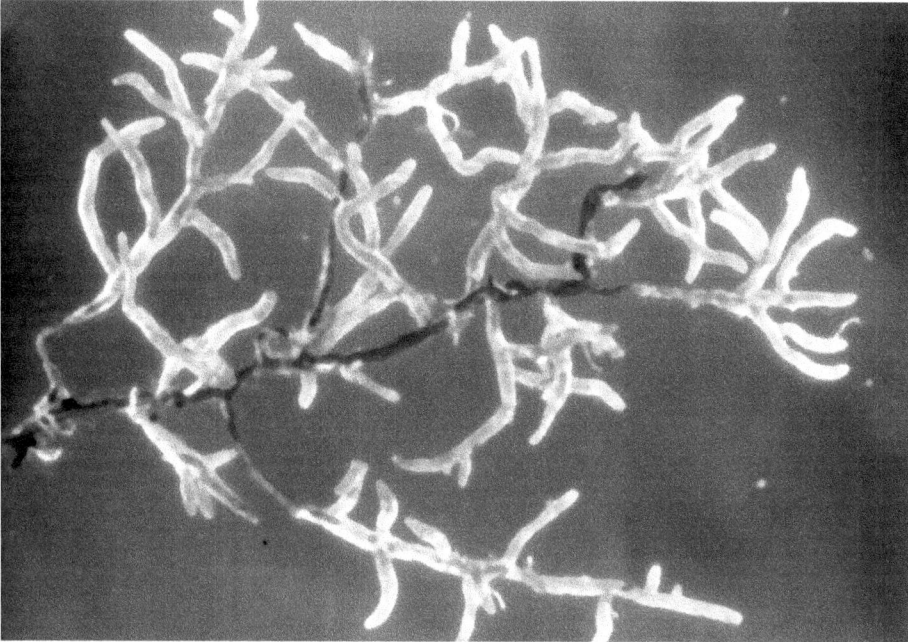

really want to understand how your bloodstream supports life. And it's the same thing when you're looking at the soil. The organisms in the soil are like the bloodstream of the human body. It's these organisms in the soil that support the growth of plants in the soil.

ACRES U.S.A. Sometimes I think we get lost in the identification of various organisms. We talk about anaerobes and aerobes. We talk about nitrogen-fixers and Pseudomonas and things like that. Is there some simple common denominator we could lay out to make this more intelligible to people?

INGHAM. What we use is something called the soil-food-web. What we've done is gone to functional groups. It's like looking at your blood, again. We can talk about the functional groups, the functional types of blood cells and make the story of all the interactions going on in your bloodstream much simpler. So we do the same thing for soil organisms.

ACRES U.S.A. We will reproduce the diagram alongside this interview.

INGHAM. In that food-web you're looking at a number of different trophic (characterized by nutrition) levels and we arrange them in columns, so all the organism groups that are in the first trophic level, the second, the third, the fourth, are arranged in a column. All of a sudden it becomes a lot easier to deal with the structure of this food-web because you're talking about organisms that are doing similar kinds of things, even though one group is bacteria, another group is fungi, another is the mycorrhiza fungi, another is the plant-feeding nematodes. They're all basically secondary consumers and they're all eating plant products of some kind. So you can talk about that trophic level as being the herbivores of the below-ground. They're all eating plant material or material that's still pretty much recognizable.

ACRES U.S.A. They form the first column in your diagram?

INGHAM. And each of the individual organism groups then is a box.

ACRES U.S.A. So by sorting our microorganisms by function, we don't really have to be concerned with the nomenclature of the genus-species of these umpteen million types of organisms?

INGHAM. Not at this level of resolution to understand how the soil cycles nutrients, how it does the job of cleaning up waste products and turning them back into something that's useful for plant growth. We don't need to know all of the species names of the organisms. It's just like you don't need to know all of the names of the different types of blood cells in order to understand how your blood system works. So we cannot get caught up in that level of detail. We can back off a little bit and talk about more of these general groups. Now when we talk about bacteria and the basic function of bacteria in the soil, we can then split that group into specific groups of bacteria that perform different particular functions. And then, within that functional group of bacteria, within that higher resolution breakdown, then we can start talking about different species of bacteria that perform that particular functional group. Now how many of those species actually occur in your soil? Have you lost some of those species because of your farm management? Have you lost all of the species in that functional group, in which case, that function doesn't occur in your soil and now, maybe, your whole ecosystem won't function because you have lost a critical group of bacteria.

ACRES U.S.A. How many distinct levels, or columns do you recognize?

INGHAM. In agricultural systems we're looking at about four to five columns of organisms.

ACRES U.S.A. What is the second column?

INGHAM. The second column are those things that eat the herbivores. So they're the predators in the system, and within that column we have things like the three groups of protozoa that eat bacteria. We have the fungal-feeders. There are two major groups of fungal-feeders, the fungal-feeding nematodes and the fungal-feeding microarthropods. Those are the organisms within that general group of predators. And then we have another level within that food-web, the fourth trophic level, or the things that eat the things that eat the herbivores. Remember that the herbivores eat the plants. So there are four columns, basically, in that food-web. The two organism groups that are in that fourth trophic level, the predatory consumers, are things like predatory nematodes and predatory microarthropods. They eat the bacteria-feeding nematodes and the bacteria-feeding protozoa, and they keep those guys in line.

ACRES U.S.A. And on and on ...

INGHAM. Right. So it's a food chain or a food web, really, because there are lots of interconnections. The predatory nematodes, for example, most of the time are eating fungal-feeding nematodes or bacterial-feeding nematodes, but they can eat protozoa—and they can eat fungi if they're really starved and there's no other food.

ACRES U.S.A. But how do they contribute nutrients to the root of the plant, such as with mycorrhiza?

INGHAM. Now we want to go back to the beginning of this whole food-web again, now that we've looked at the structure, at the levels, at who is in each column? We want to go back to the beginning and explain the function of each of these boxes as we go through this food-web picture.

ACRES U.S.A. Where do we begin?

INGHAM. Let's go back to that first trophic level; we need to go back to the plants first. The plants in the system are fixing carbon, taking the CO_2 out of the atmosphere. They're converting that carbon, using sunlight as energy, into carbon substrates for everybody else in the system. So they're making sugar and carbohydrates and lipids and all of the complex things that make life work. Plants, when they fix that carbon—and this is a big generalization—on average, 50 percent of that carbon is going to go down into the root system. So everything that's fixed above-ground, half of that's going down below-ground. Of that carbon, that photosynthate that gets moved into the root system, somewhere between 40 to 80 percent gets dumped into the soil in the form of exudates. That's a lot of carbon that the plant is essentially dumping out into the soil around it's root system. You might think that the plant's pretty stupid to take up all that energy—and it's cost the plant a lot energetically to do that—and then just dump it down into the soil. You know, it's like these plants are really stupid or something, but they're not. They're doing this for a good reason. Basically what they're doing is putting that food out into the soil to grow the right kinds of bacteria and the right kinds of fungi around their root systems so that their roots will be protected against disease-causing organisms. So when a plant grows its roots into the soil, it's expecting to find these beneficial bacteria and fungi and encourage their growth around their root systems. If we've done something to kill off those beneficial bacteria and fungi, then when the plant dumps out all of this carbon substrate, the beneficial bacteria and fungi don't grow around its root systems. Essentially, the plant's feeding pathogens, feeding things that are going to eventually kill the plant. And that's what happens in a lot of instances. A lot of the chemicals that we use, the fertilizers that we use, and the management that we perform in agricultural systems kills off those beneficial organisms. Now we put ourselves into a disease cycle that becomes increasingly more difficult to control.

ACRES U.S.A. We are creating our own problems?

INGHAM. If we can do something to get those beneficial organisms back into the soil, back into the root system, the whole system's going to function better. You have to realize that plants have done these things for millions of years, expecting to find these bacteria and fungi in the soil, and it's worked for them.

ACRES U.S.A. Well that's the exchange system then. The plant is delivering the carbon to the soil, or the sugars or the exudates, and in return getting some of the minerals and whatever it gets from the various types of life forms down there that it expects to find. Is that what you're saying?

INGHAM. We haven't gotten quite into the cycle. That's further down, in some of the higher-level trophic interactions. So in this first interaction between that first trophic level, the plants, the fixers of the carbon, the basic function, at least that scientists are coming to understand that in this first step it's really the plant getting a protective rhizosphere, protecting its roots against pathogens. Its purpose in doing this, if we can be anthropomorphic about this and give plants feelings and intelligence...

ACRES U.S.A. Well, they probably have intelligence, not necessarily intellect, but intelligence.

INGHAM. They don't have brains like you and I have, but they've been selected for a very long time and those that are successful survive and reproduce. So if you're a plant dumping out lots of carbon into the area around your roots, and you're encouraging the right kinds of bacteria and fungi, you're protecting your roots, and you're going to live, whereas a plant that dumps out the wrong kinds of exudates and encourages the pathogens—it's not going to live. You can see how selection operates in this particular system. Those things that put out the right kind of exudates and get the beneficial bacteria and fungi, they're going to survive, they're going to make seeds, they're going to have offspring. Things that don't do that aren't going to have offspring and we don't have that kind of plant anymore. Now the second thing that's going on in this interaction is that the bacteria and fungi are growing together in the soil—and there are many of them around the root system, somewhere on the order of a thousand million to ten thousand million bacteria per gram of rhizosphere soil.

ACRES U.S.A. And a gram is about the weight of a dime?

INGHAM. A gram is a teaspoon. So when you think about the amount of sugar you're putting into your coffee in the morning, that's a gram. Think about that same volume of soil; there's a hundred million, a thousand million, 10 thousand million bacteria in that little bit of soil. And then multiply that by all of the area around a root system, multiply that by all the plants in an agricultural field, and you're talking about more organisms than there are people on the face of the earth.

ACRES U.S.A. How much of this soil life is there going to be out there if you've doused the place with Roundup, though?

INGHAM. It depends. Each pesticide, each herbicide has its own particular effect on this system below ground. One thing we know about Roundup, for example, is that it does serve as a food source for certain kinds of bacteria. So when you put Roundup on, you get a big burst of bacterial growth, but it's only certain species of bacteria. The problem is we don't know which species.

ACRES U.S.A. Is that good?

INGHAM. We don't know whether those are beneficial bacteria that you're encouraging or whether those are pathogenic bacteria that you're encouraging. We have no idea. Chemical companies are not required to obtain that kind of information. There's nothing on the books that says they have to discover that information. So they don't know either. They don't know who they're encouraging. They don't know if they're improving the bad guys or the good guys. As scientists, we really can't say whether it's good or bad. We can tell you that you're getting a huge bloom of bacteria. Is it good or is it bad? We don't really know. What we're beginning to understand about soil is that in all agricultural systems—row crop, agricultural vegetable crop—those plants do much better when you're looking at a bacterial-dominated soil. The best growth actually happens when you have about an equal biomass of fungi and bacteria. That's very different from what you're looking at when you're in a shrub system or if you're in a forest system. In order to get trees or most shrubs—things like raspberries and blackberries and blueberries—we know that those soils should be fungal dominated. And you should have 10 times more fungi than you do bacteria in order to just get a seedling of one of those kinds of plants to start growing. We know that in most conifer systems you have to have 100 times more fungal biomass than you do bacterial biomass in order to have healthy growth of those trees. So different soils need very different food-web structures.

ACRES U.S.A. Soils? Soils or crops?

INGHAM. Soil, because when we're in a forest system ...

ACRES U.S.A. We're assuming a forest crop in a forest system, though.

INGHAM. If you're trying to grow Christmas trees, that's a forest crop. Those soils under your Christmas trees have to be fungal dominated or that Christmas tree is not going to grow. And there's been a number of studies now that are showing the importance of that fungal component in a forest soil. But if you are trying to grow broccoli or corn or wheat, that soil had better be bacterial dominated or you're going to see a decrease in production. You're not going to get as much yield if it's a fungal-dominated soil. Take the example where you've got a forest and you cut down all the trees. Now you're going to put in row crops. Do you have a problem? You're not going to be able to grow your row crops in that old forest soil because that forest soil is fungal dominated. So what do we do to convert that soil from fungal-dominated to bacterial-dominated? Well, it's pretty easy. It's called plowing.

ACRES U.S.A. What happens when you plow?

INGHAM. You wipe out by 10- to 100-fold that fungal biomass in the soil. You convert the soil to a bacterial-dominated soil. Now you put your row crops in and everything's fine.

ACRES U.S.A. Well, without suggesting the answer necessarily, we ran into the same thing in Brazil, where they took out coffee groves and replaced the crop with soybeans and had a very difficult time growing it. And the only correlation we could make was that the coffee would produce beautiful burdock, which is iron loving, and the soil chemistry was wrong for soybeans. But what you're saying is that there was exactly the wrong kind of life in the soil.

INGHAM. Right.

ACRES U.S.A. What they were I don't know.

INGHAM. Until you look, you don't really understand. But what we know about coffee is that it is a plant that requires a fungal-dominated soil. The things that fungi do in the soil is to produce the right kind of metabolites. They make the pH right. They alter the way mineral nutrients cycle in that soil, the availability of iron, the availability of magnesium, potassium, and they make that right for those plants that require fungal dominance in the soil. Well, when you're looking at row crops, you require an entirely different chemistry in that soil compared to a forest. Bacteria are going to push the pH more alkaline. Bacteria are going to push the form of nitrogen into nitrate or nitrite, instead of ammonia. So it's ammonia-dominated in a fungal soil. It's nitrate/nitrite-dominated in a bacterial-dominated soil. So those crops that require that kind of soil chemistry, they require bacterial dominance. You can see how the whole system works together. We just haven't known about this—because we haven't looked—until just probably the last 10 years.

ACRES U.S.A. So if the farmer is given the information as to what type of microorganisms he should have in his soil system, he can take remedial steps?

INGHAM. Right. He can do things to make sure that his management is encouraging the right kinds of microorganisms, the right kind of soil food-web structure to grow his plants and give him the best yield. And we're discovering what is that soil-food-web structure, and how do we very rapidly assess whether you've got the right organisms or not? And then how do we fix that problem if you don't have the right kind of soil food-web structure?

ACRES U.S.A. Has your investigation probed into some specifics?

INGHAM. We're discovering more all the time. It was probably only about five years ago that we had enough data to start saying agricultural crops, row crops, have to be bacterial dominated, and what the ratio of fungi-to-bacteria should be.

ACRES U.S.A. Is this always true?

INGHAM. It's always true!

ACRES U.S.A. In other words, you're talking about cereal crops, your basic storable commodities, rye, corn, wheat, soybeans, milo, that kind of thing?

INGHAM. Right. The only row crop that we found that does not do best in a bacterial-dominated soil is strawberries. Strawberries require a fungal-dominated soil, and that dominance of fungi should be around five to six.

ACRES U.S.A. Well, we wouldn't hardly consider strawberries a row crop, in the traditional meaning of the word.

INGHAM. By row crop I really mean things like corn and barley and wheat and pasture grasses and ...well, pasture grass really isn't a row crop, but it follows the same sort of pattern.

ACRES U.S.A. Have you related these findings to the type of weeds that are produced?

INGHAM. Most of the weeds in the system require the same kinds of soil, for example Johnson grass. It requires exactly the same kind of food-web structure. But, for example, the work that Ann Kennedy is doing or Mark Azevedo ... they're all working on specific bacterial species that kill weeds. We're discovering that we can apply certain specific bacterial species, and they will kill, specifically, Johnson grass or groundsel or nutsedge. The finer tuned we can be about the structure of this soil-food-web, the more specific we can be about controlling the plant species that are there.

ACRES U.S.A. By salting the soil, so to speak.

INGHAM. Right. Adding the organisms who are going to kill off the plants we don't want to have there.

ACRES U.S.A. Do they actually kill them off or do they just create an environment where they can't grow very well?

INGHAM. The mechanism of how they function is an interesting question. Now the work that Ann Kennedy is doing shows that those bacterial species specifically attack those weed seeds, so it's a decomposition process. The enzymes that those bacteria produce are the enzymes that break down the seed code, or attack the seed, the plant embryo as it germinates and starts to come out of that seed. And whatever carbon substrates that particular plant is producing, those are the ones that this bacterium needs to eat and then it decomposes that weed seed. Then, you see, it doesn't touch your wheat plant, or it doesn't touch your corn, as it germinates, because the corn or the wheat are producing very different carbon substrates. They're not producing the exact kind of food needed. Sometimes, I liken the soil system to looking at a smorgasbord. Certain people like certain kinds of food. Let's say this plant is producing the plant equivalent of chocolate mousse. And if you've got a human being that is only going to eat and consume chocolate mousse, then that human being comes along and wipes out that plant because it eats up all of the chocolate mousse. So it's gone. Well, it's the same sort of thing in soil. This particular weed is producing the bacterial equivalent of chocolate mousse and that bacteria then attacks that seed, chews it all up, and it can't survive, it can't grow.

ACRES U.S.A. How far along is that research?

INGHAM. They keep coming up with more types of bacteria that are selective against different kinds of weeds.

ACRES U.S.A. Before we interrupted our interruption ...

INGHAM. Okay, so it kind of goes back to this level of resolution that we're talking about. We're starting to delve down into the specific bacterial species and what each species does. I want to kind of back out of that, go back to a more general level of resolution and tie us back into where we were before. Bacteria and fungi are going to grow around the root systems of these plants, and they have a protective effect. I'll talk a little more specifically about mycorrhizal fungi in a couple of minutes. The other thing that bacteria and fungi are going to do in soil in general, and especially around the root systems, is to take up all of the available nitrogen. Both bacteria and fungi require more nitrogen in their biomass than any other kinds of organisms. So they are essentially nitrogen sinks. When they're using the carbon substrates coming out of the plant, when they're growing on organic matter in the soil, when they're growing on litter that's being mixed into the soil, or litter that's on the surface on the soil, they're very good at grabbing all of the nitrogen and holding it within their own biomass. They're not very efficient on carbon. They blow off a lot of carbon in their metabolism. They're going to take up a lot of carbon, but they're going to get rid of it as CO_2. They're very good at respiring CO_2. But they'll hold on to all the nitrogen. They'll hold on to most of the phosphorous that they encounter, and the other micronutrients—iron, magnesium, manganese, etc. So they're very good at retaining all of these nutrients in the soil, in those upper layers of the soil where we want to keep it. One thing we need to realize about inorganic fertilizer is, when we add that fertilizer, there's an osmotic shock all around the fertilizer you just added, and you've killed the organisms in the immediate vicinity. So now, when you wet that fertilizer, and it starts to leach or wash through the soil, you have to have bacteria and fungi alive and active in order to tie that nutrient up in the soil or that fertilizer has to hit the plants roots directly. Well, anything that doesn't come pretty much in direct contact with the root is going to leach through the soil and get into our groundwater eventually. And you and I as taxpayers have to pay money to clean up that groundwater so we can drink it. It would be better to tie up all that inorganic fertilizer that we're adding in the soil, in the bacteria and fungi, where—as I explain the rest of the system—you'll see that someday it's going to become available to the plant.

ACRES U.S.A. So this is the nutrient retention part of the cycle?

INGHAM. The bacteria and the fungi grab on to all those good nutrients and retain it in the soil, where the plants are going to be able to get to it, eventually. Again, if you're applying, for example, atrazine, it is probably a better bactericide than it is an herbicide.

ACRES U.S.A. What is the effect though?

INGHAM. It kills many more bacteria than it kills plants. If you apply atrazine year after year after year after year, what you'll eventually do is destroy most of the bacteria in the soil so you don't retain any of these nutrients. It means that the nitrogen you apply as fertilizer is just going to wash right through the soil and end up in your drinking water.

ACRES U.S.A. Your suggestion?

INGHAM. We need to stop doing that, or think about what we're doing and make sure we're adding the bacteria back in after we've filled them with these pesticides …

ACRES U.S.A. Doesn't it kind of boil down to the proposition that, in order to make the salt fertilizer work, you have to have the organic matter and the equivalent of a good compost or it doesn't work.

INGHAM. It kind of goes back to the smorgasbord idea. You need a smorgasbord with lots of different kinds of food for different kinds of bacteria and fungi. Somebody's always going to be active, somebody's always going to be available to grab on to these nutrients as they go by. There's going to be this big net in the soil waiting for those nutrients to come through in the soil solution.

ACRES U.S.A. So bacteria are necessary to keep the nitrogen from migrating?

INGHAM. Bacteria are a sink for nitrogen because they have a very narrow carbon to nitrogen ratio—a C:N ratio of around three to maybe five carbons for every nitrogen. And when you think about their food resource and what they're utilizing in terms of organic matter, it's a lot wider ratio of carbon to nitrogen. The carbon-nitrogen ratio of leaf litter material, for example, is up around 75 carbons for every nitrogen. Probably the only thing that the bacteria might be utilizing that have an equal carbon/nitrogen ratio would be proteins coming out of the roots. Well, other plants exudates are things like simple sugars and carbohydrates, which have a much wider C:N ratio. So with the food resource that bacteria utilize, their efficiency for carbon on that material is pretty low. They blow off a lot of carbon from any sugar as carbon dioxide and then retain the remaining carbon and all of the nitrogen in their biomass. Fungi are doing the same thing. The C:N ratio of fungi is about 10 to 20 carbons for every nitrogen. They might be a little bit more efficient in conserving carbon as well as all of the other nutrients. Bacteria are probably the most inefficient ones. If you are a bacterium and you took one unit of carbon, 60 percent of that, on average, would be released as CO_2 and only 40 percent would remain in the biomass or get converted into secondary metabolites, waste products from the bacterium. For fungus, it's a flip-flop. Forty percent of the carbon is blown off as CO_2, while 60 percent is retained as biomass and metabolites.

ACRES U.S.A. So both retain nitrogen?

INGHAM. Both of those organisms are basically nitrogen sinks—and also phosphorus, sulfur, micronutrients, all of those other things. But we look mostly at nitrogen because that is the limiting element for most plant growth. If they're not water-limited, they're going to be nitrogen-limited. Then you might wonder how it is that plants ever get hold of any of that nitrogen since the bacteria and fungi are a lot more efficient at taking up that nitrogen, obtaining it from the soil. And that's where the predators come into the system, because all of the nitrogen would essentially remain tied up in bacteria and fungi, unless you had something that would release that nitrogen and make it more available.

ACRES U.S.A. Earlier you explained that they never die a natural death.

INGHAM. We just don't see autolysis, or old-age death in bacteria and fungi in soil. We can get old bacteria that live in microcosm systems where there are no predators, no wolves to eat them, but you don't see it in natural soils. Some of the microcosm experiments don't mimic entirely the natural system, and so you get some results that you would never see in the real world.

ACRES U.S.A. You refer to work in the laboratory ...

INGHAM. You need bacteria and fungi to tie up the nutrients. Another function of the bacteria and fungi is that certain species are suppressive toward disease. So you want to make sure that in compost you're getting those disease-suppressive bacteria and fungi, the ones that are going to compete with pathogens, or produce antibiotics or other inhibitory compounds or that physically occupy the space on the surface of a root and prevent pathogens from occupying that space and causing disease. So there's a whole group of mechanisms by which the right kinds of bacteria and fungi will suppress disease. Basically, that's what plants are trying to do when they put out these different kinds of exudates. They're trying to encourage those beneficial bacteria and fungi that prevent the pathogens from attacking their roots.

ACRES U.S.A. Is there a pecking order of one predator for another?

INGHAM. We're going to move from the trophic level occupied by bacteria and fungi—the second trophic level—to the third trophic level, the things that eat the bacteria and fungi. Protozoa is one major group of predators that eat bacteria. And within the protozoa, there are flagellates, amoeba, and ciliates, and different kinds of soils have different amounts of those different groups. But the carbon/nitrogen ratio of bacteria contains too much nitrogen. So as all of them feed on bacteria they will always release some of that nitrogen in the form of ammonium. The carbon-nitrogen ratio of protozoa is up around 20 to maybe 40 carbons. So when they feed on a bacterium, there's too much nitrogen. It's toxic to them if they kept it within themselves, so they release it in the form of ammonium and maybe a few proteins. That nitrogen, then, is directly available for the plant to take up. Most of this process is going on right in the root system, right around the roots. And so the bacteria are eaten by the protozoa, nitrogen is released, and now it's available very quickly, very easily for the plant. In the experiments that were done in the Netherlands, research that was done by my husband when he was a graduate student at Colorado State University, work that's being done by Dave Coleman at the University of Georgia, show that somewhere between 40 to 80 percent, usually on the 80 percent end, of the nitrogen in a growing plant comes from that interaction of the bacteria and fungi with their predators. So bacteria are fed by those protozoa groups. They're also fed upon by bacterial-seeking nematodes and the carbon/nitrogen ratio for all nematodes is about 150 carbons for every nitrogen. So when a bacterial-feeding nematode eats a bacterium there's even more nitrogen released because there's way too much nitrogen in a bacterium for the nematodes. So it too releases that nitrogen as ammonium. Fungi then are fed upon by fungal-feeding nematodes. And again, there's a lot of nitrogen released because of the difference in the carbon/nitrogen ratio, there's way too much nitrogen in fungal biomass. Also, the way fungal-feeding nematodes eat is to puncture the cell wall of the fungus with a style, like a big spear, and then with their mouths they suck out the internal contents of the fungus. So really they're dealing with cytoplasm. The C:N ratio of cytoplasm is around three carbons to five carbon's for every nitrogen, so taking in a C:N ratio material that's much lower than what we see with just the fungus itself

ACRES U.S.A. And nitrogen is released ...

INGHAM. A lot of nitrogen released in that interaction. The other fungal-feeding group is microarthropods, things like springtails, and many different mites. And again, as they feed on fungi, a lot of nitrogen is released because the C:N ratio of most microarthropods is up around 150 carbons for every oxygen as well. Now we can have the situation—and this has been observed in quite a few situations—where these predators of bacteria and fungi can get real greedy and overeat the bacteria and fungal biomass. If there's nothing to keep the predators from eating and growing and producing more young, we'll have a big population burst of the predators and they'll overeat their food source. And when that happens, then we see some really negative effects in the system where decomposition slows way down, where we don't get very much nitrogen retention because these predators are overeating the bacteria and fungi, and you don't have those bacterial and fungal processes occurring anymore.

ACRES U.S.A. How does nature balance this out?

INGHAM. The next group, the fourth trophic level, becomes very important in eating the predators of bacteria and fungi. If the predators of the protozoa, of the bacterial-feeding nematodes or the fungal-feeding nematodes in the microarthropods, are kept in line—their numbers kept down to an optimal level—they'll feed on the bacteria and fungi but not overfeed.

ACRES U.S.A. It balances out ...

INGHAM. So the things that eat the predators are predatory nematodes, predatory microarthropods, insect larvae, beetles, millipedes, centipedes, things like that. So that next higher trophic level is very important as well, and people should realize that the predators of the predators of bacteria and fungi are the food sources for a lot of our above-ground organisms, birds and wolves and shrews and mice, raccoons and foxes. During certain times of the year, the majority of their diet is made up of these below-ground organisms. And if we do something to destroy this below-ground system, then we're removing that food resource for all of our favorite warm-and-fuzzies. And we see reduced numbers of the organisms that human beings like to see out in the woods or out in the real world. So this nutrient-cycling system is important in a lot of different ways.

ACRES U.S.A. So this translates into putting a big question mark around the use of chemicals of organic synthesis to combats weeds, herbicides, for instance and pesticides and fungicides and ...

INGHAM. Basically it should really call into question our overuse of those chemicals. What we see is people get into kind of a chemical dependency cycle, where you go out and you use something like methyl bromide, or you use a fumigant that really reduces all of these organisms in the soil. Another thing that we've seen from our work is that all pesticides have non-target organism effects. There is nothing that does not have a non-target organism effect. You can't really blame the chemical companies. I mean we can sort of, but it's probably really not quite fair because no one has ever asked them to look at those nontarget organism effects. They test it on the target group. Nobody has ever said, "You also have to test it on the protozoa and the beneficial nematodes and everybody else." And so we've never known that we're also killing off all of these other things when we use these pesticides and really reducing the health of our soil when we do that.

ACRES U.S.A. So if you use atrazine to take out foxtail ...

INGHAM. You're killing off most of the bacteria in your soil.

ACRES U.S.A. Killing the bacterium and setting up a fall panicum for later in the year.

INGHAM. Right. And you're reducing the ability of that soil to retain nitrogen. Look at Iowa after using atrazine for the last 40 years—they've essentially killed off all of the important bacteria in the soil, and nothing retains the nitrogen in that soil anymore. And surprise—nitrogen is showing up in all of the wells, in all of the groundwater. There's nothing to decompose the atrazine anymore. So you're seeing atrazine in the groundwater, in people's wells. And it's because they've destroyed these organisms in the soil. They've destroyed the health of the soil through years and years and years of using these pesticides. We're pretty lucky in most soils that they were real healthy when they started out, and so we have a deep savings account in that soil. There has to be a lot of years of destruction before you finally push it over the edge where none of these processes are occurring; none of these organisms are doing what they're supposed to be doing, they're just so messed up that we don't get the nutrient retention. We don't get the degradation of the herbicides or the pesticides of all of those pollutants in that soil anymore.

ACRES U.S.A. And then we see these chemicals in our well water.

INGHAM. We see the same thing in California where they have enormous levels of nitrogen and salts in the groundwater and in the rivers and the lakes and the ponds, and it's coming out of the soil. It's leaching out of the soil because we've destroyed the soil-food-web, and those processes aren't occurring in the soil anymore. And so what we have to do is get these organisms back into the soil. You have to do something to put them back in there. How do you do that? Compost, compost teas, cover crops, making sure you get the food back into the soil for these organisms and you re-inoculate the right organisms. So we've identified what the problem is. We've identified the disease, if you will, the problems that are occurring out there, and now we need to fix them. That's right where we're at, at least in my program. We've wiped out this part of the food-web, or we've reduced this part of the food-web significantly. Now, how do we get 'em back? And how do we get the right guys back into that soil so the soil starts functioning the way it's supposed to? When your plants put their roots down into that soil and they're expecting to find this in-house nutrient cycling and they don't, there are real problems. And it's called disease.

ACRES U.S.A. Well, it seems to come down to this. The pesticides and herbicides, those kind of materials, are really not good food for microbes.

INGHAM. It selects for the wrong kinds of microorganisms in the soil.

ACRES U.S.A. Is the reason we don't know because we haven't asked the question, or has the research been stonewalled?

INGHAM. For a long time we didn't know that these organisms were down there in the soil or we vaguely knew that they were down there, but we just didn't really know what they were doing or why they would be important. In the last 15 years we've figured that out. Ten years ago, we started telling the EPA that there ought to be some testing to find out what all of these pesticides are actually doing to these organisms in soil, what they're disrupting. And there really wasn't any interest. Now, I don't know whether you call that stonewalling or lack of foresight or what. It seems really short-sighted to me.

ACRES U.S.A. We've long objected to the "make-it-to-the-door" test for toxicity.

INGHAM. My viewpoint is that some government agency ought to say to the chemical companies that we have found that these pesticides are doing things to soil organisms that are really important, and we should go back and test all of these chemicals and see what effect they're having. Not so much that you would ban these pesticides because they're affecting something else, but farmers need to know that when you apply methyl bromide, or when you apply atrazine, you're killing off all of these organisms, and now you have to do something to get them back or you're going to be very dependent on that chemical forever in order to grow your crops.

ACRES U.S.A. When these chemicals break down, do we know what the breakdown products are?

INGHAM. No, we don't. Sometimes, the breakdown products are in fact more toxic than the original chemical. Sometimes, certain species of bacteria are present and they will take a toxic material and break it right down to carbon dioxide and biomass, with very few metabolites that are toxic. But most pesticides are never assessed for those kinds of effects. Different soils have different organisms in them, and so decomposition of the pesticides is going to be different in different soils. And that's never been tested. People tend to think of soil as just being soil, ad soil in California is the same as soil in Tennessee. And it's not! All you have to do is go out there and pick up a handful of soil in California and a handful of soil in Tennessee and you can see that they're very different—completely different species, all of these soil-food-web organisms. And the processes that are going on are very different as well.

ACRES U.S.A. The worst that we've ever encountered was in Tasmania, kind of an adobe soil where there were only 20,000 microorganisms per gram. How far down do methyl bromide and those kind of chemicals take the organism count in a gram of soil?

INGHAM. It depends on whether you're talking about the very first time you use methyl bromide or whether methyl bromide's been used for a series of years. The first time you apply any chemical—and this is a generality—you're only going to drop the organism counts by maybe a hundredfold.

ACRES U.S.A. That's pretty significant in itself, isn't it?

INGHAM. It is significant. You are reducing the numbers of organisms. But, in general, they're going to come back pretty quickly, as soon as the bacteria and fungi have degraded that pesticide—or in the case of methyl bromide, it's volatilized out of soil, gone up into the atmosphere so it can do fun things to the ozone layer—they'll start coming back pretty quickly. So in most cases it's a fairly minor decrease. But the second time you use it, there's a greater impact, and the third, a greater impact. You know, with methyl bromide, it probably only takes five or six applications before you've got numbers way down there and things are really getting into bad shape. In the case of the soils we've worked with in California where methyl bromide had been applied many times — at least once a year for 13 years—we saw soils that contained no fungi, no protozoa, no nematodes—except for root-feeding nematodes—no microarthropods, and the bacterial count was down to about 100 bacteria per gram of soil when it ought to have been 100 million per gram of soil. There are data like that out in the literature that document what the decreases are. We've looked at probably about 25 pesticides and documented what they actually affected and what kinds of decreases you were getting.

ACRES U.S.A. What are the worse ones you've encountered?

INGHAM. The worst pesticides ... usually the fumigants are the worst. Methyl bromide is right up there at the top of the list. Atrazine is pretty bad. And it's all relative to which group of organisms are present.

ACRES U.S.A. Is there a way to predict the impact these chemicals will have on microbial life from the chemical structure?

INGHAM. The EPA will use similar-structure activity ratios, if it's got a chlorine in this position on the benzine ring, then, in general, it has these specific types of effects on this disease or pest. What we have discovered is that you can't use that information to predict the effect on the other parts of the soil-food-web. There is no predictor that we know of.

ACRES U.S.A. In the scientific community are they still using the LD50 approach at all of rating toxicity?

INGHAM. There's a significant community of microbiologists and soils people that use LD50 readings. The problem there is LD50s are determined in the laboratory, and you're dealing with species-by-species effects. Think about the fact that in a gram of agricultural soil in Michigan we're working with probably somewhere around 11,000 to 17,000 species of bacteria in just one gram of soil. To have useful information based on LD50, you'd need to know what the effect of that pesticide was on each of those 17,000 species of bacteria.

ACRES U.S.A. So what you're saying is, getting down to a microbiologist's point of view, it's really rather useless.

INGHAM. We need a better assessment of the effect of these pesticides on the soil-food-web than do species-by-species LD50 testing. We really need to get the community of people who work on soil organisms together and say, "What's really going to work? What's the testing that we need to do in order to come up with indicators that are really very, very good." The EPA has called us for those kinds of things, and every time we have a meeting we tell the EPA folks, "We don't have enough information to tell you what's going to be a good indicator because we haven't been funded. Wouldn't you folks like to support a research program where we could really tell you what the effects are? And we could tell you which were the really bad ones and which ones are kind of intermediate, and when a farmer applies a pesticide what is really being killed and what does he need to add back."

ACRES U.S.A. Politically, we have not been able to get our agencies to really take this that seriously. Is that what we're coming to?

INGHAM. I think you've put it in a nutshell. They don't perceive the importance of soil and what's going on in soil. They don't think it's important enough. People at the EPA have said to me, "Well, these soil organisms, they're really not very interesting. The electorate doesn't care about these organisms. You can't get them excited about saving species of bacteria." Most of us look at them and say, "We're not talking about saving a species of bacteria. Get real! We're talking about soil health. We're talking about growing food."

ACRES U.S.A. Well, aren't we also talking about human health in the final analysis?

INGHAM. Absolutely!

ACRES U.S.A. And they have not made the connection, or apparently strong enough?

INGHAM. Strong enough! I'm sure you and your organization have put that out for them to understand. The information is out there, but they just don't perceive the importance. Do you call it stonewalling or ignorance? The information's been out there for them to see, and it's just not risen to the top of the heap for some reason.

ACRES U.S.A. Where does your research take you from here now?

INGHAM. Because we've identified what these organisms are in soil and who's important doing what and we've got a lot of the background information behind us on how to fix the problem, what we really need to do now is go out to real-world situations and work with farmers in the field that are having a lot of disease problems. How do we move away from the instant chemical fix and manage these soil organisms so that they don't have to use, or at least don't have to use as much, chemical, applying the chemical only when it's critical, not using it as a preventative? How do we actually get this system back in place and get it functioning for the farmer? And really, the driving thing for the farmer is to show that all of these pesticides cost money. It cost money for you to put it into your tractor, get it in the field, and drive back and forth across your field. How many times do you have to do that in a year? Working with people down in California, in conventional agriculture, they cross their fields with their tractors applying side dressings and fertilizers and pesticides and herbicides, somewhere around the order of 22 times for each crop that they produce. Sometimes that's only three months.

ACRES U.S.A. Has your research taken you into the composting art to any great extent?

INGHAM. Quite a bit. Because we've demonstrated that the main problem is they're lacking some part of the soil-food-web and the system is falling apart, the obvious question is how do you fix it? Well, the answer is to put back into the soil these organisms that used to do the job for you. So how do you put the organisms back into the soil? Well, compost, compost tea, cover crops. Those are the answers. A lot of the time what we've done is mined—with the plowing, with our management of agricultural systems—out all of the organic matter. And that's food. That's the food the bacteria and fungi live on. Without the bacteria and fungi you don't have the predators, and the whole rest of the system doesn't work. So you've got to get the food back in the soil and that means replacing the organic matter. The wider the diversity of organic matter you can have, the more types of organisms and the more probability you're going to get the disease-suppressive organisms that you need in the system. So you don't have to apply fungicides and insecticides. It's not an appropriate habitat for all those disease-causing organisms if you have a good diversity of these other soil-food-web organisms, and that's been demonstrated in quite a few studies now.

ACRES U.S.A. Have you examined any of the people or looked at any of the people who've tried to deal with this pragmatically, one at a time, farmers, for instance, who've worked out some rather remarkable systems on their farm without having all this knowledge you're talking about?

INGHAM. We've been working with the people on sustainable agriculture in California, and they've worked out ways of making good compost. Now we can go back in and discover what is really going on with the organisms in that compost. They have empirically discovered how to manage to get the right kinds of organisms. But they've gone through 20 years of trial-and-error kinds of testing. We can come along and say, "Well, when you do this, you get more of these kinds of organisms, when you do that, you get something you don't want." And it really improves their ability to rapidly make a good product. We don't know if it really worked. So it reduces the time involved in coming up with the right system to get the right organisms back into that soil. We've had a fair amount of success. We were doing work with strawberries down in California, and by putting back into the system, in methyl bromide-treated soil that really had no soil food-web to speak of, we were able to get the soil-food-web back into the system and get better yields than with methyl bromide.

ACRES U.S.A. You also suggest that farmers take a good, hard look at what type of organisms they need if you're going to grow, ginseng, for instance, and remove it from the canopy of trees out into a field, you've got a completely different problem than a lot of the people who are growing ginseng think they have.

INGHAM. We worked with one ginseng grower, and they had no idea what the right kind of food-web structure should be in the soil to grow ginseng. We can tell you what organisms are in the field where you're growing ginseng, but are those the right organisms? How do we find that out? We have to go back to China and find out what the typical soil-food-web structure is in China, or as he grows his ginseng, those parts of his fields where he's having good ginseng production versus areas where he's having real problems. We need to compare those two areas so we know what the target food-web structure ought to be and then what's wrong over here where he's having problems. We know what to add back in.

ACRES U.S.A. Unfortunately, too many of them are still reasoning from an N-P-K platform and they're not dealing with mycorrhizal food-web that they need.

INGHAM. Right, right, and when you look at N-P-K recommendations, those recommendations are made to grow corn. Well, you're not growing corn. So what's the N-P-K that ginseng really needs and are you managing that system by putting the organisms that are going to cycle that N-P-K availability that those plants need. It's a lot cheaper to let the organisms in the soil do that work for you than for you to have to go out and put on fertilizer a bunch of times or pesticides or whatever is needed in that soil. So we're talking about a system that is a lot less work for the farmer.

ACRES U.S.A. In China, we see great degradation of the soil, particularly after the introduction of Western agricultural chemicals.

INGHAM. They're getting massive decreases in harvest and yield and it's just not working at all. And what I suspect is that their soils have been farmed for so long that they've been putting back in the food for the microorganisms, but now that they've stopped doing that, they're soil's are very rapidly losing the soil-food-web, and the chemicals just can't cut it.

ACRES U.S.A. ... because they don't inject the life in them. They have a lot of soils that are a wind-blown, loose type of soil.

INGHAM. So those are things that were right on the edge to begin with. No wonder they're going over the hill so rapidly.

"We've Come a Long Way"

An Interview with Fletcher Sims

Originally Published: February 1997

In 1997, when Fletcher Sims, 1969-2020, showed up in the pages of *Acres U.S.A.* magazine, it marked 23 years since his first interview. By then, the student of Dr. Albrecht and World War II vet had become better known as the dean of American composters, a title well earned as the author of *Fletcher Sims' Compost*, which is a landmark book still available through the Acres U.S.A. bookstore. In fact, Fletcher Sims took the art and science of compost making out of the backyard and installed it near the feedlots of the high plains. In so doing, he turned a disposal problem into a utilization achievement.

Most of the details attending this development have been made a matter of record. And it was on the basis of this record that Sims received the second annual Acres U.S.A. Lifetime Achievement Award in 1995.

ACRES U.S.A. Mr. Sims, it has been some 23 years since Acres U.S.A. asked interview questions and published your answers, some of which have made their way into *An Acres U.S.A. Primer.* We didn't cover developments in composting equipment simply because the evolution of equipment was hardly underway. How was it that a fish biologist came to end up developing a style of compost making and the tools needed to complement this activity?

SIMS. I was a fish biologist, but I was also a student of the late Dr. William A. Albrecht. I graduated from Missouri University and from service as a navigator during World War II. After the war, I went to work for the government, first in Montana, and finally in the Southwest.

ACRES U.S.A. You graduated to private enterprise?

SIMS. Yes, but not very successfully. At first I opened a fish hatchery in the Texas Panhandle. At exactly the same time, America's affluence turned lakes into water skiing and boat havens rather than fishing ponds. The business was not a roaring success. I drifted into package barn building which lasted for a long time. My barns looked like traditional barns, gambrel roofs and all. But the 1960s was an era in which steel farm and shed packages came to the farm country. Not long after that time Acres U.S.A. came into being and we did our first interview.

ACRES U.S.A. From your chair just what has happened, not only to the organic movement, but to composting?

SIMS. Well, when I first met you I was grasping for any source of information that might enhance the movement of making compost and getting it accepted. I had come to the conclusion that this was the most practical way to address the deteriorating soil in the high plains because every time you'd till you would lose soil. And the soil had taken some 500 years being developed according to the Land Institute up in Salina, Kansas. I could easily see that it would be a very long, slow process to redevelop the soil, which had developed under a short-grass prairie ecological climate. During the last 40-something years that I've been in the high plains of Texas, I've seen probably half of the soil that we have now cultivated broken out of sod. I was greatly disturbed over the fact that ever-increasing amounts of nutrients were being taken from the sod and put into tillable agriculture, which was a climate that could not endure indefinitely under the conditions we have. Then, of course, the humus deteriorated, and the water absorption and retention capacity of the soil, which is part of the soil structure, was also deteriorating. The crops that we had grown on dry land for so long ceased to be viable crops for many years of lower rainfalls. Then they discovered that they could pump the Ogallala Aquifer and irrigate the land. And of course, this is hastening the depletion of the aquifer.

ACRES U.S.A. So what you really saw was the destruction of the ecosystem of the high plains?

SIMS. It's definitely true. I couldn't say exactly, maybe in 50 years or a hundred years, but sometime in the future, when this irreplaceable water from the Ogallala Aquifer is gone, and when we've burned up the humus in the soil and employed it all in the organic residue, which so far has not been adequately utilized, we will have created desert conditions. We are working to remedy that. We are seeing improvements in the acceptance of compost. Twenty-five years ago, if you said, "Compost," people would look at you with a quizzical look on their face, saying, "Composé?" They thought you were talking about a product—I think it's a sedative named "Composé." They had never heard of compost.

ACRES U.S.A. Yet compost has been around in Egypt, China, in all the old civilizations, and the United States had never developed much further than a few people using it in the city garden.

SIMS. Necessity is the mother of invention. You know, in the past people didn't need to use compost because if they wore out one farm, they could move west and get some new land. Some people pride themselves on having worn out three farms.

ACRES U.S.A. What was the turnkey that led you to composting? You were a fish biologist. You ran a fishery in which you grew minnows for transport to lakes and so on. Then you built barns, but what was it that broke you out of those time-honored professions and made you take on what has to got to have been one of the most monumental jobs in the history of mankind—convincing agriculture that it should recycle and use compost?

SIMS. I guess two things had some influence on it. I had been a fish biologist for the state of Missouri. I participated in the first statewide collection of all of the freshwater fish in Missouri—there were 247 minnows recognized at that time. Then I worked for the fish and wildlife service doing environmental impact studies over about 12 states during a period of five-and-a-half years, and frankly, I detested working for the government. This put me out into the free enterprise system. In my first attempt to try to utilize my scientific training and some of my experience, I analyzed the market for bait minnows. The market was good right after the war so I built a fish hatchery to grow bait minnows, but the business didn't last. The desire to go fishing had been met by most people, and then people begin to acquire speedboats, and waterskiing and such recreational things became popular. That is why my market began to fall apart. A chance acquaintance got me into the farm-building business, which I enjoyed because I like to build things. I like to create something rather than just shuffling papers, and farm-building satisfied that need in my life. It also kept me in contact with the people with whom I best identified, and that was and is the farmer. So I was grazing out some 300 miles in all directions from Amarillo, packaging buildings, which I did in my own warehouse.

ACRES U.S.A. But the farmers didn't teach you about composting or even approve of it. It's been an uphill fight, has it not?

SIMS. Well, true. Partially the barn business played a part in this because you build a rather substantial relationship with an individual when you make an improvement on his homestead. If he doesn't like it, he's stuck with it. So I had a clientele of some 300 or 400 people for whom I had built buildings. When I hit upon the idea of upgrading and processing manure through composting, a technology that had to be developed because no one had ever done it, I had a list of farmers.

ACRES U.S.A. Specifically, how did you hit on that idea? Did it just fall out the sky like a strange letter over the transom or what?

SIMS. Well, almost. It fell in over my car radio one night. I was still in the barn-building business and the challenge of that was beginning to wane. I was listening to the radio one night and I heard Dr. Joe Nichols, who started Natural Foods Associates in Atlanta, Texas. I observed that he sounded like Dr. Albrecht. So I wrote a letter, which they published in their journal. I began to get correspondence from people here and there who confirmed that there was a lot of concern for the soil and for the kind of products that were grown on the soil. I had just chipped the 50-year mark, so I was beginning to think about my health and the impact that the quality of food you consume has on your life, your vigor, your longevity. All of these things put together led me to think about compost.

ACRES U.S.A. How long did it take you to actually meet up with Dr. Joe and his entourage at Natural Foods Associates?

SIMS. I don't know. It wasn't too long after I began to get the correspondence that I learned of some product sold out of Iowa. I went up there to investigate it, and then I called Dr. Joe Nichols and he said, "You need to come down here to Atlanta." And of course, anybody who ever knew Dr. Joe knows that one of the first things is a visit. He's kind of like I am, people need to come to Canyon, Texas to see for themselves what I do. I made a trip down there and he took me out to the back of the office there, which was a nice—I guess it was a residence on a farm they'd acquired—and he showed me the compost that Joe Francis had made. That was the thing that gave me the assurance that there was something that could be done with the organic material that could be beneficial for the soil.

ACRES U.S.A. And Joe Francis, was he helpful in instructing you?

SIMS. Yes, Joe Francis was, at that time, still in the composting business down near Atlanta, Texas. And he was blending materials. He was making compost and blending it in with Hybro-Tite and some other things and making a real good product. I contacted him and he sent me some of his product, which I used an inoculant for my first compost. In fact, he came out there for my first attempt.

ACRES U.S.A. Which is approximately the time that Acres U.S.A. interviewed you in the first place, some fragments of which actually appeared in *Eco-Farm: An Acres U.S.A. Primer.*

SIMS. Yes. I would call Joe every time I would see something. He would say, "Well, do you see any steam coming up or do you see any condensation developing around the compost?" And I would say, "I see this or that." He would say, "That's a good sign. Dr. Pfeiffer would say so-and-so." You know, Dr. Pfeiffer was already dead then. I began to consume everything I could get my hands on that Pfeiffer had written.

ACRES U.S.A. Of course, your statement of your life's experience has been brought up to a certain point in the book, *Fletcher Sims' Compost.*

SIMS. Right.

ACRES U.S.A. I think perhaps we should look ahead and answer for ourselves the question of where this whole movement is taking us? What has really happened?

SIMS. Well, during the last almost 30 years, we've come a long way from people not ever having heard the term compost.

ACRES U.S.A. Well, how well has this whole compost idea, which has been pioneered by you and embellished by others with some successes and some failures, how well is it starting to blanket the country?

SIMS. I think not only the country but the world is interested now because I get regular calls from Australia, South America, Mexico, Canada, and Europe. I think the idea of compost has now been established. I think anyone who takes any time to look into the matter will find that it is economically feasible to make compost and to sell it at a price that is competitive with other means of producing a crop, and that it doesn't take enormous quantities to get a result. The things that I used to read in some of the garden magazines turned out not to be true. They were always talking about putting an inch or a half-inch of compost on the soil. Whether or not you could use smaller amounts was another thing that we had to determine. We found that the amounts of compost that have an effect on the soil are so small that you can't even see it unless it is a real light soil, like a light, sandy soil. In that case, you may be able to see a dark tinge where you've applied the compost as compared to where it hasn't been applied. This kind of information is now easily found out by anyone who is interested. I think that is playing a part in the fact that composting is becoming so widespread. I hear from people who now have compost products that they had never dreamed of composting, but once people try it they are quite pleased when they are done. Another thing that I've found is when a person reluctantly starts composting they are worried about whether or not they can get enough material to make it worthwhile. The first thing that happens is that stuff just begins to appear on the scene that you've never heard of. In a strawberry-growing area of Oxnard, California, the Driscoll strawberry fields are using compost. I made the first compost with Tom Driscoll in about 1973, and he was seen as the prodigal son. The older generation almost had to die off before it became acceptable to use compost.

ACRES U.S.A. In ranging around the country, we find compost operations on a massive scale just about everywhere, north of Sun Valley, Idaho, moving across the country into Wisconsin and the Dakotas, Washington and Oregon, in the South, in Texas, and in Kansas. Everywhere we see great big operations, not one of which was even dreamt of in the philosophies of the teachers who were trying to teach future generations. Now, how has that happened?

SIMS. I don't know. You don't know what all has gone into the mix. As much as I am opposed to all the limits that the environmentalists are trying to—and are—imposing upon us, this probably had some effect on people looking to alternatives instead of taking their poultry manure out and dumping it directly on their land. When they saturate their soil with it, the raw poultry begins to cause problems. They forget to consider composting. When they do try composting, they're amazed: "Look, it works."

ACRES U.S.A. Your form of windrow composting is not the only form that can be practiced successfully, isn't that correct?

SIMS. That's right. It's like you achieve composting in the field. You can put your organic materials there and if you have the moisture and you have the time to aerate the soil it will compost. People have composted in cubes and bins and static piles. One of the more unique things that I ever saw was a little movie that Joe Francis made in time-lapse photography. The movie showed how poultry droppings could decompose right before your eyes.

ACRES U.S.A. How did Joe Francis get this decomposition right before your eyes?

SIMS. They got a photographer, I don't know how professional he was, but it was a good, believable movie that Joe showed me. They took a poultry dropping and the bright lights that they had to have to make the film would dry up the poultry dropping. I think it was Joe Nichol's idea to take a hypodermic needle and squirt a little water on the poultry dropping. It was like the time-lapse photography you have seen of a rose opening or a leaf growing. They kept this poultry dropping under observation with time-lapse photography you could actually see the first dropping slowly disintegrate.

ACRES U.S.A. And this bacterial action was being accomplished even while the chickens were growing?

SIMS. That's right. Joe would go in there and spread his Hybro-Tite compost mixture with the Pfeiffer culture in it on the litter, wash down the dust and so forth, bring the moisture up a little bit, clean the building, and then take the tiller and go in there and till it up. The thermometer would register, that is, the dropping would generate quite a bit of heat. I would judge—all of this is from my recollections of his movie—that it was probably close to a foot deep. Then there was Dr. Bob Howes, who grew up on a potato and beet farm in England, got a bachelor's degree in Canada, and his PhD in Florida. He worked in South America and different places. He was down at Texas A&M. Before he came there, he was at Clemson University, where Dr. Senn was.

He was the poultry pathologist at Texas A&M and his work so alarmed the other people at the poultry department that they made him go out to an old air base about 20 miles out off of the main campus. It was there that he conducted this experiment in an old officer bunk. He built cages and put birds in there and inoculated them with all the diseases that have plagued the poultry industry. He let these chickens go until they dropped dead there in their litter. He took this out and inoculated it with Hydro-starter and tilled it to add the air and, of course, to control the moisture. Just before Bob died, which was in the Spring of 1970, he called me and told me the results of this experiment. He had grown baby chicks on this composted litter and produced healthier chickens with less disease than the control group of chicks that he had grown on virgin soil. I have since tried to get some information on this experiment, but Bob's widow told me that the poultry pathologist who followed him was not interested in that work.

ACRES U.S.A. The experiment that Howes performed was composting these diseased birds plus their litter?

SIMS. Yes. This has led me to the conclusion that the big outfits, like Purdue in the eastern United States and Foster Barnes in California, could revolutionize the way poultry is produced. Yet they all seem to be reluctant to step out. They're like people that had the feedlot manure worm when you and I first met years ago. They are more comfortable doing what they've done than to branch out to some new area.

ACRES U.S.A. Fortunately, there are a number of people—this correspondent included—who have the gut feeling that we really shouldn't be producing anything that can't be recycled. How do you feel about that?

SIMS. Well, generally I'm in agreement with that. I wouldn't say to the exclusion of anything else because every once in a while there's something you can't recycle.

ACRES U.S.A. Well, let's say 90 percent.

SIMS. Yes, to the extent that it's possible I think this should be a top priority.

ACRES U.S.A. Regardless of where you are, a little bit of rumination would probably turn up all types of materials that really could lend themselves to the composting process?

SIMS. Oh yes, of course. I've said many times, anything that's ever lived can be composted. You know, I believe John I has written in the Bible, "All things were made by him," meaning God. With Him we've got anything made that was made. That's an all-inclusive statement and an all-exclusive statement. It stands to reason that God wouldn't create living things here on the

Earth—plant and animal—that would not be recycled back into the scheme of things when they die. We would be beyond our ability to wade through the dead vegetation and dead animals if there weren't natural decomposition. Of course, Petrik has said that. Vaclav Winthrop Petrik said that everything has an organism within it which will consume it when it's dead. These are the saprophytes.

ACRES U.S.A. Except maybe plastic.
SIMS. Well, but that is not a living thing.

ACRES U.S.A. No, that's a man-made molecule. And pesticides, those are man-made molecules.
SIMS. Yes.

ACRES U.S.A. You basically developed an entire industry, and developed a good portion of it dealing with feedlot manure, some of which came to you about as hard as big chunks of pavement. Can you bring us up to date on how this effort was met by the feedlot industry, and has their attitude mellowed in any degree?
SIMS. Well, they tolerated me as long as I was paying them for the manure. They could have been helpful in a lot of ways but they didn't seem interested. Now, when the environmentalists are breathing down their throats they begin to show a little bit of interest. I have to frankly say that I have just a little bit of contempt for them because I can see how things could have been taken much further. You know when I began in business, working under your premise of capitalizing from savings and planning on income, my savings were soon depleted. I had such limited capital that it made it very slow going at earlier stages. If I had a little more help, for instance, if the feedlot would have offered to let me use their $156,000 loader instead of forcing me to acquire one myself, this would have been a big help. But you get what you pay for, and the little jibes that I've received through the years, they kind of roll off of my back like water off a duck.

ACRES U.S.A. What were these?
SIMS. Well, people would make derogatory remarks.

ACRES U.S.A. Poke fun at you in the cafés and so on?
SIMS. Yes. It was never a problem with me. I'm sure that a lot of people would sort of turn back but I just went on.

ACRES U.S.A. The average commercial farmer tends to look for the university, taking the term university to mean the generic university regardless of where it is, as an intellectual advisor. What kind of help did you get from the university in those days?

SIMS. Well, Bob Howes, as long as he was alive, was very helpful. Through Texas A&M I could send him soil samples and he would get bacteria counts which I could have never afforded with my resources at that time.

ACRES U.S.A. But there are a lot more universities than the one that Dr. Howe worked for. There is Kansas State, there are the land-grant colleges, there are the sea-grant colleges, there are agricultural schools of various stripe from community colleges on up.

SIMS. Well no, they weren't helpful. In fact, I wanted to get my work documented by the recognized people at the experiment station. I am convinced that they set up the experiment on tracts of land that probably had some terrible herbicide or something else on it that leveled the soil. In fact, I had an agronomist named Gary Meyer working for me then. He said that his professors laughingly told him how you can direct the outcome of an experiment by what had preceded the experiment on that plot. I'm convinced this was done. These people had so little confidence in honest science that they wouldn't let him deliver compost to them for use in their experiments. They came and took it at random from his compost site. That shows the lack of confidence they had in us, and at the time they were publishing all these articles saying we couldn't return to organic farming. They would show how terrible your crops would be if you didn't have fertilizer and so forth. But I think that is changing. They've had the evidence crammed in their faces often enough now that they're slowly changing. Now there's a new generation. Just as in earlier times, some of the older generations had to move on before they let the next generation try new things.

ACRES U.S.A. The generation that we encountered at San Juan Pueblo, prior to this particular interview, gave some indication of that. Would you care to recount what happened there?

SIMS. Yes. They advertised on the internet for a soil plan. And of course, the San Juan Pueblo board—or whoever makes the decisions—obviously were amenable to this idea that they would get somebody who had some academic background in sensitive soils. Here they have a very rigid young scientist who got his graduate degree from Maine.

ACRES U.S.A. And they only have about a 3-year-old history of an agricultural program.

SIMS. He told us that one of the first things they did was to compare so-called conventional—meaning toxic and salt fertilizer agriculture—with organiculture, and the results were fantastically weighted on the side of the organiculture. They have now published some of their findings in the professional literature around the country. Now everybody will know it's so.

ACRES U.S.A. Linwood Brown, who has been a consultant at San Juan Pueblo in Española, New Mexico, has been hard on the hump for any kind of a program that would return some income—actual earned income—to the tribes. He's reached out to the compost idea in a very harsh country, has he not?

SIMS. Yes, you know, the transportation costs for the compost they've procured from me has been greater than the cost of the compost. Their soil needed help. These are crops that have been farmed, some of them for centuries.

ACRES U.S.A. Well, since the Spanish Conquest.

SIMS. Yes, and they respond to the addition of this humus.

ACRES U.S.A. What precisely does compost do? We know that we put it in the soil. We can band it in. We can literally do away with toxic chemistry in most cases. What does it actually do?

SIMS. This is a complicated question. I don't know all that it does. I know that it always improves the structure of the soil—in other words, the ability of the soil to absorb and retain moisture. It obviously has a lot of nutrients in it. I believe, again, that these nutrients are complex in a manner that puts them in a state of balance that makes the plants acceptable. Man has become pretty wise with formulating ideas about compost, but I don't think that man's finite knowledge of the thing has gotten there yet. Those are some of the first thoughts that run through my mind concerning what it does.

ACRES U.S.A. Well, we do know that compost makes it possible to farm without these chemicals, some of which, at one part per trillion, are cancer-causers and Alzheimer's-givers if you want to excuse that term. That much we do know.

SIMS. Yes, but I wouldn't say it's a cure-all for all ailments. I think that it probably addresses more than any one other concept of soil improvement because of what it gives back to the land.

ACRES U.S.A. We have a unique problem today, do we not? We have to build back the soils. Previous generations merely used them up and destroyed them.

SIMS. That's right. The humus. Historically, the most prevalent tilled acreage was in the ecological climate of the prairie or sod. It's under the grass. The grass is the best healer of the soil and developer of the soil. Every time you turn that soil, till it, expose it to air, it is sort of like stoking the coals in the fireplace. If you take a poker and poke them, you're opening it up to more air and it will glow and burn faster. Well, that's kind of the way with the tilling of the soil. The longer we till it, the lower the humus gets.

ACRES U.S.A. With that continued tillage we spend, so to speak, we consume the wealth of the nation. Now, as you near age 80, you seem to still be as active as a 50-year-old. You are still operating, not only in compost-making, but in developing machinery that has been adapted worldwide to this problem that we're talking about. Can you give us a quick run-down on the growth of the machinery that's been used in composting once we get beyond just a high loader or the manure spreader and so on? Of course, any big job requires tools. When you took on the development of this whole composting game, you really didn't have any tools, did you?

SIMS. No, the first thing I did was to see what agricultural tools could be put to this task, and I found nothing. I tried the Howard Rotovator, and it would chop up manure but it was not deep enough or thick enough to maintain the heat that you need. I went through road equipment. I used a Travel Loader, a little too cumbersome and too slow. We needed to develop something that could handle big quantities at low costs so that the compost would end up at a price that could be used in agriculture. For a year or two we used a manure spreader, which will make good compost, but, again, you get as much labor in picking up the manure and dumping it in a manure spreader and modifying the blades on the spreader so it will make a windrow instead of scattering it. So, I worked with an earlier customer and friend, Louis Weick, who had a blacksmith repair shop that served the farm community in Umvarger, Texas. We knew that we had to travel at a very, very low speed. He knew where there was an old combine that had a torque converter on it. So we used that and built a frame that would straddle a little windrow that was effective. It wasn't as sturdy or fast as the ones we later developed, but it was a starting point. My kids laughingly called it a tumble bug because it even looks like it's a little backwards. I dignified it by calling it a Scarab. Since then the name has been Scarab (the Egyptian symbol for dung beetles). It's been perpetuated by another manufacturer who has taken it.

ACRES U.S.A. Basically, how many generations of machines have you gone through from development to the point where you are now?

SIMS. I'm not sure that I can remember all of them, but I would say on the order of 10.

ACRES U.S.A. There have been other makers who have picked up a lot of these ideas. There is not too much that is patentable on these machines is there?

SIMS. No, most of it is public domain. John Mattingly, years ago, was a customer of mine. I shipped him compost from the Panhandle of Texas up beyond Denver to his farm. John was the one who invented the Waterpik, and I don't know what all else. But he said, "I just know that a machine can be made to be pulled behind a tractor." Of course, the ones that we built for a long time were all self-powered. There are still people that I've licensed to use my designs and patents. There is a company in Midland, Texas, that continues to offer self-propelled machines of my design.

ACRES U.S.A. Didn't General Motors, at one time, get into the compost-turning machine business?

SIMS. Yes. That was sort of humorous. When I look back on it and think of the giant General Motors building a giant machine ... The thing was so heavy that they could only move it on the public highways with a special permit. It only turned a little bitty windrow, and it didn't really aerate as well as the self-propelled machines. It pushed the compost over and rolled it a little bit from side to side. I'm aware of three of them that they built. I haven't heard of anymore being built in 25 years.

ACRES U.S.A. And then, of course, a lot of people have made homemade machines.

SIMS. Oh, yes, I've worked with many.

ACRES U.S.A. It's pretty hard to protect a patent, such as you have on your drum, is it not—considering our law systems?

SIMS. Well, it is not an automatic thing that just because you have a good, solid patent, that people won't try to copy your idea. They do.

ACRES U.S.A. What kind of progress is the Midland, Texas, operation making in putting these machines into the field now?

SIMS. Well, they were unfamiliar with composting. I licensed them to use my design and my patent, and I've added designs since then. The freedom from the responsibility of producing and selling the machines has left me to pursue new ideas I have for this pull-type machine. They've expanded on that by developing a self-powered, pull-type machine, which means that you can pull this machine with a small piece of equipment that will barely move the machine. This is because the power of the machine is lifting and removing the material from the front of the machine. All you have to do is keep the machine up with the material that it moves.

ACRES U.S.A. So you can pull it with a tractor or maybe even a sturdy pick-up?

SIMS. Yes, but you might end up like I did. I just hooked it up to my El Camino and sat there and kept a tension on it and moved it out, but it cost me a transmission. That might be the case with a pick-up, but I think that my four-wheel-drive pick-up could do it. A farm tractor certainly could. You'd have to throttle it way back and pull it because you are sometimes going as slow a speed as 10 feet a minute, 10 to 50 feet a minute will cover the rate of travel of most machines under most conditions. Of course, the machine varies.

ACRES U.S.A. Well, now that you're freed from the actual business of shoveling compost from one truck to another, unloading trucks, creating the product, and even from designing machines, except when you feel like it, how do you fill your time?

SIMS. Well, I'm not all together free of this. I just loaded Linwood Brown. I sent him, I think it was three semi-loads of compost—a 25-ton load. I loaded most of that, not all of it, from my little experimental site. I'm not pushing that because to have produce it on in a big volume would require application equipment, in addition to transport.

ACRES U.S.A. You're a citizen of Texas, but you do spend a considerable amount of time in New Mexico. Is that not right?

SIMS. That's right. Not as much as I'd like, sometimes.

ACRES U.S.A. You have some unique guests to your mountain house.

SIMS. Yes, Phil Callahan was up there with me the last of December 1994, and he was using his devices to measure energy. He shared with me the fact that this was in an area of good, strong paramagnetic rocks. When I told him I was going to build my house out of adobe, he told me that it's a good paramagnetic material. So I went whole-hog with paramagnetic materials—brick on the floor, tile on the roof, even a paramagnetic partition wall in the bedroom—because I wanted to get all that energy I could. And I find that, when I go to the mountains—it's up at 9,000 feet—I have as much energy or more as I do down at my home, which is a little under 4,000 feet. As Callahan and I walked along my land, I got to where I felt he was such a good observer. I didn't feel the need to point things out. We were walking along this Tranch Mountain ditch, which was built naturally 115 years ago by man and animal energy. There was a lot of oak on this place. There was a grove of oak trees that run eight to 10 inches in diameter and made a complete 90-degree sweeping bend from where they come out of the ground. Actually, the main trunk of the tree is just a little beyond horizontal and this had always puzzled me. All of the theories I had offered—it was sort of like the theories that are offered on the towers of Ireland—they just didn't make sense to me. But when Callahan saw the oaks he said, "Wow!" and pulled out an instrument and began to check the energies. He said, "This is a regular healing chamber here. If you know someone sick, bring them up here, give them a stack of books to read and this will heal them." I don't know how long it would take, but I know that it would enhance anyone because I'm always invigorated by being there.

ACRES U.S.A. Well, nature knew how to compost and how to heal long before man ventured into either of these services. Would you agree to that?

SIMS. Oh yes, I certainly would.

"The Virtue of Elegance"

An Interview with Jerry Brunetti

Originally Published: August 1997

If the Acres U.S.A. community ever voted to give sainthood to contributor, Jerry Brunetti, 1950-2014, might be first in line to receive such honors. The career-long animal health professional worked in the eco-farming field long before Acres U.S.A. came into existence. He came by his insights and talents quite naturally, having farmed in West Virginia. Jerry made it north during the heyday of NFO milk organization, finally emerging into the eco-farming field in command of Agri-Dynamics, which now serves the trade. He co-founded Earthworks, a manufacturer of ecological products for the golf course industry.

The insight, exposition and analyses he accounted for in this taped exchange from 1997 revealed the level of sophistication eco-farm management obtained under his tutelage. He was a rare individual who has successfully integrated the common dynamics of nature in his mind and in his practices. Jerry was even been invited to the Mideast to consult with the high-dollar racing camel industry. Today, the eco-agriculture community remembers Jerry's kindness and contributions, and are longing for the lessons Brunetti taught to become commonplace.

ACRES U.S.A. One of the things we observe, and which we draw a lot of comments from people about, is the filth and the dirty agriculture we have out there. Dirty in the matter in which animals are kept. Dirty in the way we grow crops and in the way we fertilize. How do you assess the way we are going with this type of technology?

BRUNETTI. I think we are going to see—as we are starting to see now—epidemic disease problems in unprecedented proportion. This is no surprise when you hear people like L.J. Taylor and the *National Hog Farmer* magazine saying, "The breeding sow should be thought of and treated as a piece of machinery whose function is to pump out baby pigs like a sausage machine." When you've got that sort of mindset, and you see an animal as nothing more than a machine, doomsday is just around the corner. In contrast is an article by Craig Holdridge I had read a few months ago in the magazine *Biodynamics*. The article was titled, "The Cow: Organism or Bioreactor."

It is the discussion as to whether or not a cow is this machine, or whether it is an organism. It argues that every creature has its own reason to be. All parts have a direct effect on one another, a relationship to one another, thereby constantly renewing the circle of life. This is the spiritual perspective that I think ancient farmers understood—that they weren't merely janitors in some sort of factory, perceiving animals as nothing more than bioreactors or sausage machines. Now, they've got these ideas about farming where they are going to have extremely high volumetric productivity, low operating cost, unlimited productivity.

ACRES U.S.A. Where did this point-of-view originate?

BRUNETTI. This whole perspective came out of using animals as commercial bioreactors. Just to show you the absurdity of this mindset with animals, let's examine this problem. Since cattle are ruminants, they have an intrinsic desire to have enough roughage in their rumen in order to stimulate the fermentation of cellulose and the conversion of the crude plant source raw materials into the kinds of microbial finished products that the animal ultimately consumes. In our need and greed to get these cattle fattened out in a short period of time, we have been feeding them as if they are monogastric animals. So we've been pumping a lot of grain into these animals. The absurdity of this thing is that because they are pumping so much grain into these animals, it's causing all kinds of health problems. It's creating acidosis, liver degeneration, etc. There is a scientist by the name of Stephen Loerch who came up with a simple method to circumvent the need for roughage in steers bred for beef. He suggested feeding plastic pot scrubbers that you buy in the supermarket instead of roughage. This creates a net in the bottom of the rumen, allowing the stimulation of the rumen to constantly regenerate itself.

ACRES U.S.A. And the plastic is eternal ... it will not break down?

BRUNETTI. Right. As a matter of fact, he was apparently getting some calls from butchers wondering what the heck was going on. When they were slaughtering these animals and butchering them they were finding these pot scrubbers coming out of the rumen. The whole point again is to look at the cow as a total organism. The cow has a true need for smelling and tasting grass, chewing that grass. The cow actually has five stomachs, the four internal stomachs, and the mouth is the fifth stomach because it regurgitates its feed. The cow chews its cud and produces a lot of saliva, 35 to 40 gallons per day. The saliva then produces not just some other digestive enzymes, but also the buffering agents that keep the pH of that rumen at the proper ecological balance so that the microorganisms that are indigenous there can thrive and feed the cow.

ACRES U.S.A. What are we feeding the cow? We hear stories and see some evidence that cows are being fed everything from bin sweepings on up to solid grains with very little roughage and cattle cake, which is really previous generations of the bovine species and other animals ground up.

BRUNETTI. The original dairy belt was predominantly situated in the Northeast and the upper Midwest. Cattle being all pretty much the same species, all have similar requirements except that dairy cows are a lot more complex than beef cattle. When you look at the original dairy belt, you see that it was in the areas that were most suited for forages to be grown—pasture and hay. Now we've gotten into the selling of this bill of goods that you need heavy-metal agriculture in order to be economical. This means a lot more cows, a lot more land, a lot more machinery, a lot more buildings. The whole point being that everything is moving into the southwestern states. And this is continuing to be the case. Commodities are being sold there.

ACRES U.S.A. In the southwestern states, one of the biggest dairies we've seen is at Phoenix.

BRUNETTI. In fact, I understand that New Mexico and Idaho are two of the fastest growing dairy states in the United States.

ACRES U.S.A. It does indicate that we're trucking all types of feeds into strange out-of-the-way places that are not natural habitat for dairy animals.

BRUNETTI. Consider also the fact that there is no water there. When you factor in the cost of producing milk and beef in these areas by hauling in all of the commodities and having to subsidize all of these mega-corporation farms with water, it's pretty astounding. The cost of producing milk and meat that way is ultimately costing all of us something. Then there is the health cost. I think the biggest cost in the dairy industry today is the fact that you don't have the longevity of the cows. There are incredible records that look at the length of the average life span on these cows. You hear statistics that are published in the dairy journals where you have cow longevity amounting to less than four years on the average nationwide! You're talking about 9 million to 10 million cows. That means there are a lot of cows that are lasting a lot less time than that. Of course, we know that dairy cows don't hit their production peaks until they hit 5 or 6 years of age. So most of these cows never even see the genetic potential that these guys are spending all this money in first place to obtain. It also means we are doing a pretty poor practice of keeping these animals healthy. The reason they are being culled out is because their livers are going, their legs are going, their reproductive systems are going.

ACRES U.S.A. If you didn't cull them, they would cull themselves in another year or two.

BRUNETTI. Yes, or they just would be so expensive to keep around that they would end up being pets instead of productive animals earning their keep. Again, that goes back to the fact that milk production and meat reproduction is all in terms of quantity. Everybody is looking at bins and bushels, they are not looking at quality. What kind of amino acid profiles are coming out of those grains that are coming in these huge quantities off the farm? And how many calories are we expending in terms of caloric input on that acre of corn or wheat versus how many calories we are harvesting? It doesn't add up. The same is true with the animal business. That is why we are seeing this. I was at a meeting on the West Coast and I was listening to a number of experts talking about the future of dairy farming. Essentially, the message that was being given was the same old party line message: In order to survive in the economic crunch that's affecting the dairy industry, the only answer that everybody can see is to get big or get out.

ACRES U.S.A. That is because they have all been sold on the idea of economy of scale.

BRUNETTI. But when you factor in the longevity of the cows as being one of the critical costs ...

ACRES U.S.A. If you factor in the capitalization required for this economy of scale, and the fact that most of it is borrowed money and that cuts somebody else in on the enterprise, it's understandable why people keep going broke by the numbers, decade after decade, isn't it?

BRUNETTI. Yes. And then you factor in the fact that a lot of the western herds have the luxury of having inexpensive labor, Mexican labor. When you factor in that the industrial model works because from a labor perspective, they could afford to hire laborers willing to work at those wages.

ACRES U.S.A. Of course the social costs are not picked up by the enterprise. They're remanded for the government to handle.

BRUNETTI. Exactly. They're passed on to the rest of society.

ACRES U.S.A. Did you grow up in the dairy industry?

BRUNETTI. No, I did not grow up in the dairy industry at all. My interest in the dairy industry started when I was down in West Virginia. There was hardly a dairy industry down there at all at that time. But I left the farm there when a partnership was dissolved and ended up moving back to Pennsylvania. I started working with the dairy industry through the NFO and saw what was happening from an economic perspective. This was back in the mid-'70s. The message that was given then was, you are going to have to become more mechanized, going to have to have more cattle, more capitalization. Within a decade, it was obvious that many of the guys who took that particular advice went out business. The milk check was basically covering the cost of the new silos that were put up, the extra ground that was purchased. It didn't compensate for the unpredictable weather patterns we were seeing more frequently. We were having more dry spells every couple of years, so that coupled with low milk prices spelled doom. It annihilated the farm. Everybody was living on projections back in the late '70s and early '80s on the same prices you are getting on milk today. They were expanding and making economic commitments to their enterprise based on future milk prices. Prices did not hold up in the '80s and through the '90s the way they had anticipated they would.

ACRES U.S.A. Yet, what you are describing as the dairy industry is the bright part in animal husbandry.

BRUNETTI. Right. That was the one place that most people had some predictable expectation of having some income they could count on. In many situations, the dairy farm was primarily a cash cow to subsidize the rest of the operation. There was a distinction between a dairy farmer and a dairyman. Dairymen were into the cows. Often dairy farmers had cows around because they helped subsidize the rest of the field operation growing grains.

ACRES U.S.A. On a scale of one to 10, how would you compare the manner in which dairy animals are treated compared to beef animals in the feedlots?

BRUNETTI. That's a good question. It depends on which dairy, whether you are talking about a Lancaster County Amish dairy farmer or a 10,000-cow operation. I don't know that you would see much difference except that in the feedlots a lot of those cows aren't around long enough to be too uncomfortable. The dairy cows have to endure some pretty uncomfortable conditions.

ACRES U.S.A. Nevertheless, the regime for the beef cow that's served on the dinner table of America is not that palatable to even discuss, is it?

BRUNETTI. I would be very suspect of putting the kind of food that's coming from these large consolidated operations on the dinner table. Regardless of whether they are poultry, or beef and, sad to say, even the milk. You just have to wonder.

ACRES U.S.A. Basically your consulting firm deals with maintenance of health for farm animals.

BRUNETTI. Yes, though it always boils down to economics. Everybody in the agriculture business, in mainstream agriculture, is pretty well strapped economically, so they are always going to make these decisions based on economic input and consideration. That's what has to be laid out. By making some changes, you have to see the bottom line improve.

ACRES U.S.A. If you can prevent a turkey epizootic from taking place, if you can prevent a die down in the birdhouse, if you can prevent a kill-off of hogs sweeping through the farrowing sheds, then you've really accomplished something compared to the type of technology that is being spewed out of the land grant colleges. Is that correct?

BRUNETTI. Well of course. But it's still a symptomatic approach. One of the concerns I have when being asked to deal with those kind of potential catastrophes—without referring to some so-called expert or being involved to some extent myself—is that the reason we are having these die downs is because of the nature of the business being the way it is in the first place. There is a friend of mine who passed on a number of years ago, who was a art history professor. He came from Germany and was a student of Rudolf Steiner. His name was Johannes Gaertner. He founded the music and art history department at Lafayette College in Easton, Pennsylvania. He also wrote a book called, W*orldly Virtues: A Catalogue of Reflections*. One of the virtues that Johannes talks about in the book that I thought was especially appropriate to clean agriculture was the virtue of elegance. He wasn't referring to the elegance that consists of being fashionable. What he had in mind was the kind of elegance that arises where the effort that it took to achieve what one has done becomes invisible. Art needs its audience and so does elegance. And I think that is a relevant definition of what clean agriculture is. When you go on to a farm, and you see the kind of sustainability that really is defined further by the evidence that there is a true holistic system intact, you can see it in the color of the animals' coats, and you can see it in the smell of the ground that's being cultivated. You can see it in the overall management of the operation. You can see it in the well-being in the farm family that lives there. That is the elegance that he refers to.

ACRES U.S.A. And of course that's not available when the town that's surrounded by feedlots becomes a veritable stink sink.

BRUNETTI. Yes, how could it? All that is a matter of completely divorcing ourselves, once again, from this relationship we need with animals. One of the things I talked about at the Acres U.S.A. Eco-Ag Conference was the fact of simple observation. That is what I consider myself to be, just a generalist and an observer. I am not an expert, I am not a scientist, and I am not an R&D lab man. What I do is basically observe things, and I try to connect the dots—the similarities between the seemingly related and sometimes unrelated issues. For example, there are lot of similarities between the digestive system of an animal and the root zone of a plant. When you look at it microscopically, you can see the dynamics that work in the villi of the small intestine, and the dynamics that work around the microscopic root hairs of a plant. What you see is nature always trying to, in effect, increase its diversity and increase itself internally. We are always looking at things macroscopically, but when you look at things from an internal universe, and you look at how nature simulates that over and over again, you realize that it is these minute kinds of things that nature does. A real enjoyable interview that I read in Acres U.S.A. was with Elaine Ingham about the soil-food-web and it explained just that. This is not a linear thing. It is not a vertical column. The point that I discovered is that when you start looking at all these different kinds of dynamics, whether its horses, cows, dogs, fish, hogs, or a corn plant or a pasture plant, you find a lot of seemingly similar dynamics at work. One of the things that I found interesting about Buckminster Fuller's definition of synergy is that you can't necessarily isolate, or actually come to conclusions, or make predictions about the whole by isolating the part or studying them in isolation of one another. A classic example made was that of a water molecule. In a water molecule you have hydrogen and you have oxygen. There is nothing about the nature of hydrogen or oxygen studied by itself in isolation that would give you any clues that the characteristics of these two separate elements combined in the right proportions would give you water. The same is true of anything your dealing with in holistic principle. There was a gentleman who was an industrial architect by the name of Victor Papanek. He became somewhat of a rebel and broke away from his colleagues by recognizing that since the late 1920s manufacturers and their industrial designers have managed to sell longing and dissatisfaction, side by side. He wrote a book at that time, *Design for the Real World*. He later wrote a sequel to it called, *The Green Imperative*. He started studying indigenous peoples, like the Balinese. He felt that they were the ideal designers because everything for the Balinese is temporary. And then, it is life that becomes lasting. By concentrating on goods that don't last nearly as long as we hope and don't age well, we've lost our sense of quality and the temporary.

ACRES U.S.A. Can you describe for Acres U.S.A. readers exactly what happens in the poultry house, in the broiler house?

BRUNETTI. I am not a poultry expert by any means. But it becomes kind of an obvious situation. A number of years ago, back in the mid-'80s, we had avian influenza in Pennsylvania that clobbered the poultry industry. It was mostly the layer operations. What was interesting at that time was that there were some people that, even though they had some confinement operations, were trying to do some things differently. For example, they were using full-spectrum lightning in the houses. They figured they would simulate sunlight since birds typically receive a diet of light the way the sun provides it. They felt that it would make a difference. Some of them were paying a lot of attention to the water quality that the birds were receiving, and there were others who even went so far as to make sure these birds were getting a variety of other kinds of grains, such as flaxseed meal, as well as some greens because they were trying to go after a particular niche market. They were going after a market for birds that were being produced, if not in a free-range circumstance, at least compensating for that as much as they could. A couple of those people that I talked to didn't cave in with the avian influenza. They had the antibodies to the influenza, but they didn't have any disease outbreaks. Of course, their birds were exterminated because of the USDA quarantine that requires that to be the case.

ACRES U.S.A. They were exterminating everything. They even exterminated one farmer who was raising various exotic birds.

BRUNETTI. In fact, I heard him at a meeting speaking about the fact that when they came to his farm he felt like it was the Gestapo showing up with all these poison tanks and gassing equipment.

ACRES U.S.A. These growers that you were talking about were not feeding their birds previous generations of birds and feather meal and things like that.

BRUNETTI. That is one of the things that I observed. They were a handful of them that took a detour from the conventional approach to feeding and raising these animals. They seemed to have a beneficial consequence for making that effort. Unfortunately, they were grouped with everyone else simply because they were in the wrong geographical place at the time. But when I've gone into some of those other operations, I wondered how those birds could last more than a few days because the ammonia vapors were so fantastic that I couldn't understand how they lived.

ACRES U.S.A. Even though the fans are pumping away?

BRUNETTI. Yes. And there is no sunlight in there. There is all this ammonia and a lot of dust. And you also have the mold problems. Of course the mold problem is a soil-related issue, because we know that the mold epidemics that we have are being exacerbated by the pesticides that we are using. The salt fertilizers and the destruction of the humus is in effect creating a complete upset in the microflora that inhabit the soil and keep the pathogens in check. A lot of the people that are looking at the mold problem are looking after it gets into the grain bin instead of out in the field. I know even at this moment people are still scratching their heads wondering why we have these huge outbreaks. For example, Texas, Oklahoma, and parts of Kansas are having this huge aflatoxin problem. I was talking to a gentlemen at Texas A&M who said that this is a relatively new phenomena to the degree that it hit that area. Nobody to my knowledge is looking to the soil practices as making any contributing factor toward that.

ACRES U.S.A. John Whittaker used to publish a lot of analyses and expositions on that problem. Usually he would nail it right down, but they kept right on ignoring it. He'd turn it out to the aflatoxin laboratories and they wouldn't even listen.

BRUNETTI. Again, its always after the fact. Right now they have an *E. coli* outbreak in the turkey operations down in North Carolina. They got salmonella outbreaks that are resistant to drugs all over the place and then they've got these molds.

ACRES U.S.A. What are they doing about the salmonella and the *E. coli*?

BRUNETTI. The latest news I've heard is that because they have had some of these nasty salmonella outbreaks over in Britain, and I think there has been another outbreak of some other organism in the Far East, and of course the outbreaks in North Carolina and Virginia, the FDA approved the use of formaldehyde in the feed to try to kill these organisms. One *E. coli* organism has a cell wall about four times thicker than a normal *E. coli*, and it has evolved to produce a capsule around itself which protects it from disinfectant or antibiotics. Now, to tell if the formaldehyde will have an effect on this organism remains to be seen.

ACRES U.S.A. Have you visited the swine producers in North Carolina?

BRUNETTI. No, I've just heard the horror stories that are coming out of there. One of the interesting things I've read about is that the North Carolina hog wars had a lot to do with water contamination. First of all, keep in mind that a hog excretes about three gallons of waste a day—two-and-one-half times the average human. So if you have a 6,000-sow-and-hog factory, you are going to have 50 tons of raw manure a day. When you are dealing with an operation the size of Premium Standard Farms, you are talking about over 2 million pigs and sows. They will produce as much as five times as much sewage as Kansas City. The problem is they don't have any processing facilities for it because they are considered a farm.

ACRES U.S.A. And of course those animals are pumped full of everything ...

BRUNETTI. A hog, given the choice to free range, will derive 60 percent of its diet under the soil surface going after roots, insects, earthworms and soil itself. Good topsoil is a very nutrient-dense substrate. Hogs get the microorganisms that they need to improve their digestion from the soil. And they get the B12 and the trace minerals that builds good blood. Now when you have hogs on concrete in a confined operation, one of the problems you'll deal with is that you have to give them shots for anemia which usually includes B-complex and iron shots. Hogs raised on grass follow the sow around and drink the milk the sow is producing based on the diet that she is getting in her natural environment. These hogs don't come down with anemia because everything is there. We've taken them off the grass and thrown them on a concrete pad, force-fed them corn and soybeans, vitamins, minerals and drugs. And then we wonder why we've got a problem. They think they are giving the pigs a complete diet. According to experts in the hog industry, they have enough protein and they have enough energy. They have enough calcium and other vitamins and minerals.

ACRES U.S.A. They've given it the slide-rule approach.

BRUNETTI. Exactly. The same is true when you look at dogs. One of the companies that we make a supplement for that goes into all of their dog food asked if we could provide them with a supplement that would curb coprophagia, which is dogs eating their own stool. This has always been a mystery to many people, why the dogs eat their own stool or other dogs' stool or cat feces. It's because, in my opinion, most of these dogs are being fed this grain diet. Dry dog kibbles are predominantly grain with corn, wheat and some meat products in there.

ACRES U.S.A. So we take carnivores and turn them into herbivores, and we take herbivores and turn them into carnivores.

BRUNETTI. And as you say, taking the slide rule approach, it all pans out because everybody that is investigating is saying, this is protein, this is carbohydrate; it should work.

ACRES U.S.A. They don't even know what protein is. They don't know the difference between "funny" protein and "real" protein.

BRUNETTI. Even this moment, everyone is using the nitrogen extraction method to determine whether or not they are growing protein on the farm, or whether they are growing funny protein on the farm. They don't know. They are assuming it's protein. That gets into some of the André Voisin material. You're feeding these dogs this dry kibble that's cooked at high temperatures so there are no enzymes—digestible enzymes or food enzymes—in this product. It's loaded with grain. As the animal is eating this stuff, the assimilation has been compromised. A lot of what is going in is going right back out again. The animal now is going after its own stool or other animals' stool because one of the things that is in that stool are the enzymes its organs have secreted into the alimentary canal and passed through the animal. So it is recycling its own enzymes to help break down foods or feeds that it hadn't been able to digest the first time around. With hogs you will see this. Hogs will go after cow manure. People think that the reason that a hog will go eat a cow pie is because it is after the grain that's passed through the cow. Which is true, they will definitely devour the corn that passed through a cow. But the other thing that it is after is the B12, because the ruminant produces B12. The ruminant produces an awful lot of microorganisms and digestive enzymes. That is how that pig, which has a digestive system very similar to a dog, operates. When a carnivore kills its prey, the first thing it eats is the digestive system, besides the tongue. If it's not ravenous, it will bury the prey and come back later after that meat has started to predigest because the enzymes that are in the tissue start to decompose or actually predigest the meat so that it's digestible to that carnivore. This is a major thing in terms of the enzyme question. The book on that is *Enzyme Nutrition* by Edward Howell. To me it is one of the most important books written on nutrition. It goes into the research showing how indigenous cultures and animals have survived by following the precept of enzyme nutrition over the centuries.

ACRES U.S.A. If we ask what the future is for the meat protein industry as constituted commercially, what would you have to say on behalf of the industry or on behalf of the people consuming the product?

BRUNETTI. I don't think the picture is anything but bleak because it is creating way too much waste. Just from the surface of things, it is creating some very obvious environmental problems that are affecting all of us. The fact that the waterways and the aquifers are getting contaminated, and the air is being contaminated. If you are around these large operations it's not fit to sit outside. The regulatory agencies are going to be more and more coerced to make people obey the laws that they were intended to obey.

ACRES U.S.A. Are they going to be coerced enough to require these operations to break down into smaller units, the type of a norm we had when we had broad-spectrum distribution of average, well-managed family farms?

BRUNETTI. I don't think that is going to happen because it is just not politically a likely scenario. I think that what is going to happen is that you're going to have some major epizootic outbreaks that are going to be like trying to contain a forest fire after a season of dry tinder in a western state. I think that what is going to happen is that eventually there are going to be hopefully more and more growers who are going to realize the niche markets that exist. It will encourage them to produce clean food that's been environmentally produced, ecologically raised and that deals with all of the issues from animal welfare to having the aquifers not be contaminated. And on top of that, I think these large operations are going to run into the fact that they waste so much water. Agribusiness is a huge consumer of water. Hog factories are water hogs. You're talking about an 80,000-head finishing unit consuming 200,000 gallons of water every day. That's 73 million gallons a year for just one complex. Multiply that by dozens of these complexes, and then add over the 365 million gallons a year at the slaughtering plant and the feed mills and the concrete plants and the pump down operations for the lagoons, and you end up with not enough water.

ACRES U.S.A. How much space is there between a fantastically large epizootic and a human pandemic?

BRUNETTI. That is the question a lot of people are starting to encroach upon asking and answering. I just picked up Virgil Hulse's book. He is an M.D. out of Oregon who wrote a book called *Mad Cows and Milk Gate*. Basically, he suggests that the line between species is not necessarily a large gulf. And, more importantly, since we don't know so much, why are we acting as if this is merely a nickel-dime gambling casino in which we are playing this game?

ACRES U.S.A. Have you evaluated what genetically engineered feeds will do to the livestock?

BRUNETTI. I don't know. But I will say this, I read gleamings of this all over the place. I have real concerns when you start taking genes from plants and impregnating them into animals and visa versa, and suggesting that what we are getting out of the food chain really is irrelevant. This is the same kind of argument that they had with BST. They were saying in effect, that it doesn't matter that we are injecting cows with BST because its naturally found in the milk anyway, as far as they know. But now certain experts are suggesting that the recombinant BST and the BST that the cow produces are not the same. What kind of implication does that have? We don't know.

ACRES U.S.A. Is it the same implication that we found when tryptophan was imported from Japan and it was genetically engineered and killed 35 or 36 people.

BRUNETTI. And as a consequence, a very valuable amino acid was taken off the shelf. The real story is that it was genetic engineering that should have been indicted. Instead a natural food substance has been indicted. So now everybody is afraid of tryptophan when in fact, they should be concerned about genetically manipulated tryptophan. I found that John Whittaker for years was talking about the mold and mycotoxin issue, and that the correlation between poisons that we are applying to kill diseases is just making matters worse. This was pretty much scoffed at. And now, there is a three-year study that was done by Dr. Eric Nelson, a plant pathologist at Cornell University. It was funded by the golf course industry to find out what kinds of alternatives exist to fungicides, and whether or not they would work as better or worse than conventional fungicides that are used in wholesale applications on golf courses. The interesting thing about the study was that what Whittaker had been saying for years was corroborated by a different industry even though it is still horticulture or agriculture. That is that fungicides don't remedy the problem; they make them worse. When you go in there with fungicides, you basically indiscriminately kill a lot of beneficial organisms. When the organisms you are trying to eradicate come back after the smoke clears and the body count is tallied, it is the pathogens or the undesirables that come back with a vengeance.

ACRES U.S.A. Isn't this the story with all of this medicine and all of these approaches that have been used? They don't answer the problem, they merely exacerbate it?

BRUNETTI. The same with the formaldehyde. Right now, instead of trying to deal with why are we having an *E. coli* that is so veracious and tenacious, we are going to try to find a nuclear weapon to eliminate it.

"That's the Ultimate Form of Sustainable"

An Interview with Alan Nation

Originally Published: June 1998

The son of a commercial cattle rancher, Alan Nation grew up in Greenville, Mississippi. When he spoke with Acres U.S.A. in 1998, he had spent the last few years traveling around the world studying and documenting grassland farming systems. He joined *Stockman Grass Farmer* magazine as editor in 1977 and served in that role for decades. It is still in circulation today.

In 1987, Mr. Nation authored a section in intensive grazing in the USDA Yearbook of Agriculture and six years later in 1993 he received the Agricultural Conservation Award from the American Farmland Trust for spearheading the drive behind the grass farming revolution in the United States. He is also the author of several books on the subject.

His interview from 1998 still resonates, which featured his powerful take on the real and mythical challenges inside the livestock industry, and how following convention led many in the industry to bankruptcy.

ACRES U.S.A. Alternative agriculture pundits always speak of the great paradigm shift. What do you think about this?

NATION. One of the things that Richard Farson wrote of in *The Management of the Absurd* was the fact that the opposite of any profound truth is also true. One of the things that we have tended to see in the United States is a single dominant paradigm for agriculture at any one point in time. The 3,000-sow confinement hog house has become the dominant paradigm; the 100,000-brooder henhouse has become the dominant paradigm. The 100,000-head feedlot has become the dominant paradigm. California confinement theory is now becoming the dominant dairy paradigm. At all points in history, the dominant paradigm is never the only paradigm. It is never the only profitable paradigm.

ACRES U.S.A. So your point of view is that the so-called conventional agriculture is not necessarily wrong, but that there are other equally right or even superbly right methods ...

NATION. It depends. The dominant paradigm right now is set up to maximize the use of capital and minimize the use of labor. It's for people who have lots of capital and hire all their labor. Of course, the labor they hire is minimum-wage, non-thinking-type labor—all the thinking is done in the home office. Just because confinement theory is profitable doesn't mean that a grass-based seasonal dairy is not profitable. That a 3,000-acre wheat farm in Oklahoma is profitable does not mean that an 80-acre Amish farm can't be profitable. What we are finding is that every paradigm is a whole unto itself that runs under its own rules. You can't take the rules for a confinement dairy and put them in a grass dairy ... or a 3,000-acre wheat farm and put them in an 80-acre management-intensive farm. So everything is a whole unto itself. It appears to me that the big problem that we always have is that we are trying to make the dominant paradigm work on a smaller scale.

ACRES U.S.A. So is the problem not in the direction the farmer chooses, but the fact that he is trying to straddle the fence and is pulled in multiple directions, failing in each?

NATION. He is trying to make a big farming whole or paradigm work on a small acreage, and you can't do that. A small-acreage farm is put together entirely differently than a big farm.

ACRES U.S.A. How, so? Could you illustrate this?

NATION. Big farms are set up on small margins spread over large units. Small farms, on the other hand, have to spread big margins over fewer units. A confinement dairy in California makes a net profit of about $200 a cow. If you have a thousand cows, you've got a pretty good income. An Amish dairy in Ohio runs on a net profit of about $1,800 dollars a cow. If you've got 40 of those, you've got a pretty good income. But 40 cows at $200 dollars is not the same thing. You've got to have big margins per unit of production to make a small farm work. There are two ways to do that. One is to direct market and get a premium price for a product in a small, specialized market. What generates the premium price is the fact that you are pursuing a small market that the big guys can't cover their overhead serving. The other way is to lower your costs. Find an unfair advantage that your knowledge, your climate, and your markets allow you to compete on. For any commodity-based product, you've got to have an advantage of about 30 percent over the average producer to succeed as a small farmer or in a startup situation. In other words, you've got to produce at about 30 percent less than the statistical average of your industry.

ACRES U.S.A. So how do farmers gain this 30 percent advantage? Can they do it through applying the advice of the conventional growers and extension?

NATION. You've got to know something that the other guy doesn't know; you've got to find the advantage. You can build a farm around organic matter rather than weed control, for example. That would produce a 30 percent yield advantage. Lower cost—it's all a knowledge game pretty much. The paradigm has been to let the chemical companies do the thinking and let the universities to the thinking. There's no thinking being done on the farm. You've got to get the thinking back down at the farm level.

ACRES U.S.A. Does it go deeper, to the actual sources of information?

NATION. Well, it goes to the point that most farmers don't basically understand how nature works. The closer you can get to, "Well, how does nature make that work?" the closer you are getting to gaining advantage. When we understand that nitrogen, which is the driver of the system, comes from the breakdown of organic matter, you can see that if you want a higher yield the first thing that you have to concentrate on is building organic matter. But you have to understand how nature does that. For example, everybody's talking about the problem of coyotes eating lambs. Well the reason the coyotes eat the lambs is primarily because we were lambing in a time of the year when there was nothing else for those coyotes to eat but lambs. All of a sudden, some producers started looking to nature. They said, "Well, maybe we ought to lamb and we ought to calf when the buffalo and the deer and the rabbits and the birds and all the things that coyotes eat also are out there, which is May and June." What we found out is that we can produce a lamb or a calf about 30 percent cheaper if we get in sync with nature. That's the 30 percent advantage I was talking about. If you can understand how nature would do something and then get as close as you can to that natural way, that will usually give you about a 30 percent advantage over modern agriculture.

ACRES U.S.A. The techniques of management-intensive grazing often call for playing to the natural tendencies of the production system. For instance, keeping cattle in tight herds fits in the their natural tendency, instead of spreading them out over thousands of acres. How does this fit into the profit picture?

NATION. Let me give you an example of that. When we started putting cattle in tight herds, it helped us to survive the predators. The only animal in nature that will penetrate a herd of animals is man. A wolf won't do it. A coyote won't do it. A dog won't do it. Herding animals get into a tight herd because nothing will penetrate the herd for fear of being trampled on. The big cost in Western ranching—labor cost—is in finding the cows. If you've got 200 cows spread out over 100,000 acres or 50,000 acres, finding those cows becomes a tremendous labor cost. All of a sudden when we kept them in these tight herds, there goes that high labor cost.

ACRES U.S.A. The smart grazier simply gives a whistle, sends out a dog, and the work is done.

NATION. Once you don't have to spend half a day looking for a cow, all of a sudden your labor productivity goes through the roof. What we are talking about is that you've got to learn to think in whole systems. Once you start getting in sync with nature, its just like pulling a train; everything is hooked together, and it all works for you. Or it will all work against you if you start going against nature.

ACRES U.S.A. So what advice would you offer the farmer or rancher who has gone down the path of conventional thinking of the last 50 years? How should they start to move closer to nature to find this 20 percent to 30 percent advantage?

NATION. The first thing you have to do is to quit struggling. Whenever you find yourself in a difficult situation, it's just like being stuck in quicksand. The harder you struggle, the faster you sink. You've got to stop struggling and start thinking. You are going to have to buy some books and read. Ask yourself, "Where does this come from?" and "What is that?" You are going to have to stop relying on a salesman who comes to your farm to give you all your solutions. You are going to have to come up with some on your own.

ACRES U.S.A. So farmers should turn down the freebies—and advice —from the pesticide company?

NATION. Yes, turn down the free steak dinner. You've got to understand, if you get a free magazine in the mail, somebody paid to send that to you. There is no such thing as a free lunch. There are no free steak dinners. There is no free extension service or anything else. Somebody somewhere paid for that. And they are going to get their money back out of you somehow. You just need to understand that you need to start doing some thinking on your own. One of the big problems that we see is that most people get interested in sustainable agriculture the day before they go bankrupt. They've totally worn all of the organic matter out of their soils. They've put all their money into rusting, rotting things, things that are losing value. And then the day before they go bankrupt they call up Acres U.S.A. or *Stockman Grass Farmer* and say "Okay, I am ready to do it your way now." The problem that we've got is that nature takes time to heal. We've got all sorts of technologies that will cut a wound faster, but we don't have any technology that will heal it faster. Ninety-nine percent of wealth production in agriculture is not in getting out of difficult situations, it's in not getting in difficult situations in the first place. Once you find yourself up to your neck in quicksand, you've got a pretty slim chance of pulling yourself out. You had better pray for a miracle. The most important thing is to not get into that situation. I see so many people that—even though the trend is set and they are heading toward bankruptcy—will not do anything until the absolute last minute.

ACRES U.S.A. Is it your belief that in various states where we see dairy farmers falling like flies—where they are forecasting half the farmers will be gone in a few years—that it is not a problem of the moment's markets, but is a continuation of a long-term trend?

NATION. The biggest reason that we are seeing dairy farmers go down has nothing to do with the cost of production of milk. It has everything to do with the fact that there has not been a re-placement of dairy farmers attracted to that industry. There are just as many dairy farmers going out of business every year regardless of what the price of milk is. The reason is age. We've got the oldest dairy farmers in the world. And they are 10 years younger than the average beef farmer.

ACRES U.S.A. So the mega-operations are by necessity filling in the gap?

NATION. Yes. One of the things we have done with CRP—and all of the government pro-grams—is that we've skewed every government program to benefit people who are currently in agriculture. That was the voting constituency. What that does is effectively ensures that the next generation can't get into agriculture. And now this generation is all retiring and there's nobody coming along behind them. The beautiful thing about what the New Zealanders have done with their dairy system is that they've developed what they call a share milking system where a kid right out of high school can start milking cows, and then buy a few of the cows on the string, and then buy all the cows in the string, and then turn around and buy the land. Then he hires a 17-year-old share milker and it all starts again. A guy gets in at 18 and he's out at 50. It makes a beautiful generational transfer. Those are the kind of things that I think we need to start taking a look at. If the young can't get into farming, the old can't get out.

ACRES U.S.A. One of Joel Salatin's favorite messages from his farm pulpit is that we all are lost if we fail to romance the next generation toward agriculture. No matter what we do to clean up our farms, remove chemicals, and grow good food, it's useless if the next generation doesn't follow.

NATION. That's right. One of the things that my Dad always told me—and it's pretty much true—is that all your extension efforts have got to be directed toward the young. Once a man gets to be 50 years old, he is more interested in getting a payback for something he already knows than in learning something new. That's a major problem in agriculture where the average person is 58 years old. That means that half of the people in agriculture are over 58. Twenty-five percent of the cow-calf producers in Alabama are over 75 years old. How would you like to go create change in that situation? You are just not going to do it. It's really difficult to see that what's really important is what's not here yet, but what's coming. We want to put all the effort into what's here now, but what's here now in 20 years is gone. It takes about 20 years to really create any change because you actually turn the people over. I've got a saying that it's actually easier to change the people than to change the people's minds. The mindset that we are in is when we get to a certain age, we quit learning. We just have to accept that.

ACRES U.S.A. There is a small layer of people that do keep learning, do keep innovating, but the vast majority have no interest in learning something new.

NATION. If you look at the statistics you'll see that there are about 2 percent who everybody else is following. Those 2 percent are the cosmopolitans. The average American farmer is not interested in anything that is happening outside of his own community. That really limits your perspective. I like to follow and cover what's going on in New Zealand, what's going on in Argentina, what's going on in Africa, Ireland. Most of our farmers don't want to hear that. I tell them, "But that's your competition, you better hear that." If you are going to be in a commodity-priced market, you've got to keep up with cost. And if somebody's come up with a lower-cost way of doing it, you need to take a look at that. So many say, "No, I want to read about my little part of the world and people like me." It's tough.

ACRES U.S.A. When new technologies do come on the horizon, have you found a cycle in the rate of acceptance?

NATION. Let's take the Internet for example. Everybody said that the Internet was going to replace print technology. Now anybody in the world can have a site and their own publication to the point that there is now one Internet site for every three subscribers. What are the odds of that working? The Internet first went through this huge hype cycle—it was going to revolutionize the world. Now it's coming back down, because there are more sites than there are viewers. With any new technology, after the hype cycle there is a trough of disillusionment. Only then do you enter into a slow period of incremental change. This is so slow that to the average person it appears that it was a failed technology. But it's not. It's putting the pieces together and growing slowly. In 20 years it emerges and people say, "By golly, there really was something to that after all."

ACRES U.S.A. But before that evidence of true success, you are surrounded by neighbors who are quick to criticize and slow to acknowledge any success.

NATION. Yes. One of the things that Richard Farson pointed out in his book was that it's very difficult for people to learn from other people's success. He said we usually learn by watching failures. We learn what not to do rather than what to do. It appears to me that the only time people will learn that they should have been watching successful people is after they have become successful themselves. Everybody, Joel Salatin, a lot of the people written about in Acres

U.S.A., always say that the first thing a person needs to do is to enter an apprenticeship, really learn a trade. I bet there is not 1 percent of the people in the world who do that. Most of us jump into whatever we are doing and try to learn it while we are doing it. Boy, that's an expensive education. What happens is that most people dig themselves into a great big hole right on into the front end of their life. A lot of them don't ever get out of it. The ones that are successful say, "I wish I had done an apprenticeship. I wish I had learned a trade." I think that's the success of the New Zealand system. Think about the average kid in America that thinks that he is going to farm. He's going to first have to create the capital to own land. That just puts it totally out of his perspective.

ACRES U.S.A. It's difficult for them even to inherit the father's farm because of the taxes on inheritance.

NATION. What nobody is telling that kid is that there's a whole lot of land out here that is not being utilized. If he would ask people, "Can I do something with this? Can I run some sheep out here in this wheat field?" You'd be amazed by the replies. Most of the people who own land are elderly people who would love for somebody to do something sustainable with the land. Everyone is waiting for some salesman to come knock on his door and say, "Hi, I'm from the land leasing company. Would you like to come lease a farm?" The beauty of the New Zealand system is that it's an establishing system. When you are 18 years old, you've got a lot of energy. Well, go milk some cows and start building capital. And you'll wind up with your own farm. We've got to find a system where young people with that energy can find a way into farming. What's happening now is that people leave the farm to work for IBM and create money. When they come back and finally buy their dream farm, they don't have any energy to do anything with it. They are too old. So we have to look at the structure of these situations. If you look at the average full-time farmer in America today, there is not but two to three percent of them that are farming land that they own, or entirely land that they own. Usually it's 60 percent leased. You've got to do that. You can't sit here and try to buy every acre that you farm. It just won't work out.

ACRES U.S.A. In your books and journal you look to New Zealand often for success stories beyond the social model they've created. What sort of technologies have come out of New Zealand, and why there versus from our own farms?

NATION. I have learned that you can learn from any farmer in the world who is selling at a lower cost than you are. What you have to do is to say, "OK, who is selling at a lower cost and making a full-time living doing it?" New Zealand, for dairy, is the ultimate country to take a look at, I think, because they are making a living on $6 or $7 milk.

ACRES U.S.A. How does a New Zealand dairy operationally compare to an American dairy?

NATION. Well, there's nothing there but cows and grass. There's no machinery. There are no buildings. There are no barns. The milking parlor is just a flat-roofed, open-sided building with a ditch dug down into the middle of it. The whole thing is a livestock operation. If you want to make money in livestock, then 90 percent of your investment has to be in livestock. It can't be in dirt, machinery and buildings, and things that rust, rot and depreciate. And that's what they've done. If you want to look at a beef paradigm, Argentina is a great place to start. How do they produce grass-fed beef that's as good as any steak that you've ever had? You can go down there and see what it takes. It's simple. I don't think you are going to learn anything going to Germany or somewhere where they are selling it for a price higher than what we are getting, because those farmers tend to let their expenses rise to their income. The thing to do is to see how the guy is making a living at half the scale that you're getting. What you want to do is to keep the American price and produce at their cost. Then you really knock a hole in the wind. The only way I have found to learn is you've got to go somewhere where you can see a whole system. That's the problem with research stations. It's all presented in little bits and pieces and disconnects. It's not put together into a whole.

ACRES U.S.A. So this reductionist science mentality of isolating production factors to one element or one tillage method or one of any single factor is fundamentally flawed?

NATION. The major paradigm that we have in the United States is that if you produce more of something, it is more profitable. If you produce more milk per cow, it is more profitable. If you go and study the least-cost producers in the world, that's not what you are finding out. The New Zealanders make about half as much milk per cow as we do. But they are making three or four times the income per cow that we do. The total cash production cost of a dairy cow in New Zealand is about $100 a year.

ACRES U.S.A. And they don't need to buy BST?

NATION. Right. No BST and all of that stuff. If I can get you sold on the idea that you can produce yourself into wealth, you're a sucker that will buy whatever I've got that'll increase production. The thing about a commodity business is it's all least-cost production. It's not production at any cost.

ACRES U.S.A. So the two options you see for American farmers are either to get into higher profit specialty goods and services or lower the cost of the commodity production?

NATION. Or do both at the same time.

ACRES U.S.A. Who has the most to gain in changing over to a more natural system?

NATION. The person who has the most to gain in any new technology is the person who's not in the old technology. Remember the lesson of the steam locomotive. The major cost of a steam locomotive was not the locomotive itself, but it was the attendant ash pits, water tanks, coaling towers, and roundhouses. All of these things had to stay as long as there was one single steam locomotive left on that railroad. They didn't get the benefit of converting to diesel until it the system was entirely diesel. Only then they could deal with that high-cost, depreciating infrastructure. And that's what's hard. If you slowly transition your farm, you've got all the costs of the old and all the costs of the new. So your costs don't go down—your costs go up. That happens in any transition situation. You go through a period where there is actually more cost going toward the lower cost, because you are running both technologies at the same time. The quicker you can get through this stage, the better off you are because you can drop the other costs down. A lot of farmers say, "I just want to put my toe in this, my toe in that." They are not seeing any benefits and won't see any benefits until they have made the complete transition.

ACRES U.S.A. Do you see farmers that are conventional feedlot cattle operators, for instance, who try to tip-toe into rotational grazing or HRM methods?

NATION. The biggest problem we see is people going into rotational grazing who don't have their calving season right. If you are calving in the middle of winter when there is no grass, there is nothing that rotational grazing is going to do that is going to benefit you at all. It has all got to be put together as a whole. The other thing a lot of people don't understand is that what we are trying to do in humid environments with rotational grazing is to create a variable stocking rate to maintain a variable grass growth rate in order to keep that pasture vegetative. If you don't know what the goal you are trying to create is, you can't figure out how to use the technology. We've seen people start stringing electric fences who don't really know what they are trying to do. They are trying to keep that grass vegetative, and that might mean that you have to drop half of it and chop it for silage, or bring steers in. The goal is to keep that pasture without any seed heads and stems on it and keep it vegetative. When you start improving quality, you have to go up in the class of animals that you are using. You've got to go to steers, not just a cow-calf operation. And then, of course, the ultimate animal is a dairy animal. It's all hooked together. You can't just take one piece of anything in agriculture and hope that it will work.

ACRES U.S.A. You've mentioned that there is a sort of a biological imperative, that we have to look to nature and emulate nature. But at the same time, a farmer can take a natural system synthetic overnight. But by converse, a synthetic system doesn't heal itself back to natural overnight. How is this shown on the farm?

NATION. If you go back and look, it was in the '50s when we started moving away from using rotations and legumes and manure, away from the systems in place when we had a lot of animals on the farms. When we started moving away from these proven systems we were starting from a very healthy ecosystem. It's like when you start smoking—gosh, you don't feel bad for the first 10 years or so. If you decide to quit smoking two days before you die, it's probably not going to do you any benefit. People don't understand that there is a transition time, a pretty lengthy transition time coming back. Things don't simply heal overnight. I think the farther south you are and the wetter it is, the faster it heals, however.

ACRES U.S.A. Do you feel that any transition needs to be to a complete, holistic system, not just picking up a new product, substituting a natural fertilizer for a chemical fertilizer?

NATION. For example, if you had a weed-filled field and you were wanting to bring it back, you need to look to the law of the opposite. What I would do to bring this cornfield that's totally gotten away from you back is to take it to grazing first. Every major weed in row-crop agriculture is a high-quality forage plant. That's why you don't find them in pastures; animals graze them out. Every major weed of pastures is quickly removed with tillage. If you could come up with a system like Argentina has, where you have a planned rotation of pastures and crops, all of a sudden you have solved lots of problems on both sides.

ACRES U.S.A. There are growers who recommend this as a method to transition from conventional to organic, spending the three years of transition time in grazing versus trying to fight the weeds that result from a cold-turkey quitting of chemicals.

NATION. Research last fall at Uruguay found that you can grow just as much grain in three years with a grass rotation as you will in six years in grain.

ACRES U.S.A. So the rotation is not lost time?

NATION. It's not a loss. It's a win-win situation. That's what I am saying. If you want to look at what drives yield, it's organic matter. If you want to see what builds organic matter fastest, its pulsed grazing of cool-season plants. With rotational grazing, because the plants grow up tall, the roots die back when we graze it off. It sheds those dead roots in the soil where they form organic matter.

ACRES U.S.A. So the mistakes that modern agriculture is paying for date back to the

'50s when animals were pulled off of the farm so we could plant fence to fence. But farmers believe that manure that wasn't spread on fields anymore could more easily be handled in a lagoon …

NATION. A lot of people don't understand manure. If analyzed, there is only 5 percent of cow manure that has mineral content to it. What we are putting back on the land is organic matter. The benefit of spreading manure on the land is organic matter. The minerals content that you are putting out there is minimal; you can buy that out of a sack. It's that organic matter that we are putting back in the soil that we put out there with manure that people have forgotten. Nobody's selling that in a sack. Everybody's selling the minerals, but they are missing that organic matter.

ACRES U.S.A. Or they try to "buy it in a sack" in an almost bizarre way. For instance, some California organic growers are applying up to ten tons of compost three times a year on fields trying to undo the damage of the last generation.

NATION. I have not found a sustainable agricultural civilization anywhere in any history book that did not have a ruminant animal at the centerpiece. That's the ultimate form of sustainable. This whole idea of creating bare ground is not the way. You don't find that in nature. The only way we can get away with that is to do it infrequently. The rest of the time you've got to put that scab back on and either heal or till the ground. I know a lot of people don't want to hear that. We are divided up into this culture of crop or cattle and there's this big wall between us. The interesting thing is that I have found that same wall in Argentina, too. What they do is rather than a cattleman trying to learn crop technology, he leases his land out for three years of cropping. Or the row cropper leases his land out for three years of grazing. They cooperate with each other rather than each trying to learn two technologies.

ACRES U.S.A. So two neighboring farmers should essentially switch farms every three years?

NATION. Sure. Just switch farms. There is something in it for both of them to work together rather than working in opposition to each other.

ACRES U.S.A. Ohio State University suggested that if dairy farmers switch to intensive grazing they can temporarily increase their profits, but they cannot battle the inevitable that they are going to have to expand to survive. Is this your experience?

NATION. What is happening is that the price of whatever you are producing is trying to stabilize at the cost of the average producer. That's why it's important that you know what the average cost of production is. The other thing that was driving it is that we were narrowing the margin because we were spending more to produce more. Everybody was saying if we went back to non-grain-fed dairy or non-grain-fed beef agriculture ... well, go back to look at what happened to the price of milk, back there in 1996 when we had $5 corn. It went through the roof. That's what everybody's looking at. If you get off this high-production treadmill and start putting some profit margin back in production, then you are working with the market, rather than against it. If everybody produced a little less for a lot less cost, you'd have the price going up while you are increasing the margin.

ACRES U.S.A. The only people who make money on the market shifts are the traders, not the farmers, because the farmers get hurt on selling and they get hurt on re-stocking.

NATION. It's a mark-up deal. The grain traders don't care what the price of grain is, as long as it's moving. They make their money on moving grain. That's their business. And you just need to realize, well, that's fine, that's how they make their living.

ACRES U.S.A. You believe farmers should accept this, rather than shout how unfair it is?

NATION. It's not unfair. It's a guy just making a living.

ACRES U.S.A. So farmers must design their operations so they can survive and thrive in the midst of these unfriendly market forces. There are many dire forecasts for what's happening to agriculture. Do you see any bright lights? Are there people making more than a decent income in farming?

NATION. I think that you can make as good a living in farming as in anything else. Gordon Hazard in Mississippi teaches schools for cattleman. He is 71 years old, runs a ranch all by himself, and makes almost double a doctor's income doing it.

ACRES U.S.A. Can you profile his operation?

NATION. He's got an all-grass beef-stocker operation and does not own one piece of machinery other than one little beat-up old pickup truck. He runs it all by himself, with one part-time cowboy that works about 12 hours a week. When he sells an animal, he buys an animal. He doesn't fight the spread between when you sell an animal and when you buy; in other words, you buy a profit in an animal. You don't buy it and sell it; you sell it and buy it. He says that his entire capital investment is that he's got two hammers and a fence stretcher. When times get really tough, he sells one hammer. And it sounds like a joke, but it's all for real. His net return on investment including land cost is 20 to 30 percent a year. He's bought all of his land, all 3,000 acres, all just from retained earnings. But he had to get away from tractors and machinery and all. He doesn't have any barns.

ACRES U.S.A. Yet he's selling into the commodity market. He's not selling individual cuts of meat himself organically at the farmers' market ...

NATION. He analyzed the market and found out that for where he lives in Mississippi, you've got to be selling big, heavy 900-pound steers. But it's not every month of the year that people in Nebraska want a 900-pound steer from Mississippi. He analyzed the market and found out that there was a hole in the market in August when those people did want those animals and that they could be fed and dead before it turned cold, which is something you've got to consider with a Brahman-cross animal—so he went for that target and it's worked fine for him. He didn't moan and groan and say, "Isn't it a shame?" He looked for the hole. That's the unfair advantage I was talking about. Where's the hole? Everywhere anybody lives there is an unfair advantage to where you are.

ACRES U.S.A. What does this lack of thinking that you sense going on in American agriculture result in? When American farmers do try to change over, what do you see?

NATION. If I was trying to heal land as a crop farmer, the first thing I would do is go to grazing for several years. Because all those weeds are wonderful cattle feed or sheep feed. In everything in nature there's a prescribed sequence to change. The most financially successful dairyman in Ireland, a guy named Michael Murphy, has made a million dollars in the dairy business over there buying up bankrupt farms and putting them back to a natural system. And he said that the necessary sequence to change into a seasonal dairy is to get your grass sward up to maximum speed with seed, lime, phosphorous and drainage. Second, change to cows selected for higher reproductive reliability on grass. Third, go to seasonal production. What we have tried to do in the United States is to take guys that have a confinement dairy and make the least change possible. They've got hybrid Holsteins and a weed patch and they are trying to go seasonal with that. We've got the sequence out of kilter. A beef cow is a scavenger of bulk grasses, low-quality bulk grasses. She's there to improve that pasture for a higher class, for a steer, for a finishing lamb, or for a dairy cow. In other words, she is a means to an end, not an end to herself. It's important that you understand these sequences. Every niche in nature is creating a sub-niche. Algae growing on a rock is creating soil which is going to create a grass plant, which is going to create organic matter which is going to allow a higher succession of plant life to grow. You need to understand that there is a successional series in creating positive change. You've got to start at the right end of it.

ACRES U.S.A. And when you are looking at the massive note just on that new tractor, you are locked in. When you are paying for blue silos you've just eliminated your wiggle room.

NATION. I always say that capitalization is like an Australian water gate, the gate that's like a V and the cow can push through one way, but then it swings shut so she can't get back out the other way. That's exactly the way capitalization is. That's why 99 percent of building a successful farm management scheme is making it flexible. Don't get locked into anything. A lot of people say, "How do I get back?" And I say, "Boy, I wish I knew." It's tough coming back. Peter Drucker says there are only three ways out of debt. One is inflation. Two is bankruptcy. Three is put your nose to the grindstone.

"We Have to Start with Education"

An Interview with Gary Zimmer

Originally Published: December 1999

This is a fact: Listen to Gary Zimmer speak for 30 seconds, and you'll learn something. Or a lot of things, most likely. The father and entrepreneur is a dynamic educator and speaker who has helped countless diverse groups, including farmers, financial groups, ag consultants, environmental groups and extension offices. He is best known for the quick pace of his talks, and as the father of biological farming for the books he wrote with his daughter, *Biological Farming* and *Advancing Biological Farming*.

Mr. Zimmer was actually raised on a dairy farm in northeastern Wisconsin, and after receiving training in dairy nutrition and earning a bachelor's degree from the University of Wisconsin, he later earned a master's degree at the University of Hawaii. He then spent five years teaching a farm operation and management course at Wisconsin's Winona Area Technical Institute.

Over time, he has pursued an open, responsible approach to farming and taught tens of thousands through conferences and field days. In the early 1980s, he helped establish Midwestern Bio-Ag, a sustainable agriculture input supplier and consulting firm that works with thousands of farms and has dozens of trained consultants. Today, he continues to farm with his family in Wisconsin, and travels to speak and consult wherever he is needed most.

ACRES U.S.A. It seems strange sometimes that so many of us happened to start in at just about the same time. You started your company the same year Acres U.S.A. was founded, in 1971.

ZIMMER. It is amazing. There was quite a bit of talk at that time, and it was not always well received in the countryside. Everybody would argue against it. At that time, no one had foreseen all the problems that would come about through conventional farming.

ACRES U.S.A. In the process, you've established a rationale for biologically correct procedures on the farm.

ZIMMER. Yes, when we got started here, we began by looking at the procedures governing this type of farming, and we found that a lot of the materials were missing. There was not a well established, ordered structure or system to follow. Ultimately, we want to educate, to give people practical ideas and guidelines, so we came up with what we call our "Rules of Biological Farming. "

ACRES U.S.A. How many rules do you have, and what are they?

ZIMMER. We've got a total of six rules, and I cover them in detail at the start of the book. The first rule is to test and balance all the essential soil nutrients. You want to get a complete soil audit and look at all the places where your soil is short of any nutrients, because any nutrient in short supply becomes a limiting factor. With a good soil test, you know the condition of your soil, and then you've got a starting point. You can go after soil correction and begin moving the soil toward a balance, keeping in mind the soil's excesses as well as the deficiencies. When we balance the soils, we want to use as many natural rock mined materials as we can get hold of to do the job. We use a natural rock phosphate and right on down the line. Rule number 2 is to use non-harmful, life-promoting fertilizers. Here again, we prefer to use elements that are mined, with minimal processing. It's important to know that different soils will respond better or worse, depending on the type of soil we're looking at. Sometimes the cheapest source of an element is not the best source. I prefer materials like Idaho phosphate, North Carolina reactive rock phosphate, high-calcium lime, gypsum, potassium sulfate, and those sorts of things. We can't, of course, forget the trace elements—the sulfate and chelated forms I find are the most effective. On the flip side, it's important to avoid harmful fertilizers which may be cheaper per-unit, but they are not cheaper in the long run, because they can degrade soil structure, harm soil life or injure plants. As before, fertilizers will perform differently on a variety of soils, and some otherwise harmless fertilizers could do quite a bit of damage on some crops.

ACRES U.S.A. Can you give us some examples?

ZIMMER. Sure, take for instance dolomitic lime. It's a calcium and magnesium source, but in some areas, soils are already naturally high in magnesium. Adding more magnesium to the soil doesn't contribute to balancing it and can interfere with the uptake of other elements, especially potassium. A magnesium level that is too high makes some soils more compact and tight. So fertilizers like dolomitic lime, potassium chloride, anhydrous ammonia and certain dry fertilizers and oxide-form trace elements should be eliminated if at all possible. Rule number 3 would be to use herbicides and pesticides only when absolutely necessary. Put them in the category of necessary evils, and do everything you can to reduce them. Biological farmers who are just starting out find that the need for herbicides and pesticides will decrease as the soil comes into balance and is more biologically active. When they do encounter problems, they should meet them with non-toxic methods such as maintaining a short crop rotation, keeping soils balanced and, in some cases, rotary hoeing. Biological farmers should also invest some time to learn what weeds grow and what cycles are and the kinds of things to better help them fight those kinds of problems. Rule number 4 is to run a "tight" rotation or a short rotation. Some people ask, "Is that absolutely necessary? Can I grow corn on corn?" If you did do such a thing, we would want that corn crop interseeded with a clover, so we would have a clover/corn rotation. Always growing crops in a tight rotation is one way of reducing your insecticides fairly easily. Rule number 5 would be to use tillage to control the decay of organic matter and control soil air and water. We've gotten a lot of good advice from Don Schriefer, beginning way back when he took a look at what the purpose of tillage was. He told how best to shallowly incorporate our residues, how to do deep tillage for aeration and avoid flipping the soil upside down. The soil life wants its food on top and wants to be left alone. We watch tillage fairly closely.

ACRES U.S.A. But you don't necessarily pursue the idea of no-till?

ZIMMER. No we don't, because no-till, to me, is not a viable method of farming. It's a decision whether to till or not to till. If you opt to go no-till, you are really limiting your options, particularly if the soils are wet and tight and you've gotten them compacted. This year in Wisconsin, we had an extremely wet year, so we'll do a lot of subsoiling this fall. If you say you're going to be a no-till farmer, I think this really limits your production and limits what happens on your farm, biologically.

ACRES U.S.A. And then there's the added factor of having to insert so many chemicals into the system.

ZIMMER. That's why that would violate our rules, because we couldn't follow rule number 3, which says to use herbicides only when necessary. In this case, you would be totally depending upon them. Our last rule, number 6, is to feed the soil life. We take care of the soil biology, and that means using something besides corn stalks and raw manure. We want to grow green crops whenever we can. We were planting rye on our farm today, following corn silage. We interseed our corn. If you're a dairy farmer or have livestock, there tends to be a lot of forages on the farm, and you're always going to do something to feed the soil life. Again, the soil life wants its food on top and it wants to be left alone, so we would be incorporating green manures and putting on compost—those kinds of practices, as often as we can.

ACRES U.S.A. Where do you start? Do you take the soil audit first, or do you do a little educational work? Because if people don't understand the grammar of the subject, they're not too receptive, are they?

ZIMMER. Because it is such a different method of farming, we have to start with education. That's why we've always put out a lot of informational materials and we've put on winter meetings and do educational seminars. They have to understand the concepts. Taking a soil test and not understanding what to do with it would be a waste of time. We like to go and do a complete farm evaluation. What's the farmer doing now? This isn't a radical jump from one extreme to another. We ask, what is being done now, and what practices can be done that don't cost any more. Some of them may actually save costs. One example might be reducing tillage or changing a rotation. So we want to evaluate the whole farm and figure out what the farmer is already doing—sources of fertilizer, when he does it, crops he is growing, what he is getting per yield—and then we want to pull the soil test so we know the chemistry. We want to know what fertilizing and liming practices have been used. Very few people have looked at the soil in detail, as far as providing all of the minerals necessary to grow a good crop.

ACRES U.S.A. Touching on your move into non-conventional farming, how did you overcome that rampant hostility we faced in the early 1970s?

ZIMMER. It was difficult in the beginning, but we certainly found farmers who were receptive, and we traveled all over the countryside. We had to find farmers to whom these ideas made sense; but at the same time, we had no examples, nobody to point to and say, "Here's somebody who is doing it and who is quite successful." But we did find a few farmers who were always cautious about using chemicals or just felt something was wrong with the present system of agriculture, that there was a better way. We would get those guys going. Part of the reason we're so scattered is because of that—not every farmer down the road was receptive. We shook a lot of bushes, and now we see the success of the farms. I truly believe that is what is going to change agriculture. It isn't going to be by some government policy or some magic research, but by the success of the farmers who are farming and looking at different methods and different ways of doing it. Take for example Louis Bromfield, way back in the 40s, he had several books out, *Malabar Farm* and *The Farm*.

ACRES U.S.A. And *Pleasant Valley*.

ZIMMER. Yes, he was quite a prolific writer. I look back, and some of his books were really interesting to read because the concepts were there. You are talking about 50 years ago, and people began to get in tune to this. In the 40s, that's when Albrecht did his work. That's when *Plowman's Folly* was written, and Bromfield was out here. Chemicals were just coming in to full swing at that time, and even though people were getting smarter about how they were farming without them, the chemicals were cheap and easy, and we didn't see all the problems that were to come forth. Bromfield's thing was to have field days where he talked about building six inches of topsoil in five years. It was really a program of taking a soil test and mineralizing, all the way down to copper and zinc and manganese and boron, using different sources of lime, using a hydrated lime or a kiln lime, which we have found to be quite effective today. He would plant oats and clover, then cut it and mulch it on the ground. He actually had rotavators like we do today, and he would shallowly incorporate that into the soil. Within five years, they could build six inches of topsoil.

There were field days in the 1940s when 8,000 farmers showed up. I still think his ideas and concepts are valid and should be followed if we are going to see agriculture change to what he called "pilot" or "model" farms. I always challenge everybody to go out and look at the farms we are working with, the really successful ones, not that they're going to copy what they're doing, but it seems like a logical way to get ideas. There isn't near the risk involved today, and that's why it now grows so rapidly. There are many, many guys who have been doing this successfully for a long time.

ACRES U.S.A. Do you have problems finding qualified consultants to do this fieldwork?

ZIMMER. Our biggest challenge is finding good quality consultants. It is extremely difficult to find qualified consultants because they have not necessarily been educated in this method of farming. We either have to re-educate or, certainly, they have to be receptive. Today, there are more and more of them, just like there are more and more larger farms that are looking at this type of farming. Many of our consultants came from farms that we have worked with—maybe a brother from one farm, or a son, or the farmer himself. They've seen it work on their own farm, they've been practicing it, and they are really convinced how well it can work, so they often make the best consultants. Once they commit to becoming consultants, they go through some "basic" training and then commit to what amounts to a lifetime of continuing education. We have monthly training sessions and also send them to outside training. After two years, those who qualify can go through our Certified Consultants Program and get advanced training. Throughout all this, they are supported by a staff of individuals who have technical expertise in areas ranging from nutrition and veterinary medicine to agronomy and other areas specific to biological farming. This has really been working well so far, and I hope that in the future we can continue to attract more qualified and motivated candidates.

ACRES U.S.A. Which is more successful, the rural meeting or getting a farmer to actually visit another farmer who is making a success of his system?

ZIMMER. We combine those two. As part of the rural meetings, we will usually have a farmer we are working with invite his neighbors, or we have what we call "round-table meetings" at people's houses or restaurants where there are just a half a dozen farmers. They're usually friends of someone we're working with. These people have been watching their neighbor and now they come and get the whole picture painted for them of what is involved. They're already quite receptive when they come, so now they have both things working for them. That's been real successful in getting people to understand this.

ACRES U.S.A. But the actual walking in the field, is that part of the process?

ZIMMER. Absolutely. We do a lot of the meetings and presentations in the wintertime, of course. In the summertime, though, we take it out in the fields. We look at both the crops and the livestock. We also try to get across that there's no reason that because you're going to reduce nitrogen and reduce chemicals you have to settle for less. In fact, it actually works in the other direction. Midwestern Bio-Ag also operates a farm where we do demonstrations. It's not pure research, really. My concepts are that we had better demonstrate it first, and if people want to get all the statistical numbers, then do the research after we're pretty sure we can demonstrate a method of farming. We look at the whole system, not just one piece, when we do our demonstrations. Every year, we have a two-day field day, both with livestock and organic agriculture, looking at soils and crops. We have educational booths and demonstrations, and we bring in about 800 people. We also have field days all over the countryside and try to get back to every farmer we work with at least twice during that summer to look at crops and show them some new things and give them ideas of what to look for. They should see the soil structure change, they should see weed pressure change, and they should see earthworms start to appear in their soil. Overall health and root system development in plants should be improved. For some of these things, it depends upon where the farmer is coming from, as far as what shape his soils are in—if they're really hard and dead, obviously, it's going to take longer. He has clues and things that he can see along the way that keep him encouraged. You must realize, of course, that when you start this program, an imbalanced soil didn't get the way it is overnight, and it's not going to change overnight. Many times you are looking at a three- to five-year process to really get this thing working. If there are major soil chemistry corrections to be made, the budget may stretch it out longer than that. It might take the grower 10 years to get the mineralization that is needed into the soil.

ACRES U.S.A. But this is a foundation process you have to go through to deal with the livestock end of it, correct?

ZIMMER. Yes. I'm a dairy nutritionist by training—and we work with a lot of dairy farms; that's the only kind of farms we worked with in the beginning—and I wouldn't do a farmer's dairy nutrition unless I could work with his soils and get him involved with manure management and soil nutrients. On the dairy farm, the farmer gets paid twice, compared to our conventional farmers who grow corn and soybeans. Except for higher test weight and maybe a little cleaner crop, they don't get rewarded for the quality of what they have produced, for having more minerals in their crop and having a healthier crop. But a livestock farmer gets to feed that. He goes through the same process. Today in the livestock industry, there is a lot of milk produced, but they are always supplementing and adding more—meat and bone meal, animal fat, cotton seed, chelated trace minerals, a pound and a half of minerals. They keep adding more and more things to the cow to get her to produce, which has certainly been successful in getting production up in conventional agriculture, but we have not been successful in keeping the cow healthy and around for many years. There's a real shortage of dairy heifers in the countryside. We've gotten the production, but we've really destroyed our cow. We have to get back to what our cow is supposed to be eating, which are forages and grasses. It isn't made to eat grain and animal fat and meat and bone meal.

ACRES U.S.A. Yes, and if you burn up the cow in the process, then getting that extra hundred-weight of milk isn't that important, is it?

ZIMMER. That's right. One of the difficulties of dairy we call the "dipstick mentality." The dairy farmer looks in his bulk tank and says, "Yeah, I've got more milk. I must be doing something right." But the cow is standing there looking like she's dying from trying, and she's not reproducing. She doesn't last long. Biological farming isn't just getting the soils right; it is also getting the quality forage so we can produce the milk off of forages. If you put the cow on a high-forage diet that has lots of balanced minerals, then she will produce really well and she'll last a long time. I'm not saying it's the highest-producing herd in the county that some of our farmers have, but they're very profitable and they have extra females to sell off their farm, and their cows aren't dying from trying.

ACRES U.S.A. They're not dying from the use of bovine growth hormones either.

ZIMMER. Yes, and that's a real interesting one, the bovine growth hormone. Again, if you consider a cow, she has plenty of natural hormones she produces in her body, but if you put her on a high-starch, high-corn silage, high-protein supplement diet, the level in her bloodstream is actually going to be quite low. There was a test herd in Michigan that worked with BGH, and they actually saw a negative return and didn't see any response in milk production. They were using a placebo to measure against, but people still thought somebody was cheating. What the people who did the test failed to understand was that these cows were on a high-quality forage diet and they didn't need any synthetic BGH.

ACRES U.S.A. Isn't there something wrong with the idea that the cow should produce by having a needle stuck into her every eight or 10 days?

ZIMMER. That's what we like to say. Whatever gave people the idea that a needle in the tailbone was a limiting factor on that cow. A good-quality feed is what they need, more forages. Some of these cows could use a little dry hay for some cud chewing. We like to do everything possible to get the soils healthy and to get a balance of minerals and do everything we can to get the livestock healthy. Then you can't stop them from producing. To get livestock healthy, we use kelp and probiotics and yeast and high-quality minerals. We give them comfort and fresh air—there are a lot of factors involved, just like there are in soils. It's never just one thing. That's why there are a lot of clues and ideas that come from research, but it's very difficult for a researcher when you have so many variables to follow, particularly when research doesn't capture a method of agriculture that really works. It's kind of interesting to see that the new disease in dairy cattle out here—it's not a disease, really, but something we make fun of—it's called "potassium toxicity." It's really nothing more than dumping straight potash on hay ground and getting piles of tonnage, and then, with all that manure, all that potassium, we've got the soils so out of balance that we've actually created a new disease. To me, it's the most logical, straightforward thing in the world. People's reaction to it was to feed more corn silage, because corn is a plant that doesn't take up very many minerals. Our reaction to it is, let's go to the hay fields and alfalfa, being crops that if you fertilize properly, you can get a balance of those minerals. Get your calcium and potassium in line, and don't fertilize with potassium if you have plenty in the soil. Use a balance of all minerals when you fertilize.

ACRES U.S.A. And how effective has this been?

ZIMMER. We've been able to demonstrate a 50-percent increase in mineral uptake in the alfalfa plant, and we give a lot of the credit to soluble calcium. It isn't a matter of just putting lime on the soil and saying, "Now I've got my calcium taken care of." We want to balance that soil, but we also want to feed that crop a balanced diet. We add soluble calcium to a soil. We look at gypsum as one source, and processed kiln dust or what we call our Bio-Cal, which is a highly soluble calcium source with sulfur present and also boron. These products really increase the mineral uptake and completely change the fiber within an alfalfa plant. If you only fertilize with potassium on hay and you go look at alfalfa fields, the stems are coarse and hollow. Traditional thinking is that you have to cut that hay in the bud stage or cut it early, that's the secret to getting quality. Well, early cutting of alfalfa that has an imbalanced mineral content will result in a loss of energy, and that's one reason why farmers have to keep supplementing all the energy to the cow. If you get the minerals balanced properly, you'll get a solid stem in that alfalfa. We give a lot of credit to calcium for changing the pectins in the plant and giving us more soluble carbohydrates and providing more energy, although it may not even be noticeable on the feed tests. Feed tests and soil tests, as well as research, all give you clues; but a feed test doesn't tell you the whole story. People sometimes get hung up on these numbers.

ACRES U.S.A. Is anybody still using *Morrison's Feed and Feeding*?

ZIMMER. *Morrison's Feed and Feeding* is a book I always have on my desk. It's the good old basics—and the cow, I tell people, has not changed. There are a lot of wonderful ideas. In regard to calcium, *Morrison's Feed and Feeding* says that alfalfa hay is supposed to have up to 2 percent calcium. The average in the state of Wisconsin is only 1 percent.

ACRES U.S.A. Is that meaningful, considering the disparities that we have between types of agriculture out there?

ZIMMER. Yes. There's no question about how we fertilize and how we've done things with the system of agriculture; we've really changed the plant. What the old Morrison's *Feed and Feeding* guide recommended we feed cows years ago was a totally different kind of balance. However, if you go back in those days, what was missing was still an understanding of what biological farming was. There was also a lack of understanding of all the minerals necessary to grow this mineralized crop. Then, of course, we didn't have all the tools or equipment to get it harvested properly and stored properly. Making dry hay in Wisconsin when it rains a lot is very difficult. It's kind of interesting comparing the 1940s or even the 1970s to the present because today there are so many farmers who have figured out how to obtain mineralized biological soils. That's why I can see such a rapid change coming with the biological farming movement. There are a lot of successful farmers that are figuring this out. Our potentials and our possibilities are far beyond where we are today. I am convinced of that.

ACRES U.S.A. What you are saying suggests that a dairy farmer who wants to operate by buying his feeds in the market is really playing Russian roulette with his herd.

ZIMMER. That's right, and today, dairy heifers are selling between $1,500 and $2,000 to replace an animal that sells for $300 for meat. You're paying an awful lot of money for an udder and reproductive tract. And that's the part we can't keep going. That's right, they're playing Russian roulette, saying, "Oh, we have to have $13 a hundred for milk or we lose money. I say, "That's not true for every farm." Not that the farmers are getting overpaid, but I guess farmers are in competition. In reality, they do have to be able to make a living through the market that exists. There are a lot of farmers doing really well once they figure out this biological system, because they're not buying all those inputs. They're not buying the problems; they're not buying all the feed. They're growing it, not buying it. I think that's the same thing, too, for a crop farmer. He can grow clover and produce nitrogen and get minerals exchanged from the soil microbiology. He doesn't have to buy all of those inputs.

ACRES U.S.A. You've encapsulated most of these ideas in a book. What are you calling that?

ZIMMER. We struggled quite a while with a name for the book, because what started out as a training manual for our consultants, over the years turned into the book. I had a lot of help from Dr. Harold Willis with the technical chapters. But as far as the name goes, we went through everything, even the "ABCs of Biological Farming," to try to get a name that would fit, because it's a "how-to" guide.

ACRES U.S.A. Do you take specific crops by chapter, or do you more or less generalize on the foundation principles, like what Albrecht talked about?

ZIMMER. In the book, I go through the basic concepts—soil tests and understanding fertilizers and the different sources and how they perform. Then I go through case studies and experiences on my own farm. I follow up with soil reports, what we put on and how the soils and crops change. Along with that, I've got sections that go through individual crops—corn, soybeans, alfalfa, small grains. I even have potatoes included. There are many crops that concern a biological farmer, and accordingly, our study has spread into every crop imaginable. We do a lot of work with potatoes in Wisconsin. We started working with potato farms after this one particular farmer rented a field that had been on our fertilizer program. The farmer grew potatoes, which are paid on quality—and calcium is so critical to get a quality, healthy potato, as is as a balance of all minerals—that when the farmer rented this field from a guy that had been on a biological program with mineralized soils, he contacted us and said, "We haven't grown potatoes like that since we broke virgin land. What are you guys doing?" That's how we got into that industry. There's a chapter in the book specifically on potatoes. The potato growers are as close to what I call "precision farmers" as I've seen out here. They monitor those crops. Actually, they are getting a lot of pressure to reduce chemicals in the crops that they produce. We've had some fairly dramatic results. I think as far as getting rid of some of the fungicides, I give copper quite a bit of credit. We give calcium credit for producing a healthier potato. We're looking at green manure crops and growing brassicas in rotation to get healthier soils. There has been pressure for potato farmers to reduce chemicals, and also, they get paid for quality, so our type of farming is really the right approach. What's more, potatoes are having a hard time storing the way conventional farmers have been raising them.

ACRES U.S.A. Yeah, they turn into soup by about April, don't they?

ZIMMER. That's absolutely right, and the buyers are getting smart to that. So the potato industry, since they get paid on quality, really took to biological concepts and principles because they could see them happening. Today, I work with cranberries and cherries also. I always say it's really quite simple, we're always going after balance, and then we put together a fertilizer that is specific for the crop. I say to the farmer, "What are you doing now?" He typically responds, "Well, I'm using N, P and K." To me, it seems very simple. What about sulfur, what about calcium, what about zinc, manganese, copper, boron? What about some of these natural mined materials we use and Idaho phosphate—that's quite high in molybdenum. All those minerals are necessary. We use as much science as we can, but on the other hand, we go as close to nature as we can because we realize science doesn't have it all figured out and maybe never will. Maybe never needs to. We use natural rock, for instance, that might have trace minerals in minute amounts that we haven't yet figured out. Then there's kelp, which we like to use in our fertilizer program. We get a positive response from observing farmers, identifying what they're doing and locating the limiting practices. Whether the crop is cherries or apples or cranberries, we seem to get a really good response. We've actually done real well in the cherry industry. I don't know a whole lot about the cherry crop, how it is harvested or anything else, but it's really quite simple. I ask, what are you doing now and what seems to be limiting factors.

ACRES U.S.A. Once those basics are in place, you can practically take the crop system any direction you want, can't you?

ZIMMER. That's absolutely right. Out here, we've got some of what we call the "myths" of agriculture (things commonly believed that sometimes should be questioned) and one of them is putting calcium on potatoes. The sentiment is, "We can't get our calcium out there because we'll get our pH too high and we'll get scab." I'm not saying that is not true, but the plants still needed calcium. It didn't need lime to change the soil pH. If we go to blueberries as an example, there are blueberries being grown on high-pH soils. A blueberry wants certain minerals, and if your soil is high in pH, those minerals, like manganese and iron and aluminum and some of those heavier minerals, would not be available. We have to get that mineral balanced, no matter what the pH of the soil is. But like you said, those base concepts need to be in place, and some of the specifics, then, just need to be added. I think that's what's really missing in our food chain. We really notice dramatic results in livestock farms when we mineralize those crops. The livestock really perform. I always say our foods today are not grown on balanced, mineralized soils—that's why I'm quite an advocate of people taking food supplements—we know our foods aren't fresh, and we know they aren't grown with a balance of all minerals. So unless you have control over what you grow, you'd better supplement your diet with something. I tell the farmers that we take better care to provide healthy food for cows than we do for humans.

ACRES U.S.A. And then we're adding to the problem with genetically modified organisms in the food supply and on the grocery store shelf.

ZIMMER. It is amazing that some things are allowed to happen. To me, it was the same when they came out with BGH in dairy. Someone would ask, "Well, what do you think about it?" And I would say I don't know, but the researchers said it was totally safe and good, and I'm not doubting their research, but I would suggest that those guys take milk from cows with high levels of BGH injected for 20 years and see what impact it has on the next generation. Then maybe we can reevaluate it. To jump right in without knowing enough, I think, is our difficulty with any technology.

ACRES U.S.A. This is the point of view that's been adopted by the Japanese. They say that they are going to watch the American children and see what happens to them over the next 10 years.

ZIMMER. Right, and it's the same with research. It's no easy task, watching downturns of health in humans and trying to pinpoint exactly what did it, especially when everybody's stress level in life is different. It's easier with livestock. We at least have them contained on one farm and we can control their diet. In humans, it's impossible. Society in general seems to have quite a few health problems, and there are a lot of people today who are quite concerned about these things; even more so, in some of the European countries. At this point, we don't need to put pressure on to produce more food. We ought to be putting pressure on to grow it better and to take care of our soils. Our system of farming today produces an abundance of food, but what about its sustainability? What about five generations from now? Will we be farming that way? I have a feeling we won't. There are a lot of restrictions coming in now, and it seems kind of sad that we have to farm based on restrictions. Why don't we farm based on common sense? We have to ask ourselves why we overuse nitrogen, why we let our soils erode away. If we get a healthy, loose soil, it won't wash away. If we're producing this cheap food, where is the sustainability in our system? I think we're going to have to pay the price. Right now, everybody has to make the change, and I'd love to see agriculture give the farmers a carrot for change instead of keeping a stick over their heads. Give them big incentives to reduce their chemicals and grow green manure crops and use proper minerals. Even if we did take a reduction in supply (which might only happen in the beginning, since once they figure this out, they do better), we could use the surplus food we have on hand now. Right now is a wonderful time to make a progression to change agriculture.

ACRES U.S.A. Change it to the quality that the American people demand.

ZIMMER. Yes, exactly.

ACRES U.S.A. Now, this farmer who wants to do something, he's been mulling it over for a long time, so he's a little sheepish about approaching you because of the gossip at the coffee shop. What can he do? Does he just get hold of one of you guys and say,

"Hey, we need to get together and have a little meeting here"?

ZIMMER. Yes, that would be one way, if he doesn't know a farmer who is doing it. There are various ways he can get hold of information. We've got tons of educational material available. Attending an Acres event would be a good starting point for some of these farmers. You need to get yourself familiar with the ideas and visit farms that are doing it so that you feel comfortable with it. You had better know where you are going before you start out on the road. I think it's a process of self-educating. Another thing I find on our field days is that the attitude of the farmer is changing. We're putting fun back in the farming again, making it a challenge for them while giving them more control.

ACRES U.S.A. Are we putting any young people back in agriculture?

ZIMMER. That's the other question. We've driven many of them off the farms, saying, "This is a terrible way of living. There's no money in it. Go become something else." My son is 21 years old, and I strongly encouraged him to go to college. He wanted to farm, but I wanted him to first gain the experience that college provides in opening up the mind to different ways of thinking and living. After three years, he came to the conclusion that he was not learning what he needed to learn in order to farm the way he wanted to and started farming full time. Today, he's farming about 400 acres certified organic. He's also going to put in an organic dairy and milk around 100 cows. He's got an excellent herdsman helper, and he's excited about farming. If you put a pencil to it, the young guy is going to do extremely well at something he likes to do and finds challenging. And he's proud of what he does. He doesn't have to apologize for how he farms to anybody. He gets a lot of compliments from the local people around here about how he farms.

ACRES U.S.A. He must be a pretty smart youngster, because he's figured out something that took us most of a lifetime to figure out, and that is that most education is self-education in any case.

ZIMMER. Absolutely right. I always tell people that I drive between 50,000 and 75,000 miles a year across the countryside, and if they were to ride with me and follow me for a year, they would farm differently. They couldn't do the things they keep on doing if they saw all the things we see out here in the countryside. My son had the opportunity to be around many of these successful farmers and hear them; it is a matter of self-education. He can see and observe. It's a wonderful way of making a living. I don't think you'll find a happier kid out here. He loves what he does. He puts in long hours, but he's enjoying them, and he's proud of what he's doing. About getting around the people that are successful, if you have to go to the local coffee shop, sometimes that conversation about the newest, latest chemical doesn't fit what some of these guys that we work with are trying to do. Some of them say they don't like to go to the coffee shop any more; they almost have to make new friends. I think that biological farming starts with what's going on between those two ears. You have to reevaluate and think differently, you have to open up that mind. The farmer has to get the basics and the logic behind any new ideas, the ones he picks up at the coffee shop included. He needs to sift through all the rhetoric and, using common sense, come up with the method of farming that will work best for a sustainable future for all.

CHAPTER

18

"It is Really Fundamentally a Different Philosophy of Life"

An Interview with John Ikerd, PhD

Originally Published: December 2000

Many might not know that John Ikerd, the author of *Small Farms are Real Farms*, began his career as a "traditional" economist. His area of special attention was livestock marketing, market outlook futures, and all the nuances of farm marketing reality.

"I really began to question a lot of the things I was doing during the farm crisis of the 1980s," Mr. Ikerd said in this interview published in 2000. This led to an in-depth exploration of farm financial management and how to help the trends of consolidation swallowing up family farmers around the country. Facing this hardscrabble reality forced Ikerd to move in the direction of sustainable agriculture, and for years taught the economics of such practices at the University of Missouri. When he sat down with Acres U.S.A. in 2000, the year organic certification rules were finalized in the United States, his answers suggested a new frontier for eco-farming.

ACRES U.S.A. One of the bankers at a recent meeting of that fraternity made a remark to the effect: "Let the cartels have agriculture. Who needs it? We are getting along fine without the individual farmer." Without trying to serve you up a loaded question, how would you assess the pros and cons of that statement?

IKERD. I think from the pro standpoint that he is talking about what economists have been talking about for years; they are saying that the only thing that really matters with respect to agriculture is that we produce food and fiber as cheaply as we can possibly produce it in terms of purely economic cost. The generally accepted premise is that the most efficient way to do anything in this economy is to do a kind of industrialization, which is specialization, standardization, and eventually, centralization of decision-making. That is the process that we have been going through in agriculture throughout my professional career and before that for at least the last 50 years. We have been trying to industrialize agriculture. We have been centralizing first by going from diversified family farms to the specialized crop and livestock and now specializing in a particular crop or livestock or even in a particular phase of livestock production.

ACRES U.S.A. Such as your hog operations and dairy operations?

IKERD. Hog operations are a good example. The crop farms in this area of the country are corn and soybeans, and in some cases, only corn or only soybeans. We have specialized and then we have standardized the whole process to make those systems work. We tried to get it down to the point where we could farm by recipe as much as possible. We came out with recommended practices or best management practices or the best variety—whatever it is. We are trying to make the same thing work on every farm by standardizing every process and by standardizing through grades and standards so that we only sell basic, generic commodities. Where we are now is really in the final stages of that third stage of industrialization, which is the consolidation or centralization of decision making. We have really been going through that for some time from the standpoint of moving to larger and larger farming operations, which mean fewer and fewer decision makers.

ACRES U.S.A. This is monoculture, basically?

IKERD. Monoculture really operates with the farm as a factory and running a field or a feedlot as if it were an assembly line.

ACRES U.S.A. Doesn't it rely to some extent on a certain type of accounting too?

IKERD. It is basically just the economic accounting in terms of measuring production in terms of output per person, per farmer, or in some cases, production per acre. Those things assume that the limiting factors are the number of people that are farming.

ACRES U.S.A. We have often pointed out that in terms of capitalization per acre, and even in terms of production per acre, smaller units seem to measure up a lot better.

IKERD. Smaller units that are more intensively managed, that have more managers per acre, more people, more farms, if you will, have the capability certainly of increasing production per acre over and above what you can do on a large farm. The output per unit of capital in agriculture, and the output per unit of purchased input in agriculture, have been going down over the years. But if you measure simply in the generally accepted industrial economic terms, then back to your original question, the big corporate operations would tend to move you toward a more efficient kind of agriculture from a strictly economic standpoint. That is what people are going by. I don't think that you can even justify where we are now in terms of the corporate takeover of agriculture from the standpoint of economic efficiency, although that is the economic argument. I think that now the consolidation and corporate takeover is being accomplished by purely market power strategies where the independent farmers that are out there, even if they are more efficient and have lower cost of production than the corporate operations, are losing access to the markets. In other words, the corporations are using their market power in the processing and distribution sector to eliminate markets or to at least impede market access on the part of the independent producers.

ACRES U.S.A. Doesn't that extend to management of public policy as well?

IKERD. Yes, I think public policies that we have had over the years have reflected this bias in terms of what we have concluded was the way to achieve the most efficient agriculture. I think you can look straight down the list of all the government programs that we have had in agriculture and those programs have really promoted or provided farmers with an incentive to become more and more specialized, more and more standardized, and more and more centralized, or consolidated, in terms of the management and operation of the farming operation. For example, take the price support programs that have been popular with farmers because they look at them as way to prevent having to accept low prices. But what the commodity price support programs do is take the risk away of having an extremely low price. In other words, the government makes up the deficiency in price. That really allows farmers to move away from the diversified farming systems and to specialize in a specific commodity. What they don't realize when they do that is that the increased incentive to specialize by removing some of the downward price risks is driving farms toward becoming larger and larger, which means as they move in that direction, some farmers are going to be forced out of business.

ACRES U.S.A. It hasn't prevented a lot of farmers from consuming their own capital, has it?

IKERD. No, generally that is part of the process by which the farmers are forced out of business. You have something come along, usually a technology that comes in, and even the technology that we have developed through the USDA and the public institutions—the land grant universities as well as in the private sector—that technology invariably promises that the farmer, if they adopt it, will be able to produce at a lower cost per unit—at a lower cost per bushel or per hundredweight. But to gain full advantage of that new technology, they have to expand their production or their output. When we have a new round of technology that comes in, the early innovators adopt it and they expand their output as more and more of them adopt that new technology; then, of course, production increases and the prices drop. You get into a situation then where the people who don't adopt the new technology quickly enough eventually have to adopt it just to survive. Then you are left with people who for one reason or another, because of their financial situation or because of a bad crop year or because of not being in a position to adopt a new technology, they are the people who are forced out of business. That process of forcing them out of business is something to look at. They don't go out of business over night. They get into a situation where they are not competitive because of something new that has come along.

Initially, rather than cut back or go out of business, they will simply, as you say, consume some of their capital. They will live off the capital appreciation as they try to stay in business.

ACRES U.S.A. The one thing they don't seem to want to do is think the thing through and come at it from a different angle.

IKERD. They get caught up in this race of saying this is the way agriculture is; we are going to have to find some way to continue to get larger and larger and more and more specialized. They don't realize that if they play that game, there are going to be fewer and fewer winners every round. The system is forcing you to fewer and larger farms, which means there are fewer and fewer farmers that are going to survive each round. The losers, so to speak, use up the capital and whatever they have got in agriculture, and eventually they either go bankrupt or they see the handwriting on the wall and realize they can't continue to farm this way and they sell out, while they still have some equity, to the neighbor who is still on the treadmill and still trying to get larger and larger.

ACRES U.S.A. Who thinks he is going to be an ultimate winner.

IKERD. Yes, he thinks he is going to be an ultimate winner without stopping and thinking that in the end there will be fewer and fewer winners each time. This has been going on for years and years, but what is different this time is that we are in a process now where we are not just consolidating farms into fewer and larger farms with those farms continuing to be under family ownership or independent farm ownership. What we see now is that corporations are moving toward agricultural production in a big way. I think they learned from the poultry model. They see, in particular, an opportunity in agriculture where if a few enough of those big farms can gain a large enough proportion of the total market, then they can stabilize supplies and they can stabilize profits over the long run.

ACRES U.S.A. But is that going to be possible? I am thinking of Steven Blank's treatise on the end of agriculture in the American portfolio. The big fish will swallow the little fish and then the foreign fish will swallow the American fish. What is your take on that?

IKERD. I think that is a logical next step with what is happening now. We are in a situation where the corporations are taking control of American agriculture. The farmers that are left out there become contract producers for the corporations rather than independent decision makers. More and more those corporations are going to be multinationals in one sense or another. They are going to be a subsidiary of some foreign firm, or they are going to be a U.S. firm that has subsidiaries in other countries. When you get into that situation where total agriculture is really under the control, either contract or direct control, of just a handful of giant multinational corporations, then those corporations have no sense of place. They have no commitment to a particular community because they are not a person; they have no particular commitment to any country. They are simply trying to make as much profit as they can and grow as fast as they can. That is all a corporation can do; it is nothing more than a piece of paper. What Blanks says, and I agree with him, is that when you reach that point, then you are going to have agriculture located wherever in the world that resources are the cheapest, particularly the land and the labor resources. His argument is that land and labor are always going to be cheaper somewhere other than in the United States as long as we have an economy where our level of living exceeds most of the rest of the world. Agriculture will be moved out of this country to someplace where the land and labor are cheaper.

ACRES U.S.A. An associate who used to be an editor of the *Tulsa World* newspaper once made the remark that in closing down the independent farmer, we are closing down one of the most successful institutions of learning that this country has ever had. As an educator, how do you see that?

IKERD. I agree with that. I think that statement comes from the standpoint of seeing the real potential productivity for the future, and actually the real future of human civilization depends upon having people who are productive as opposed to just having technologies or factories or whatever the means of production might be. One of the big concerns I have now is that we are moving toward a system where fewer and fewer people are doing the thinking and making the decisions. Ultimately the benefit in any society or economy goes to the people who do the creative thinking and who make the decisions.

ACRES U.S.A. And that has always been the farmer, hasn't it?

IKERD. Historically, in American agriculture—and this is not just true of agriculture, but of a lot of other sectors of the economy that have been virtually destroyed up until now—but in agriculture it was the farmers who were doing the thinking, it was the farmers making the decisions. Each one of these farms out here was a unique farming system, a unique operation; they had their own particular set of enterprises which for many of them were very diverse—hogs and chickens and cattle and whatever. Up until the chemical age in agriculture they had to understand how to keep their particular piece of land fertile and keep their crops growing. They had to understand how to rotate crops to at least minimize the pest cycles and keep the productivity up. In other words, the farmer had to have a knowledge and they gained this knowledge either by growing up on the farm or working with someone else who had the knowledge. I think that is what a lot of people, including Thomas Jefferson and various leaders through the ages, have referred to as the real solid foundation for society and the economy among rural people and among farmers. We had that tremendous stock of knowledge then in the heads of these people. Going back to this industrialization of agriculture, what we have done is transferred this thinking and this knowledge off the farm. These new technologies that we have developed over the years, these industrial technologies, have not been developed by farmers; they have been developed in a university experiment station or by someone in an agribusiness laboratory somewhere. That information is simply transferred to the farmer and all the farmer is doing with that technology is just applying it. They don't understand what is going on because they are not doing the thinking. As we move into these contract production operations, the farmers out there become contract operators, or at best franchise holders, and they are not doing the thinking; they are not making the decisions. What we are doing is really depleting this intellectual resource by concentrating the thinking and the decision making into the minds and hands of fewer and fewer people.

ACRES U.S.A. We are getting diminishing returns in the resources themselves, in the landed resources, let's say. For instance, out there in the Imperial Valley there are thousands of acres off duty because they have been salted to death by that kind of label-reading farmer who didn't understand the system.

IKERD. That's the downside, going back to your original question of how would you assess the pros and cons of agricultural industrialization. What we have talked about until now has been primarily the pros in terms of simply pure economic efficiency. But pure economic efficiency basically ignores what we are doing to the inherent productivity of the natural resource base of the land and the water. It also ignores what we are doing to the inherent productivity of the people, such as the farm intellectual resources. In the process of gaining economic efficiency, we have really been depleting that basic resource base that ultimately we are going to have to depend upon to feed and clothe ourselves.

ACRES U.S.A. Is the sustainable model a viable alternative?

IKERD. I think the sustainable model is the common sense alternative if you look at it in the generic sense. To me sustainability is nothing more than saying we are going to look at the long run; we are not just going to look at the short-run economics as we have with the industrial model. We are not just going to ask the question, can we feed and clothe ourselves today, but can we sustain the ability to feed and clothe ourselves indefinitely into the future and can we do it in such a way that it is good for people? To me sustainable agriculture just simply asks, how can we meet the needs of the present and leave equal or better opportunities for people of the future? It is farming for the long run and farming that will last.

ACRES U.S.A. It is also farming for the community, is it not?

IKERD. Right. It is farming in the sense of not only meeting our needs, but meeting the needs of society into the future. I like to think of it as applying the Golden Rule. The Golden Rule says, "Do unto others as you would have others do unto you." We go about farming and it is important that we are able to take care of ourselves as farmers or people involved in agriculture because we can't do well by other people unless we do well by ourselves. You have to have respect, confidence and some integrity yourself or else you can't do well by other people. But it goes beyond the economics of self-interest to be taking care of ourselves. We have to see that it is really in our benefit to take care of other people, and that is people within communities, within families and within societies today. It is developing the kind of agriculture that is not just good for the producer or the consumer, but that it is good for society in general. Sustainability really extends that and says it is not just about doing for people now, but it is doing for people of the future as we would want them to do for us if we were going to be a member of that future generation.

ACRES U.S.A. Joseph Schumpeter, the economist, once made the remark that it is the innovator that launches the assault on the establishment. Are we going to have enough innovators in sustainable agriculture to turn this situation around?

IKERD. We have thousands of innovators now. Nobody can foresee the future with absolute certainty, but we have the innovations all across the country. I am one of the privileged people that really has the opportunity to go around the country to attend and speak at various conferences where we have hundreds of farmers at any given conference, with thousands of farmers all across the country, who are really following this general approach in terms of sustainability. They are farming in such as way as to meet their needs and be profitable while being sustainable.

ACRES U.S.A. They are able to do this with a niche type of approach?

IKERD. They are doing it mainly by being more intensive managers, by making more of the decisions themselves, which allows them to reduce the costs and gives them access to a particular high-value niche market such as organic and hormone- and antibiotic-free meat and things of that nature. It is recognizing their particular abilities as producers, so they do the things that they feel strongly about and the things that they do best. They also work together and build relationships, and they are able to make these work. They have an acceptable economic level of living. I am not going to say that these people that I am talking about are necessarily going to be the wealthiest farmers in the county. I will say that in most cases they have a better level of living if you include the whole realm of how much they are contributing to their net worth, whether or not they are living off their capital, and whether or not they have a sense of economic security that will enable them to go on year after year without government programs and things of that nature. In other words, they have an acceptable level of economic returns and they are able to do it in such a way that they are taking care of their natural resources. They are farming in a way that to them is ecologically sound. They are not risking their health with pesticides, agricultural chemicals, hormones or whatever it is that they are trying to move away from. They are not all organic, but they all have that ecological conservation or environmental dimension to what they are doing.

ACRES U.S.A. You think these two dimensions are what it takes?
IKERD. I think there is a third dimension beyond the environmental or economic dimension. These farmers have a very strong sense of community and family and their farming is very much about building relationships with other people, particularly like-minded people in the organic or sustainable agriculture community, or other people that have a different vision for the future. I used to be pretty much in conventional, mainstream agriculture and I still go to some of those conferences. Mainly what you hear in conventional agriculture, and have for years, is depression. It is about what the government isn't doing that it should be doing or is doing that it shouldn't be doing; it's about market, it's about struggling to stay alive, and it's about trying to beat out each other to see who can have the highest yield or who can farm the most land or who is going to get the next farm that comes up for sale.

ACRES U.S.A. It is all about traditional competition for resources.
IKERD. It is a very competitive, destructive, and depressing environment in which you are operating. In the sustainable agriculture meetings it is exactly the opposite. These people are optimistic and enthusiastic and they really have a lot of hope for the future. They think they are involved in something that is working and that is going to work even better in the future. They share information with each other. It is generally just an altogether different attitude. I was a little bit concerned last fall when I went out on the meeting circuit because agriculture commodity prices have been so depressed for the last couple of years. I knew a lot of what I would call sustainable producers, or alternative producers, sell a lot of basic commodities. I thought it might be difficult for them to hold on and continue doing what they are doing. But what I found was just the opposite; these people were more optimistic about the future because they realized that if they hadn't changed their operations over the last 10 or 15 years then they probably would have been in a very severe financial situations this year. I talked to quite a few people who said that the downturn in agriculture commodity prices had really encouraged them, or in some cases pushed them into going ahead and making the changes that they wanted to make anyway. In other words, it had made them even less dependent on the commodity market than they had been before because it kind of gave them a push to go ahead and do more direct marketing or niche marketing or something different that would make them less vulnerable to these economic

downturns.

ACRES U.S.A. Thirty years ago we staged a conference and it was the only one in the country.

IKERD. Yes, but now they are going on all across the country. But my perception is that attendance is falling off because there are fewer and fewer farmers left out there. I think there is growing evidence of a viable movement, if you will. I resisted the term for a long time, but I think when I talk about sustainable agriculture now, that it is a movement. I think sustainability is a social movement; it's a change in farming, it is really fundamentally a different philosophy of life. That's what I see and that's what I hear among a lot of people at these conferences. The people that come to Acres conferences have a different philosophy of life than do the people that are attending a lot of these commercial agriculture meetings.

ACRES U.S.A. To what extent will the university participate in this movement?

IKERD. I am very pessimistic about the universities participating. I have worked very hard for the past 10 or 11 years at the University of Missouri to convince the university that sustainability is a part of the future and it is particularly a responsibility of the public and land grant universities to stay in tune with the needs of the people as individuals. The corporations can basically take care of themselves. I am not opposed to technology or even the industrial technology if someone wants to develop it as it arises out of the free market, but I am opposed to taking taxpayer dollars and using them to support the development of the kind of technology that displaces people, degrades the resources and degrades the ability of people to make a living. I think that is a lot of what is going on. I said that when I was with the university and since I have retired. I am still a Professor Emeritus at the University of Missouri and I still say the same things because I believe them to be true.

ACRES U.S.A. You are writing off the university?

IKERD. I'm concerned. I am not going to write off the university in terms of working with the people in sustainable agriculture. There are some good people in the university that are working very hard to try to maintain the integrity of the program and try to continue to build it so that it will serve the needs of what I call the common people—small farmers and people in rural communities. Those are not the people who are basically going to be on corporate contracts in the future. I think there will be less and less need for the things that universities legitimately did in the past, when we were going through the industrialization of agriculture, because there will be fewer and fewer decisions that will be made on the farms out here. More and more of the decisions regarding breeds, variety, production practices, equipment, technology, buildings or whatever; all of those decisions are going to be made by the corporations. The people who are on the land, the farmers, are going to be either contract laborers or contract landlords.

ACRES U.S.A. It sounds like the opening chapter of *Grapes of Wrath*.

IKERD. Yes, I think so. It has been a long time in the making, but I think we are in the final stages of it.

ACRES U.S.A. This corporate agriculture depends largely on a peasant labor force, so to speak.

IKERD. That is how some people have referred to it. Some talk about serfdom and say we are going back to medieval times. The farmer is going to be the equivalent of the serf who is out working the land for someone else. Some people refer to the folks who are in the poultry industry today as the new serfs of agriculture. I don't think that is necessarily unfair. I think a closer relationship to what we have today is a franchise operation, but most people look at a franchise and don't understand what is involved in the ownership. People see a McDonald's hamburger franchise and think this is a real going operation. They think the manager is making money, but that is not the case. It is the people who own the franchise who make the money, and the people who pay them to put up the golden arches are barely getting a living wage out of it.

ACRES U.S.A. Certainly the people who are working there aren't getting anything but minimum wage.

IKERD. The people who are working there don't get anything, and the people who even own these franchises, if they also operate the franchise, don't get anything much more than a living wage.

ACRES U.S.A. Are you going to continue speaking on the circuit, so to speak?

IKERD. I am continuing to speak out on these issues and write about them and do whatever I can. As I say, I think this is a movement, I think it is important, and I am concerned about what has happened to agriculture during my career. I guess I feel somewhat responsible, since during most of my career I was at least a part of what I see as negative in agriculture. I wasn't doing anything in particular to move it in a positive direction. Now I want to do what I can to use what I have learned along the way to help make the situation better. In the long run, the agriculture that we have today in this country is simply not sustainable. In the longer run, we are going to lose the ability to feed and clothe people in the process of trying maximize production and profits in the short run.

ACRES U.S.A. We do have the situation where ConAgra and these other big corporations are increasing profits exponentially, but their primary suppliers are going down the tube.

IKERD. I think we can expect to continue to see that kind of split in the proportion of farms that are going to the corporate sector and the proportion of farms that are going to independents or the contractors.

ACRES U.S.A. The only answer to that is to change the system. If not the entire system, then at least your farm system.

IKERD. That's what I tell people. A lot of the farmers have already done that. But eventually, even the farmers that are out here today, for instance, those on managed grazing systems, the main thing they are working on is reducing the cost of production. Up until now what has been focused on is reducing costs, but I think the day will come when the small dairies, as the chicken people did, are going to lose access to the market unless they develop their own market all the way through to the customer. I think individual farmers have to break away because the industrial system of processing, marketing and distribution eventually is going to wring all of the profitability out of whatever kind of farming or production operations sell to them or feed to them through a contractual arrangement.

ACRES U.S.A. Is there any good news?

IKERD. I think the good news is that independent farmers are proving that they can, in fact, break away from the system. A lot of people look at it and say, well these are just niche markets, they are just selling through farmers markets or CSAs or direct marketing; these are small markets. But in reality, every consumer market is a niche market. We all have tastes that are a little bit different and we all want something a bit different from what anybody else wants. We just buy all the same things at the supermarket because that mass production, mass distribution system can only handle a very limited variety of things even though we think there is tremendous variety in the grocery store. That store still isn't providing all of the variety that all of the customers would prefer. These individual farmers are giving people the individualized service and attention they want. The place where it shows up now is in the fastest growing niche market, the organic market. It has been growing about 20 percent a year for the last eight years; it has been doubling every three or four years. You probably know that there is also a growing market for hormone- and antibiotic-free meats. It is not necessarily organic, but it is people who do not want meats with hormones and antibiotics. There is a fast-growing market for range chickens and pastured poultry. People don't like the taste or think that the chickens that come out of the broiler factories don't have any taste.

ACRES U.S.A. If they have a taste it is an unwholesome taste.

IKERD. People will pay almost anything for chickens that are raised and slaughtered right on the farm. I haven't met anybody yet who could raise and slaughter as many chickens on the farm as customers that would come out and pay anywhere from $6 to $8 a bird.

"We Wouldn't Go Back to the Way We Were Farming"

An Interview with Klaas & Mary Howell Martens

Originally Published: February 2001

Klaas and Mary-Howell Martens are New York farmers, teachers and community leaders. Both are veterans of the chemical drift that fogged America in the wake of university instruction in the 20th century. As the parents of three children, they came to the conclusion that a toxic environment suited neither the children nor themselves. An analysis delivered the stunning conclusion that toxic chemical inputs meant bankruptcy on the installment plan. These conversations in 2001 create both insight and conclusions as to why this transition made sense to them, and why it would make sense to thousands of others who followed.

Klaas' parents came to the United States from Germany and he grew up on their family farm. Mary-Howell grew up on Long Island and simply "wanted to be a farmer." Klaas graduated from the state university system in New York and Mary-Howell from North Carolina State. She then took a master's degree in plant breeding at Cornell University. They are still active teaching, farming and leading community activities today.

ACRES U.S.A. What prompted you to make the transition from conventional to the type of organic farming you now do?

MARY-HOWELL MARTENS. When we got married, the farm environment was brand new to me, but I had been working in agricultural research at Cornell. Neither of us were comfortable with the amount of spray that seemed to be required for what was considered normal agriculture. We discussed this a lot. We were good at the chemicals. Klaas was the neighborhood expert on herbicides, and I planned the grape spray program at the experiment station, but we knew that we weren't doing what we wanted to be doing. It wasn't the right thing to do.

KLAAS MARTENS. When our kids were little we knew that we were bringing these chemicals into the house that were not safe. You can wear all the protective clothing you want, but we were being exposed. The thing that finally pushed us over the edge was 1992, which was a disastrous year both price and weather-wise here.

MARY-HOWELL. That was the first year that we were out of a farm partnership with Klaas' brothers and we were farming on our own. We saw that crops and the prices for conventional products were very low and we weren't going to be successful.

KLAAS. There was no way we were going to feed our family and have a decent standard of living farming conventionally. If you take the prices the way they were, you couldn't make the farm big enough to have a decent living.

ACRES U.S.A. Isn't that the case with most of so-called conventional agriculture today? But peer pressure is keeping most farmers from making the transition. Wouldn't peer pressure, or at least peer approval, contribute to forming the type of cooperative arrangements that you have with farmers in your area?

KLAAS. There is more to it than peer pressure. We hated using these poisons. You buy a bag of Bladex and it says, "restricted due to oncogenicities." I know that means it causes cancer. Maybe they figure most of the farmers who buy that bag can't figure out what the word means.

MARY-HOWELL. They also supply protective equipment when you buy the bag so that hopefully you will protect yourself. But what they don't tell you is that there is no way that you can wear enough equipment to exclude all exposure.

KLAAS. We are not directly answering your question here, but it goes into the "big lie." The big lie is that you have to use chemicals, that this is the only way you can farm.

MARY-HOWELL. It is not just peer pressure. That is the word that farmers get from the universities, from extension, from all the chemical salespeople who come to the door, and also from the farm papers and magazines. There really is very little alternative information.

KLAAS. We are constantly being given the story that if you want to farm successfully, you have to use these chemicals.

ACRES U.S.A. There is a lot of information out there, it is just getting it broadly distributed.

MARY-HOWELL. One of the things that has really struck me within the past year is that when we read various farm journals, practically every issue has an article on organic farming.

KLAAS. Back in 1992, 1993 and 1994, there was less information out there. We didn't know about Acres U.S.A. at the time. One big breakthrough for us was to discover that our neighbor, John Myer, had been farming organically for 12 or 14 years. That kind of blew the big lie all apart for me.

ACRES U.S.A. Once you decided that it could be done, how did you proceed?

MARY-HOWELL. We decided that we needed to learn more about how to do it. We figured that back in the pre-chemical days there were smart people who were doing research on agriculture.

We thought maybe we could find some of that old research. We went down to the Cornell library, poked around the dusty stacks and found several resources that, while they didn't answer all of our questions, pointed us in a direction of inquiry that made it possible for us to understand more of what the organic system is, as opposed to organic by neglect or organic by input substitution.

KLAAS. One of the key pieces of information that we found was a paper written in 1939 by a German professor named Bernard Rademacher, promoting the idea of cultural weed control. What he wrote was that cultural methods should form the basis of all weed control.

MARY-HOWELL. It was enlightening to us to think that we have some control over both the weed pressure and the weed species. We don't have to just take what comes. By thinking that way, then we need to think about cultural methods that will enhance crop growth but reduce weed growth.

ACRES U.S.A. Rademacher and Albrecht were contemporaries. They knew about research going on as far back as the beginning of the last century. But this material became eclipsed in approximately 1949, with the beginning of the chemical era.

KLAAS. In the Cornell library there was virtually none of that old material. It was very difficult to find anything.

ACRES U.S.A. Most of it has been pitched. Did the article deal with insects as well?

KLAAS. It didn't, but the same applies. We have learned this from your writings and from our own work.

ACRES U.S.A. You intellectually made the transition. How did you do it physically?

KLAAS. My mother and father were German peasants and there was a system of farming that they had never completely given up. My father died in 1981, but he never gave up the sound rotation. This is something Rademacher wrote about, that each crop should follow its most suitable predecessor. It is the idea that each crop sets the ground up to be able to grow another crop to advantage. Rotation, just by common sense, breaks up insect and pest cycles. We set up a sound rotation and we already had practice doing that because we were using a rotation. We noticed that when we were in the healthier part of the rotation, our crops were much more profitable than when we tried to grow corn the second or third time in the same field. This was just direct observation.

ACRES U.S.A. You already had healthy practices on the farm?

KLAAS. Let me back up. I came out of college one of those smart kids that was going to teach

Dad how to farm. College had given out the recipe of how to farm scientifically. We had the "green revolution" on our farm and I watched it all in a very short time. When I came home, the first couple of years were phenomenal; yields doubled overnight on some crops because we started using chemicals. But within two years, it was like a drug, we had to use more and more inputs to get less and less back. Looking back, I know what we were doing. The first year we harvested the organic matter and the health of the soil and got this terrific jump out of the chemicals we were throwing on. Then, as the soil weakened from the continued use of the chemicals, we saw things falling apart. We saw weeds and insects coming in that we had never seen before. We had a basic body of observation that told us that the system we were using had something wrong with it.

ACRES U.S.A. The rescue chemistry failed to rescue you?

KLAAS. And we saw it happen quickly. I've always been keen on observation and keeping records. We saw that the system was not a long-term system. It worked real well for a short time and then fell apart. After that, the answer was always, "Well, you need more of this expensive chemical to make it work." Nobody is turning around and saying, "Why do we need more of this expensive stuff?" If the system works, it should work the way it worked last year.

ACRES U.S.A. Where did you start then?

MARY-HOWELL. The main impetus of starting organic was not just that we were dissatisfied with spraying, but we actually saw there was a market. The first market we got was for wheat, so that was our first organic crop. We saw a small classified ad from a company here in New York that was looking for organic wheat. We had not realized that there was a market out there for organic field crops until then. We knew that people bought organic vegetables, but not the kind of crops we grew.

KLAAS. As soon as we talked to him about growing wheat, he started asking if we grew soybeans or corn. Usually when you are farming conventionally, you are supposed to stand at the elevator and beg them to take it away from you for next to nothing, but here we had people asking us for our crops.

MARY-HOWELL. It quickly became apparent to us that there is a strong market, especially for soybeans. New York is considered one of the prime growing areas for the Vinton 81 soybeans, the food-grade soybeans.

KLAAS. It is a very high-protein variety that makes excellent-tasting tofu, at least the Japanese consider it the best.

ACRES U.S.A. But it grows best in northern climates, doesn't it?

KLAAS. Yes, it is a narrow belt of latitude that goes across New York and across the Midwest.

MARY-HOWELL. It has been shown that the protein levels are higher the farther north and east you get, to a point. New York is the intersection of the best growing areas.

ACRES U.S.A. Your competition today is the Roundup Ready which is being grown in the Cotton Belt now that cotton prices have been destroyed.

KLAAS. One thing we have seen is that the protein of soybeans is creeping downward.

ACRES U.S.A. That's your so-called conventional?

KLAAS. Yes, they are garbage. We see new varieties released every year in the magazines and they are 33 or 34 percent protein. Organic Vinton 81 has easily tested 44 percent protein. Vintons are criticized by conventional growers because they won't yield, but if you multiply pounds times percent protein, it is those conventional beans that are the dogs.

ACRES U.S.A. So, you were growing wheat and soybeans?

MARY-HOWELL. Since we were looking at organic certification, we needed to build in other rotations, so then there was corn and other small grains. We found that as buyers learned about New York's ability to grow organic Vintons, we have had many buyers come to this area actively seeking farms. The markets have come to us. That has been a driving force for bringing us and many other farmers in our group to organics. The fact that it works, the fact that our yields are as high and we can control weeds and we are successful, certainly keeps us here, but it helps that we have a good market for our crops to bring us here initially.

KLAAS. One interesting thing about this market is that we have been able to give the market away, and the more we give it away, the more we have. There is so much demand. Buyers are much more interested in coming to an area where they can find 10,000 acres than where there is only 1,000.

MARY-HOWELL. We can do that by working together. Our OCIA chapter is also a very effective educational group and encourages working together. By working as a unit, not marketing as a unit, we are able to attract buyers who are more interested in a large quantity.

ACRES U.S.A. How many people do you have in your association?

MARY-HOWELL. We will be certifying 42 this year, which represents about 15,000 acres. We have about 50 additional associate members, some of whom are certified with other organic certification agencies, some of whom are conventional farmers who are trying to feel their way into organics, some are industry representatives, and some are university people who want to come because they enjoy our meetings and want to be associated with the group.

KLAAS. Our chapter started out with five organic farmers in 1994; the growth has come from conventional farmers who have walked into the meetings and noticed that it can be done.

MARY-HOWELL. We try to have monthly meetings through the winter. At each meeting we try to have an educational issue. It might be how to harvest with high quality; the seed industry and how to obtain untreated seeds; the GMO threat and what that means to organic farmers; or soil health and how to interpret soils tests and plan your soil amendments.

KLAAS. We also had three farmers who are in the process of transition talking about their experiences—what went right and what went wrong.

ACRES U.S.A. You are expanding some of the work we have been trying to do for the last 30 years.

KLAAS. Right—and it is just amazing. The conventional farmers out there, many of them, are just hungry for a way out. They want a way to do better and a way to survive.

ACRES U.S.A. Do you take them a step at a time, for instance, teaching them decay management?

KLAAS. Yes. Right now I am consulting with one farm that is close to 3,000 acres that has decided to convert the entire operation.

MARY-HOWELL. There is another farm here that we are working closely with that is planning on converting about 1,800 acres. They are not going to do it all at once, but these are large farms.

ACRES U.S.A. You don't want to give the impression that it has to be a large farm?

MARY-HOWELL. No, not at all. About a third of our group is old order Mennonites. They are interested in the smaller, simpler way of farming and they are equally welcome and get an equal amount out of the group.

KLAAS. It is important to understand that what we are learning about is a biological system. The biology is the same on a small farm as it is on a big farm.

ACRES U.S.A. You mention rotations as being an integral part of the environment or process. What do you do if you have grape arbors or orchards which can't be rotated?

MARY-HOWELL. No, you can't rotate, but you can be careful about what kinds of crops you establish in your row middles, and you can try to establish natural disease and insect control. There are some grape growers in New York that are successful at organic production. New York has very tough conditions; we have weather that is very conducive to fungal diseases. It is more difficult when you are dealing with a perennial that is also being farmed non-organically in the immediate vicinity.

KLAAS. I think genetics is a problem in fruit. You plant a variety and you don't change it from year to year. If you have planted the wrong one, you are stuck.

MARY-HOWELL. We planted an orchard when we first moved here using varieties that were being discarded at the Cornell experiment station, and we have peaches that have a thick layer of fuzz. It is a variety that commercially probably isn't attractive, but we find that insect damage on the really fuzzy peaches is much less. We don't spray anymore. I think the insects can't get their egg-laying equipment down through all that fuzz. We always have good peaches. One of the things that I would recommend if someone were interested in growing a fruit crop organically is look at some of the older varieties, because those varieties are probably better adapted to low-spray and low-fertilizer conditions.

KLAAS. Another thing that will help in fruit is that the same Albrecht equations, with the ratios of cations, apply to perennials just as they do to our annuals.

MARY-HOWELL. You have to balance the soil in the root zone.

KLAAS. Having the cation exchange capacity loaded with the right balance of cations will improve overall health.

ACRES U.S.A. In one of your lectures you made the point that organiculture is not simply a matter of taking hostile-type materials and finding more friendly materials.

KLAAS. That is the mistake that universities make. They think that conversion to organic is merely input substitution—that we just find an organically acceptable substitution for each conventionally used input. They are not changing their management system. Doing that guarantees failure; you have to relearn the system.

MARY-HOWELL. That is one real problem with the so-called organic research that is being done at universities, this trying to compare organic production versus conventional production. People don't understand that organic agriculture is a multi-year system that requires you to think through rotations, cover crops and soil health before you can even think of comparing yield.

KLAAS. Organics has to solve a problem at its source. If we go back to the early '60s, extension was pushing people to throw muriate of potash on their alfalfa in huge amounts. A few years later we saw the weevil move in and a few years after that we saw the potato leafhopper. Then we started having metabolic problems with our cows. The cows would have a calf, get milk fever, go down and die. Along with that we began to have infertility problems because the high potassium chloride was causing the alfalfa to be full of nitrates. That caused stress on the cows that caused endocrine problems and all of these other problems. Now, if we were doing input substitution organic, we'd have to find an organically approved way to kill the weevil, to kill the leafhopper, an organically approved way of treating the downer cows, and an organically approved way of taking care of the breeding farmer. If organic farmers use the proper fertility to make a healthy alfalfa plant, then we don't have to deal with any of the other problems. What we are doing is recognizing that it is the improper fertility management that caused all of the other problems, so we are solving causes instead of symptoms.

MARY-HOWELL. This goes very close in hand with the philosophy that seems to be driving the use of genetically modified organisms, especially GM crops such as Bt corn. Because people are looking at ways to solve a symptom—that is the epidemic of corn borer—without looking at why there is a problem to begin with. The problem is that they are not rotating, they are applying chemicals that kill off organisms that would decompose corn residue, they are using a fertility program that is enhancing insect pressure, they are using monoculture, they are using no-till where the residue doesn't get tilled under for decomposition, all of these practices are actually creating the insect problem. But without looking at that, Bt corn is doomed to failure. I was at a USDA advisory committee meeting talking to people who said that the experts now are estimating that the length of effectiveness of Bt corn is about three years and after that it will be no longer economically advantageous to use it. The farmers will have to spray anyway because of resistance.

ACRES U.S.A. Have you got a handle on the enhancement of estrogen in soybeans under the Roundup Ready system yet?

MARY-HOWELL. I think that the natural phyto-estrogens taken at a natural level are healthy, but if you alter the chemistry of how the plants are being raised and alter different products that are being produced, it could get unhealthy. What I am really concerned about is that fact that Roundup itself has been linked to non-Hodgkin's lymphoma and other cancers.

KLAAS. They are spraying it more often. The first year one spraying worked, the second year they had to come in a second time as the weeds are adapting. Just like any other herbicide, it is taking more to do less.

ACRES U.S.A. We've heard the major sin of chemical agriculture is the chemicals and some of the fertilizers, but it is much broader than that isn't it?

MARY-HOWELL. One of the things that Klaas and I feel very strongly about is the breakdown in cooperation and community between farmers. The university system has been teaching for the last 50 years that the only way you can be successful in agriculture today is to get bigger or to get out. The only way to get bigger is to take over your neighbor's farm. So farmers are increasingly looking at each other as potential takeover candidates. It is very hard to be cooperative, to help when someone is having a rough time, to be friends and be supportive, if you are hoping that your neighbor is going to have an auction so you can buy his farm and he's hoping the same. This is really what is breaking down the sense of farming being a lifestyle and farming being a satisfying occupation.

ACRES U.S.A. You know that the Grange was founded in New York and one of the precepts that was required was that every meeting have a social aspect to it.

MARY-HOWELL. One of the things that we find the most satisfying in our organic certification group is the social time. We make sure that everybody brings something to share, and for about an hour in the middle of the meeting we just stand around and talk. That builds friendship. People talk about cooperating on different projects, once a year we have a picnic and we get to know each other in a way that we are not just the neighbor down the road.

KLAAS. For the farmer in transition there are enough stresses, financially and otherwise. When you start to transition you are starting something new. It helps to have neighbors that are there to pat you on the back and say, "I've been there, it will get better, you can make it, we want you to make it." That is an enormous difference rather than to have your neighbors saying, "This isn't going to work and we are going to be there to buy up the farm when you fail."

MARY-HOWELL. This year in New York it rained constantly from April until the end of June, then July was the coldest July on record, and in August there was more rain. We are having a really tough year and yet we feel strongly that we have to pull together and make sure that everybody in the chapter makes it through the year. None of us will have tremendous crops, but all of us have to make it.

ACRES U.S.A. But you have developed markets, or markets have found you. Isn't that half the problem?

MARY-HOWELL. One of the problems that a lot of the farmers face is that in order to satisfy rotational requirements, they have to spend at least some of the acreage on lower-value crops like corn, that are harder to find a good market for. One of the things we have done is start an organic feed business. This has grown enormously as the New York organic dairy industry has grown. We buy from most members of our chapter to give them markets for their out-of-rotation crops.

KLAAS. We actively look for markets.

ACRES U.S.A. In this rotation system, how many crops do you require?

MARY-HOWELL. We don't require a certain number of crops, we require a variety. We need some to be row crops, and some are soil-building crops. It is not a set rotation that people have to follow, but we have to see every year that there is crop variety and that people are planning ahead so that they are not going to put all their acres in soybeans this year and corn next year. Not only is that not sustainable, but financially they are going to have years of boom and bust and that is not going to make them good farmers.

ACRES U.S.A. How many crops makes an optimum?

KLAAS. I would like to see seven, 10 or more.

MARY-HOWELL. We try to have our rotations be four to five years.

ACRES U.S.A. What are the crops you are growing?

KLAAS. All the different small grains: wheat, spelt, barley, oats, rye, buckwheat. We grow corn, soybeans, red kidney beans, sweet corn for processing, hay, and alfalfa.

MARY-HOWELL. We underseed all of our small grains with clover that isn't harvested. That's a cover crop but it comes up after we harvest the small grains, grows until frost, and will start to grow again in March. It produces a lot of nitrogen and organic matter.

ACRES U.S.A. You need it to make your nitrogen cycle work.

KLAAS. We need it to keep healthy soil. It is a terrific food for microbes, with a lot of sugars. It is a well-balanced food for getting a real biological population set in the soil.

ACRES U.S.A. How would you counsel people who have no peers around and no one to compare notes with?

KLAAS. I would start by saying we have to bring in small grains and hay right away. What worked well for us was to start the first year of organics with soybeans. There is generally a grace period. The first year you could grow soybeans just dropping all the chemicals and spending a lot of time on weed control. As soon as we pull away the chemicals, it is like a drug withdrawal. We have the lack of biological diversity and activity that the use of chemicals has caused and we are not putting in the factory fertilizers and the salts anymore. So we do have to go through a period of withdrawal. Soybeans seem to be able to tolerate that with no yield loss for a year. We try to follow those soybeans with winter grains, and we almost always have a poor crop of winter grains because that is when payday comes. Because the poor crop of winter grains is a perfect way to seed clover, we get a terrific cover crop of clover. That would be your second year transition. I like to put corrective treatments on that clover. At that point you could use manure, but I like to use compost. You do a soil test and put on lime and gypsum, rock phosphate, granite dust—whatever corrective materials are called for, according to a soil test.

ACRES U.S.A. Can you get non-acidulated phosphate to be active fast enough for you?

KLAAS. This is the reason I like clover. The legume root system is able to use the tricalcium phosphate much more efficiently that most crops do because it is picking it up and turning it into a biological form which will feed the next crop.

ACRES U.S.A. Do you feed any of these materials into your compost operation?

KLAAS. We have always put them onto the green manure and we have done so with good results. I think we would probably gain faster if we did. If we produced compost with these materials fed into them, we might be able to lessen the yield drop in the second year transition.

ACRES U.S.A. You do have livestock?

KLAAS. We don't own animals except for a few sheep.

MARY-HOWELL. I've gotten some composted manure in the past, but we are trying to get away from that partly because we are trying to get away from purchased inputs. We also believe we can supply most of our nitrogen needs just with green manures and cover crops.

KLAAS. We do have earthworms.

ACRES U.S.A. What we used to call billions of unpaid workers.

KLAAS. Yes. When we grow clover, for instance when the spelt comes out, there is already knee-high clover in the field. That crop will continue growing and it will go flat and a new growth will come up through it in the fall. I've seen as much as five tons grow after taking off the small grain. The next spring the ground is virtually bare and that whole five tons of clover has been consumed by earthworms and microbes. All you see where that biomass was is holes and earthworm castings.

ACRES U.S.A. Do you require any special equipment to farm as you describe?

KLAAS. We are using conventional equipment—the moldboard plow and such.

ACRES U.S.A. In your area of New York, what kind of soil do you have?

KLAAS. It is a loam and it is a soil that has reasonably similar amounts of coarse, fine and very fine particles. If that soil is not thoroughly mixed, the movement of moisture will actually seal in the pore spaces and you end up with a soil that has very little air. We have a saying here, "no till, no corn." They are trying to no-till into these soils that are biologically dead, where the movement of water is actually putting the small particles in between the big ones and plugging up all the pores. It gets extremely hard. It is too high in magnesium to begin with, and the use of chemical nitrogen has let the magnesium continue to climb while the calcium has been depleted.

ACRES U.S.A. That also contributes to things like foxtail and other weeds.

KLAAS. I mentioned that when we were conventional, we had to use more chemicals for less results. We started seeing a shift in our weeds. After a few years of using these new chemical fertilizers, we saw foxtail and we'd never seen it before. Then we had to change our herbicide program. We went through this whole succession, and every time there was a new weed on top of the old one. The chemicals are bringing the problem. We have picked up a lot of land because the owners didn't want chemicals on it. One of those farms was a neighboring place. The guy who had rented it was a chemical farmer who had gotten to the point where he had nothing but weeds. The one field had such a mess of barnyard grass, foxtail, panicum, you name it, that we couldn't even put a plow in the ground. He came to us when he heard we had rented it and said, "Let me tell you something, nothing grows on that land, absolutely nothing grows on it." He was just disgusted.

ACRES U.S.A. What do you grow on it now?

KLAAS. We are farming corn, soybeans and spelt. It is not as good a land as we have here at home, but it will produce way above average yields, and it has made us a lot of money because we rent it cheaply.

ACRES U.S.A. But you are able to grow crops successfully on it?

KLAAS. Yes, but it had to heal up from what he had done to it.

ACRES U.S.A. Do you demonstrate any of these things to field day crowds?

MARY-HOWELL. We just held our first organized field day that we had widely publicized in New York. But we always have lots of visitors. People come by and we take them around. We spend a lot of time consulting.

ACRES U.S.A. There is such a thing as people pressure. Is people pressure starting to impact Cornell?

MARY-HOWELL. It definitely is. Some of the recent, younger faculty are inclined toward looking at organic systems. They may not encourage adoption of the whole system, but at least they incorporate as much of organic principles as possible into their recommendations. I heard recently that Cornell is going to officially start recommending that farmers incorporate more organic matter into their soils either through actual additions, or through crop rotations with cover crops. This would have been unheard of several years ago. Organic matter was considered absolutely worthless, nothing that you would want or need in your soil. Now researchers are beginning to recognize the value. I think this is just the beginning.

KLAAS. We made a stir that a lot of people noticed last year when we were part of the great Northeast drought. We had organic corn that six to eight weeks into the drought was still not curling badly when everyone else's corn looked like a bunch of onions or pineapples. Come fall, John Deere brought a combine in here—they developed this new rotary with a GPS and a yield monitor—and they were hot-dogging that machine. They said, "There won't be any corn in

New York that is able to challenge this combine." They had been doing conventional corn in our neighborhood at 40 to 60 bushels per acre. Robert Hall, he is our right-hand man who manages our field operations, was at the coffee shop and the guys were all asking how our corn was going to do. He said, "Oh, it might make 200 bushels." It wasn't an hour later that they were all sitting here at the end of the field because they were going to see that it wasn't true. One of them was riding along with the combine and watched the yield monitor peg 275 bushels around the field.

MARY-HOWELL. It was very interesting to drive around in July and see all the conventional corn rolled up by nine o'clock in the morning. Our corn was not. It was not really stressed. By the end of the drought it was definitely hurting some, but not like the conventional corn. Then at Rodale Institute, Jeff Moyer has shown the same thing in a more severe drought, that the organically farmed crops did much better and showed a lot less stress probably because of the organic matter in the soil.

KLAAS. We think it was all the salt fertilizer—the harder those plants pulled for water, the more salt they got in their roots. They were literally burning their own roots up, whereas the organic crops had the fungi and the other beneficials actually helping to release water to them. They didn't suffer from the salt burn. That's conjecture, but I think it is a pretty good explanation of why there would be such a drastic difference.

ACRES U.S.A. You obviously keep excellent records.

MARY-HOWELL. We do. We have a computer program that we wrote to keep our field records. We try to be accurate and honest about cost of production so that it is not just a record of the operations, but also we can estimate how much it costs per acre on a field or per unit. We can do modeling to determine whether an input is going to pay. We feel strongly that to have a good knowledge of whether organic farming is economically advantageous, we have to know what our costs are.

KLAAS. Our records go back to 1986. We have complete records of what we were doing when we were farming conventionally, and they were part of what was telling us that the system was not working.

ACRES U.S.A. You must have some records that told you what the drag was because of the chemical costs.

KLAAS. Yes, and that was really one of the things that told us we had very little to lose to try organic. One year we had two fields. One was a National Corn Growers entry that placed fairly high in the state. The other was a field that yielded a little less, but it was following sod and it had very low inputs. We didn't make any money on the field that we bragged about. Our cost of production was half as much on the one that was following sod and had low inputs.

ACRES U.S.A. It sounds as though you are slowly passing the niche market and assuming a little bit bigger role than most organic folks have been thinking about.

KLAAS. Our cost of production is less than it was when we were farming chemically. If we lost the organic premium tomorrow, we wouldn't go back to the way we were farming. The problem is that conventional prices are so poor that nobody can survive at those prices, no matter what.

ACRES U.S.A. They are consuming their equity and then one day they post the "decided to quit farming" notice.

KLAAS. And they are driving away their families.

ACRES U.S.A. Is there any last advice you would give our readers?

MARY-HOWELL. You can do it. It is not that difficult to farm organically. A lot of conventional farmers hesitate, at least a lot of the ones that come to our meetings, because they don't have the confidence to do it. It may not work perfectly the first year. You may have some rough times during the transition or even the first years of being certified. But you can do it, especially if you can find other organic farmers—either in your area, over e-mail or by telephone—to talk to, get information from and network with. You will be able to be successful and you will feel satisfied by what you are doing.

KLAAS. I think the biggest help that anybody could get is to have a group of farmers to do this together as our group is doing.

MARY-HOWELL. After our talk at the Acres U.S.A. conference, several people came up to me and said, "We would really like to have a group like the one you spoke of, where can we find them?" The answer to that is that you have to do it yourself, you have to make one. It will take some work, but the payoff, the rewards, are so great—both personal and financial—that it is well worth doing. We need to intentionally create situations where farmers build community, build trust in one another, work together, and feel the need for each other's success so that we are not fostering competition but trying to get everybody to succeed. Organic farming can be a very successful commercial-scale venture.

"The Moral Imperative is to Waste Nothing"

An Essay & Interview with Wendell Berry

Originally Published: July 2002

Wendell Berry, 1941-, is a character in the American cast built on uncommon good sense. His commentaries are those of a social philosopher anointed with the refinements of poetry and literature. His *Unsettling of America: Culture & Agriculture* and dozens of other books, poems and platform presentations remind us that a world often takes on the image of what we present to it.

"I now suspect that if we work with machines the world will seem to us to be a machine, but if we work with living creatures the world will appear to us as a living creature," he wrote.

Wendell Berry's gentle rebel with the uncanny ability it takes to unmask blunders, impale contradiction with the offender's own petard, and explain to his fellow countrymen the long range consequences of the public policy sold by spin minsters with syrupy logic. In 2022, he was still busy advocating for

Mr. Berry is a poet, a recognized artist whose awards fill several pages of small type. Included in the above is a Guggenheim Fellowship and a Rockefeller Foundation Award, Vachel Lindsay Prize, Poetry Magazine's Bess Hokin Prize, and the T. S. Eliot Award.

Reputation and credits are merely an aside. What is important today is the thinking of this much loved spokesman for America's heart, its culture, history and freedom, both economic and political. His essay "Thoughts in the Presence of Fear" is reprinted here, followed by an interview published in Acres U.S.A. in 2002.

ESSAY:

Thoughts in the Presence of Fear

by Wendell Berry

I. The time will soon come when we will not be able to remember the horrors of September 11 without remembering also the unquestioning technological and economic optimism that ended on that day.

II. This optimism rested on the proposition that we were living in a "new world order" and a "new economy" that would "grow" on and on, bringing a prosperity of which every new increment would be "unprecedented."

III. The dominant politicians, corporate officers, and investors who believed this proposition did not acknowledge that the prosperity was limited to a tiny percent of the world's people, and to an ever-smaller number of people even in the United States; that it was founded upon the oppressive labor of poor people all over the world; and that its ecological costs increasingly threatened all life, including the lives of the supposedly prosperous.

IV. The "developed" nations had given to the "free market" the status of a god, and were sacrificing to it their farmers, farmlands, and communities, their forests, wetlands, and prairies, their ecosystems and watersheds. They had accepted universal pollution and global warming as normal costs of doing business.

V. There was, as a consequence, a growing worldwide effort on behalf of economic decentralization, economic justice, and ecological responsibility. We must recognize that the events of September 11 make this effort more necessary than ever. We citizens of the-industrial-countries-must-continue-the-labor-of-self-criticism and self-correction. We must recognize our mistakes.

VI. The paramount doctrine of the economic and technological euphoria of recent decades has been that everything depends on innovation. It was understood as desirable, and even necessary, that we should go on and on from one technological innovation to the next, which would cause the economy to "grow" and make everything better and better. This of course implied at every point a hatred of the past, of all things inherited and free. All things superseded by our progression of innovations, whatever their value might have been, were discounted as of no value at all.

VII. We did not anticipate anything like what has now happened. We did not foresee that all our sequence of innovations might be at once overridden by a greater one: the invention of a new kind of war that would turn our previous innovations against us, discovering and exploiting the debits and the dangers that we had ignored. We never considered the possibility that we might be trapped in the webwork of communication and transport that was supposed to make us free.

VIII. Nor did we foresee that the weaponry and the war science that we marketed and taught to the world would become available, not just to recognized national governments, which possess so uncannily the power to legitimate large-scale violence, but also to "rogue nations," dissident or fanatical groups and individuals — whose violence, though never worse than that of nations, is judged by those nations to be illegitimate.

IX. We had accepted uncritically the belief that technology is only good; that it cannot serve evil as well as good; that it cannot serve our enemies as well as ourselves; that it cannot be used to destroy what is good, including our homelands and our lives.

X. We had accepted, too, the corollary belief that an economy (either as a money economy or as a life-support system) that is global in extent, technologically complex, and centralized is invulnerable to terrorism, sabotage, or war, and that it is protectable by "national defense."

XI. We now have a clear, inescapable choice that we must make. We can continue to promote a global economic system of unlimited "free trade" among corporations, held together by long and highly vulnerable lines of communication and supply — but we must recognize that such a system will have to be protected by a hugely expensive, worldwide police force, whether maintained by one nation or several or all, and that such a police force will be effective precisely to the extent that it oversways the freedom and privacy of the citizens of every nation.

XII. Or we can promote a decentralized world economy which would have the aim of assuring to every nation and region a local self-sufficiency in life-supporting goods. This would not eliminate international trade, but it would tend toward a trade in surpluses after local needs had been met.

XIII. One of the gravest dangers to us now, second only to further terrorist attacks against our people, is that we will attempt to go on as before with the corporate program of global "free trade," whatever the cost in freedom and civil rights, without self-questioning or self-criticism or public debate.

XIV. This is why the substitution of rhetoric for thought, always a temptation in a national crisis, must be resisted by officials and citizens alike. It is hard for ordinary citizens to know what is actually happening in Washington in a time of such great trouble; for all we know, serious and difficult thought may be taking place there. But the talk that we are hearing from politicians, bureaucrats, and commentators has so far tended to reduce the complex problems now facing us to issues of unity, security, normality, and retaliation.

XV. National self-righteousness, like personal self-righteousness, is a mistake. It is misleading. It is a sign of weakness. Any war that we may make now against terrorism will come as a new installment in a history of war in which we have fully participated. We are not innocent of making war against civilian populations. The modern doctrine of such warfare was set forth and enacted by General William Tecumseh Sherman, who held that a civilian population could be declared guilty and rightly subjected to military punishment. As a nation, we have never repudiated that doctrine.

XVI. It is a mistake also — as events since September 11 have shown — to suppose that a government can promote and participate in a global economy and at the same time act exclusively in its own interest by abrogating its international treaties and standing apart from international cooperation on moral issues.

XVII. And surely, in our country, under our Constitution, it is a fundamental error to suppose that any crisis or emergency can justify any form of political oppression. Since September 11, far too many public voices have presumed to "speak for us" in saying that Americans will gladly accept a reduction of freedom in exchange for greater "security." Some would, maybe. But some others would accept a reduction in security (and in global trade) far more willingly than they would accept any abridgement of our Constitutional rights.

XVIII. In a time such as this, when we have been seriously and most cruelly hurt by those who hate us, and when we must consider ourselves to be gravely threatened by those same people, it is hard to speak of the ways of peace and to remember that Christ enjoined us to love our enemies, but this is no less necessary for being difficult.

XIX. Even now, we dare not forget that since the attack of Pearl Harbor — to which the recent attack has been often and not usefully compared — we humans have suffered an almost uninterrupted sequence of wars, none of which has brought peace or made us more peaceable.

XX. The aim and result of war necessarily is not peace but victory, and any victory won by violence necessarily justifies the violence that won it and leads to further violence. If we are serious about innovation, must we not conclude that we need something new to replace our perpetual "war to end war?"

XXI. What leads to peace is not violence but peaceableness, which is not passivity, but an alert, informed, practiced, and active state of being. We should recognize that while we have extravagantly subsidized the means of war, we have almost totally neglected the ways of peaceableness. We have, for example, several national military academies, but not one peace academy. We have ignored the teachings and the examples of Christ, Gandhi, Martin Luther King, and other peaceable leaders. And here we have an inescapable duty to notice also that war is profitable, whereas the means of peaceableness, being cheap or free, make no money.

XXII. The key to peaceableness is continuous practice. It is wrong to suppose that we can exploit and impoverish the poorer countries, while arming them and instructing them in the newest means of war, and then reasonably expect them to be peaceable.

XXIII. We must not again allow public emotion or the public media to caricature our enemies. If our enemies are now to be some nations of Islam, then we should undertake to know those enemies. Our schools should begin to teach the histories, cultures, arts, and language of the Islamic nations. And our leaders should have the humility and the wisdom to ask the reasons some of those people have for hating us.

XXIV. Starting with the economies of food and farming, we should promote at home, and encourage abroad, the ideal of local self-sufficiency. We should recognize that this is the surest, the safest, and the cheapest way for the world to live. We should not countenance the loss or destruction of any local capacity to produce necessary goods

XXV. We should reconsider and renew and extend our efforts to protect the natural foundations of the human economy: soil, water, and air. We should protect every intact ecosystem and watershed that we have left, and begin restoration of those that have been damaged.

XXVI. The complexity of our present trouble suggests as never before that we need to change our present concept of education. Education is not properly an industry, and its proper use is not to serve industries, either by job training or by industry-subsidized research. Its proper use is to enable citizens to live lives that are economically, politically, socially, and culturally responsible. This cannot be done by gathering or "accessing" what we now call "information" — which is to say facts without context and therefore without priority. A proper education enables young people to put their lives in order, which means knowing what things are more important than other things; it means putting first things first.

XXVII. The first thing we must begin to teach our children (and learn ourselves) is that we cannot spend and consume endlessly. We have got to learn to save and conserve. We do need a "new economy," but one that is founded on thrift and care, on saving and conserving, not on excess and waste. An economy based on waste is inherently and hopelessly violent, and war is its inevitable by-product. We need a peaceable economy.

INTERVIEW:

ACRES U.S.A. In the presence of a new world order industrial model for agriculture and the perception that there is an inexhaustible supply of raw materials, where are we headed?

WENDELL BERRY. We are heading toward exhaustion. The assumption of inexhaustible raw materials is false. It has been the reigning assumption for a long time, but we know that we are dealing with finite quantities all along the line.

ACRES U.S.A. And the industrial model is exhausting that finite quantity rather rapidly, is it not?

BERRY. That's right. It is the economics of exhaustion.

ACRES U.S.A. How did we get into this lot? What happened to the United States that started out with such lofty ideals?

BERRY. I think we always had the illusion that there were inexhaustible quantities of raw materials. We stood on the edge of this huge land mass and we realized that it had been very little exploited by its original inhabitants. It was there for us to use as we saw fit and there was an enormous amount of it. We began with the illusion of inexhaustibility. Our estimates of how long it would take to settle the country were way off. We did it much faster than the predictors thought we would. Now it is no longer an illusion; it is a fantasy. The idea of inexhaustibility of resources is a consciously maintained fantasy, and it is not going to work.

ACRES U.S.A. We sort of made the transition from an agriculture based on husbandry and agronomy to one of mining, did we not?

BERRY. Yes we did. Wes Jackson has been very good on this subject. We are using agriculture, which is a potentially renewable economy, as another one of the economies of exhaustion. We are using it as a way of mining. In other words, we are using a renewable resource as an exhaustible one, and we are exhausting it as fast as we can.

ACRES U.S.A. In some of your writings you have mentioned the proxy at times. Can you explain what you mean by that?

BERRY. I am making an analogy to the idea of a political proxy. Modern consumers have given their proxies for production to all sorts of people, among them are the farmers. The modern food consumer typically does not produce any food. The proxy is given to somebody else to do it. Someone else must get down on hands and knees to deliver the calf; someone else must work in the weather. The rule about proxies is that you have to supervise them. We give our representatives in the government our proxies to represent us, and the rule, as people have been telling us since at least Thomas Jefferson, is that you have got to oversee the proxy holders or you ultimately lose your freedom. These agricultural proxies that have been given by modern consumers have not been overseen. The modern food consumer typically has been an ignorant consumer who has no idea where the food comes from or what is involved in its production or what is done to it.

ACRES U.S.A. Or even whether or not it is food!

BERRY. Even whether or not it is food. Yes, that's right.

ACRES U.S.A. Let's talk about romantic environmentalism, which is sort of the handmaiden or opposite side of the equal sign from romantic capitalism and romantic communism. That's been one of your favorite themes.

BERRY. I don't know whether it is a favorite or not, but it is one I have needed to go back to from time to time. The idea of romantic environmentalism begins with the idea that the world is an environment that is different from us and outside of us—our "surroundings." This assumes that the world is away from us "out there," and we experience it as scenery or a recreational place to "go to." This is a radical separation of human beings from the world that sustains us and with which, in fact, we have a much more intimate relationship. We eat the environment and we excrete it. We breathe it in and out. We live in a world of which we are a part.

ACRES U.S.A. In other words, we are part of the organism.

BERRY. Yes, we are. We are little organisms who are parts of a great organism.

ACRES U.S.A. It is a little easy to get carried away with the idea of that vision. In the '20s and '30s Lincoln Stephens and some of those people had this great vision of communism—the future now. In our own day, of course, we have this vision of capitalism that qualifies under your term of romantic capitalism.

BERRY. I think one part of it is the romanticism of progress—that things are inevitably getting better and better.

ACRES U.S.A. We act as if there is just not any system that could hold its own against this great capitalistic empire that delivers so much for so many people, when, in fact, a sizable section of these people are being ushered into abject poverty. Isn't that the point?

BERRY. Perhaps these romantic ideas are a mask for the fact that the big economy, what we now call the global economy, is a way of allowing a few people to plunder all the other people, and to plunder the world in the process.

ACRES U.S.A. But the leadership speaks of it as a moral imperative for the entire world.

BERRY. The moral imperative, it seems to me, is always to take care of the things that you use, that you are dependent on, and that you will pass on to generations still to come. The moral imperative is to waste nothing and ruin nothing, to keep the world as an inheritance given in trust to pass on to coming generations. There is no moral imperative to waste and ruin and plunder. That's hogwash.

ACRES U.S.A. Regardless of whether it is justified on the basis of false accounting or not.

BERRY. If you do plunder, waste, exploit and steal, that leads directly to false accounting. You do not get false accounting to begin with. I think you get false accounting after you have made the fundamental moral error, which is to assume that the earth is yours to do with as you please. I am deliberately avoiding religious language, but obviously there is a religious reference involved here.

ACRES U.S.A. It all comes back to this: you have 6 billion people on the planet and 4 billion of them are probably shared out completely. Would that be correct? If that is the case, this system of globalism can have consequences that will lead to the destruction of political and institutional arrangements across the board. It comes down to the Walmart approach to cheap regardless of what it does to the local situation is not to be ratified.

BERRY. The doctrine of cheapness at any price. This takes us back to bookkeeping because what the bookkeeping has done so far is externalize all of the costs except for the ones that the corporation in question wants to honor or thinks it can readily pay back. To permit Wal-Mart to destroy your community in the name of cheap commodities is insane, because we simply have no way of getting onto the books what the cost of a ruined community ultimately will be. We don't have any way at present to account for the value of a functioning community. Wes Jackson says that the science of the new century will be bookkeeping, and he means learning to set the books up so that the real costs of things are accounted.

ACRES U.S.A. What this all boils down to is that we have to go back to that family farm unit that was squeezed into being bigger and bigger until now it is a relief client.

BERRY. This again is a failure of our way of accounting. We have lost those little farms because we insisted on classifying them only as units of production. We have got to learn to take a more complex view of it. A small farm belonging to a family is not just a unit of production. It's a home, for one thing. It's a family trust. It's a place of entertainment. It's a place of trouble and trial, too, there's no question about that. It's a place to raise children and a place to educate children—it's a school. It has lots of functions besides just that of production. It's a place of maintenance also, and we had better say that before we go any farther. The industrial idea of production involves no idea of return. We cause the thing to produce to the point of exhaustion. But always in a natural system, as in a religious system, there is this counter movement of return, of gratitude for the thing that has been gained or given, but also there is the obligation of maintenance, of good care. To return the waste to the soil, for instance, is an obligation that we have ignored on a big scale.

ACRES U.S.A. The farm managed on an individual basis takes more of these aspects into account than the one that gets bigger and bigger until it consumes its own capital, doesn't it?

BERRY. Yes. The idea of a farm being the consumer of its own capital is a suicidal idea. A well-managed farm preserves itself, and it produces satisfaction along with an income. We always have to look for these multiple returns. When you are cleaning your barn, if you do it right, you are enriching the soil, and you are satisfying yourself, so you are not doing just one thing. I think on a good farm you rarely are doing just one thing.

ACRES U.S.A. We used to have an associate who was a farm editor on a paper in Oklahoma. He said, "The demise of the family farm also means the demise of one of the most successful institutions of learning the country has ever had."

BERRY. There is no question about it. The English agrarian writer William Cobbett in Rural

Rides of "a little hop-garden" where he worked and played as a small boy: "This was the spot where I was receiving my education." It was this education, he said, that had kept him from becoming a milksop and a fool. I think all of us who grew up farming knew that we were going to school. These one-time farm boys who get into agricultural economics and on one hand acknowledge wistfully the benefits of growing up on a farm and on the other hand justify the destruction of the very childhood that they enjoyed are ridiculously at cross-purposes.

ACRES U.S.A. Jimmy Carter's book about growing up on a Georgia farm argues this educational aspect that we are talking about now. Let's depart a little. In one of your books you talked about the Menominee Indians. Could you recast that for us now?

BERRY. We were going to have a forest conference here and a friend of mine, William Martin of Eastern Kentucky State University, asked me if I would speak. He said I should talk about the kind of forestry we ought to have. Our state of Kentucky has 12,700,000 acres of forest, mostly privately owned by people who do not know what their woodlands are worth, much less how to take care of them. I thought I needed an example. I had heard of the Menominee forest and I had seen some statistical reports on what those people had accomplished. The most significant thing, I suppose, is that they have been logging their forest for about 150 years, and it has more standing board feet in it now than it had when they started.

ACRES U.S.A. And this was through good forestry practices?

BERRY. The story is that when they were forced out of their ancestral lands into their reservation in Menominee County, Wisconsin, the tribal leaders instructed the young men that the tribe would now have to live from their forest. The young men were to start on one side and log their way across, but they had to do it in such a way that, when they reached the other side, they could go back to where they started and do it again. That is essentially the program that they have followed. It is a very sophisticated forestry program now, with computer models and all of that, but it is still essentially the same program. They go into each area of the forest every 15 years, and they mark the trees that are not likely to live for the next 15 years. Those are the only trees that they cut. The forest ecosystem in Menominee county is intact, with very old trees still standing. The foresters have left them standing because they are healthy, they are still making wood. To drive from the Menominee forest, which is still an intact and thriving forest ecosystem, northward to where the forest was clear-cut many years ago is an instructive experience, because you drive out of a healthy ecosystem into a degraded one. It is one that has been degraded for a long time and will continue to be degraded. It is readily apparent what the difference is.

ACRES U.S.A. And this same degradation goes on as we move into field agriculture using the toxic technology and so on, doesn't it?

BERRY. Yes. I know you are acquainted with the work of Sir Albert Howard. He said a half century ago that if you want to know how to farm, you have got to look at the forest. Wes Jackson came along maybe 40 years later and said, since he was a Midwesterner, if you want to know how to farm, you have got to look at the native prairie. Both of them were saying that nature makes the rules for farming, and they both have recited the rules: Farming has to be diverse; you must farm with both plants and animals; you must keep the land covered; you must build and maintain great reserves of fertility; and so on. These are the principles, and they are absolutely contrary to the principles of industrial agriculture. This is biological agriculture that we are talking about, and it takes lots of forms, but essentially the biological program says that you have to keep intact on the farm, in the domestic or economic landscape, the processes that are at work in the natural landscape.

ACRES U.S.A. And do you think this requires broad spectrum distribution of land?

BERRY. Absolutely it does, because farms have to be fitted to the land. Another great writer, J. Russell Smith, said we should "fit the farming to the farm." We all know that in even the most uniform looking landscapes, the land varies from one acre to another. In order to farm those landscapes well, we have to have enough people on them to know them intimately and use them well. They have to be people who understand the history and nature of the place, and how to preserve it at the same time as they use it. Good farming is inseparable from the idea of economic democracy. The land has to be divided among a lot of owners, as Thomas Jefferson said, and as other writers going all the way back to Virgil have said.

ACRES U.S.A. Yet we are going in the opposite direction and the worst aspect of it is, and you've been highly critical of it as we have, is this World Trade Organization, especially NAFTA and this idea that this commission of corporations can trump the United States Supreme Court. Bill Moyers had a good program on it the other night and, as he pointed out, if you allow them, they will take Canadian Lakes and water the deserts in Sonora with them.

BERRY. There are people who would do that. The proper thing to do, however, would be to go to the Sonora Desert and see how the Papago and the Pima farmed, rather than try to transport the agriculture of Illinois to southern Arizona. The argument for the proper use of the arid West started a long time ago, with John Wesley Powell. Bernard DeVoto argued it; Wallace Stegner, who was my teacher, argued it. You have got to fit the farming to the farm. What we must have is a locally adapted agriculture.

ACRES U.S.A. That was the same statement Walter Prescott Webb made. You can't expect to transport Ohio to Arizona.

BERRY. It is a mistake that we have been making all along. We go to a new place and pretend that we can behave there in the same way that we did in the old place. We have to let the place prescribe the terms of settlement to the settler. The settler can't prescribe terms to the landscape — that has never worked, and it cannot work.

ACRES U.S.A. What you have been saying meshes pretty thoroughly with what you say in your "Thoughts in the Presence of Fear" essay. What prompted you to write that?

BERRY. Wes Jackson and I were worrying about the calamity of September 11 and decided that maybe we ought to write something. My immediate thought in the face of terrorism was that we have now the most vulnerable food system imaginable. We have a national food system that is trying, and so far succeeding, to be a part of a global food system. This is a highly centralized system economically, but productively it is more and more extensive. This means we have long supply lines everywhere. These supply lines are highly vulnerable to disruption by sabotage or terrorism. Moreover, the supply lines, even without terrorism, work as conduits for diseases and exotic pests. Even before September 11 we knew that. The forests and fields of our country are being invaded by damaging organisms brought in from other places by the food transportation system. In my little essay I said that we now face a choice: we can either keep this extensive, highly centralized system with its long supply lines and undertake the very expensive and politically threatening task of defending them, or we can come to our senses and develop decentralized food systems based on the idea that localities, states and regions should be, as far as possible, self-sufficient in their economy.

ACRES U.S.A. As local as possible.

BERRY. Yes. The idea, as my brother and I have been saying back and forth to each other for years, is that we should go back to the principle of subsistence. The farm, the community, the region, the state, the nation, each should supply itself first and then export the surplus. Always take care of the local or immediate need and then the surplus is for export—for sale or for charity or however you want to use it.

ACRES U.S.A. This goes exactly opposite to what Stephen Blank says, that the big fish will swallow the little fish and eventually the foreign fish will swallow the American fish, meaning that we will end up importing all of our food.

BERRY. In the process, we will all be swallowed by a food dictatorship that will make us all the serfs of the agribusiness corporations. As I have argued elsewhere, the idea of competition doesn't propose a lot of small entities competing with one another indefinitely. It proposes, finally, one corporation. Competition ultimately destroys competition. The law of competition is the same as the law of war: finally you have one victor. When you come to that, you have come to capitalist totalitarianism.

"They are All Nutritional Diseases"

An Interview with Pat Coleby

Originally Published: August 2003

Like everyone in this book, Pat Coleby, 1928-2015, exhibited all the qualities of a global and national treasure. In this circumstance, the nation was Australia, one of the early adaptors of the organic movement. In the first few paragraphs of the wide-ranging exchange from 2003 that follows, she told her story, a mixture of biography and insight that makes her books—*Natural Goat Care, Natural Cattle Car, Natural Sheep Care* and *Natural Horse Care*—so valuable. The bottom line according to Coleby is simply this: all diseases are a consequence of inadequate and unbalanced nutrition. The range of her observations answers questions and explains answers.

Makers of public policy would do well to give room and board to her advice. In fact, her recommendation to treat snakebite with vitamin C was ridiculed by the establishment. It is now an acknowledged course of action.

ACRES U.S.A. Pat Coleby, your titles on the natural care of goats, cows and horses have arrived as something of a blessing from down under. Now you are also working on a book on natural sheep care. I think people would like to know a little bit more about you. Would you tell us how you were led into this vocation?

PAT COLEBY. It's a fairly long story. I had one passion in life, and that was to be a vet. I was qualified at 15 to go to university, but wasn't allowed to go. I was sent to work on a commercial grower's garden on the Isle of Man, between Ireland and England in the North Sea, where my father was working at the time. I was very lucky, though I didn't know it then, because this was the only commercial grower in England who declined to espouse the N-P-K theories of the time. He kept his garden natural, as he always had. My father never used these things either. I was lucky in that I grew up knowing that none of these modern inputs were necessary and that the farm worked very well without them. What I learned working in the Isle of Man just underlined that.

ACRES U.S.A. And so you moved to Australia after the Isle of Man?

COLEBY. I worked in various jobs in England, whatever I could find, including horse racing and jobs on farms and market gardens. My husband and I came to Australia in 1959 with three children, all under 3 years old, which soon became four children under 4! At that time, I had worked a lot with cattle. My husband was pretty ill, he had a very hard war. I realized that I wasn't built to manage cattle on my own, so I decided I had better make my living doing goats. I had four young children, so there was no way I could work away from home, and fortunately my landlords were very sympathetic and quite glad of what I was doing. The crunch came when we started to get diseases on the place, and that was when I began a correspondence of many years with you about the lack of calcium and magnesium being a cause of mastitis. We started to get diseases, and the vets at the university who were interested in goats—but didn't know anything about them—just said to me, "Oh, it's a germ."

ACRES U.S.A. But cattle came first, before goats.

COLEBY. Yes, because I had worked with cattle in the past, but as I said, I realized that a woman on her own with cattle was too hard for me to do. It was physically too hard for me alone. With goats I could manage on my own.

ACRES U.S.A. Which of your books did you write first?

COLEBY. A book called *Australian Goat Husbandry*. We had no books at all in Australia at that time. There was one excellent book from England written by Dr. David Mackenzie, called simply *Goat Husbandry*. He used no modern antibiotics, nothing, and it was a very good book. The regrettable thing is that it has since been rewritten to include all the antibiotics and awful things that the first edition did not contain. I used that original book as my bible. I began to think I ought to write something myself for Australia—people kept saying they couldn't understand the other book because it was too technical. That is where I began, and one thing led to another.

ACRES U.S.A. This focus on land—did that come first because that is what you first encountered in the Isle of Man and in your work in England?

COLEBY. That's right.

ACRES U.S.A. But the land in Australia is considerably different from the land in England, is it not?

COLEBY. I work now with quite a lot of farmers in Australia, America, Europe and England, and we use the same analysis I use here. In England we used the Albrecht method, and have done so from way back. The farmer I worked with in England worked it out rather like me, before he had even read Albrecht's excellent books. I didn't realize quite how different the soils could be until I met someone who is a big cattle farmer in California who had done an analysis. I was consulting with him by phone, and he said, "I will get my agent to ring you with the analysis." When the agent called, he asked if there was anything he really needed to know. I said that he should not put on trace minerals if the pH was below 5.8 calcium, because he would lose them. He said, "That's not a problem—it never is, is it?" I told him that here our pH is 3.3. He thought I was joking. That is what we start with here. He asked me why I farmed here. He could not understand how anyone could farm on land as poor as we do, but it has taught me a very great deal, because of course you can farm this land if the nutrients are right.

ACRES U.S.A. Australia is probably as old as Eniwetok and Hawaii, how would it compare to California with reference to the age of the land?

COLEBY. Australia missed the last Ice Age. In other countries you have wonderful glacial rock, for instance, right through the Americas. Australia does not have this glacial rock, the only rock we have, luckily, is dolomite lime and gypsum. We have a little bit of basalt and stuff like that, but not much.

ACRES U.S.A. How do you overcome a pH that acidic?

COLEBY. We rented our land for the first 20 years, and then the house, which was 150 years old and glorious, started to fall down, and the landlords wouldn't come to the party—so my husband, Peter, and I decided we would buy our own place. We had very little cash. When you have very little cash, you buy land at the bottom of the heap. Peter died about a year after we moved, and my youngest son stayed home for a bit. It took about 4 1/2 years to get that place on top line by improving the nutrients—then we sold it and bought another one at the bottom, worked it up, sold it, and started again. It was very exhausting, but each time that we moved to different land, improved it, and successfully farmed it, we proved you could farm wherever you were.

ACRES U.S.A. Are you still improving the land?

COLEBY. Now I am retired from big-time farming—I only have six acres, and it was the absolute bottom, it was mining shale. The man who fenced it for me said, "Pat, you have gone mad? No one can grow anything on this!" I said, "I'll prove you wrong." We started with a pH of 4.1, cation exchange capacity of 9, which wasn't bad because it is rocky soil, and the various minerals were just so low that it was incredible. The first top dressing was a ton of dolomite to the acre, and the subsequent dressing two years later was half a ton of dolomite, lime and gypsum to the acre. The reading that interested me the most was the trace mineral copper, which started at 0.2—now, six years on, we have a pH of 7 or 7.5, the CEC has gone up to 22, the copper—which is the mineral that determines good health or bad health—is 8, not 0.8 but 8. The last top dressing I did was 25 kilograms of magnesium sulfate, to which we added about a quarter-ton of rock dust and made a good mix. That land is top-line now.

ACRES U.S.A. You gravitated to goats immediately when you came to Australia?

COLEBY. I had never seen a goat before I came here.

ACRES U.S.A. Goats are grazers and browsers, are they not?

COLEBY. They are, but they also have the interesting characteristic that a milking goat requires six times more minerals than any other animal you keep on farm. If you don't get that right, you don't have a herd. I've got three of my original stock that I am milking now. We are in deep drought and have had no green stuff visible for a very long time, and I can't stop them from milking. They are made to milk all year round, but they are not stopping—it is incredible. I can't believe how well they are doing in these conditions.

ACRES U.S.A. How do you get by without green stuff?

COLEBY. In Australia you have to. I use cod liver oil to make sure the vitamin A and D levels are good, and nothing else is necessary. During the 1983 drought I was farming 50 miles north of here, and I was milking 35 animals. The great thing about Australia is that drought is never universal—on that occasion I was buying semi-trailer loads of feed from the next two states. This time I have bought a truckload of beautiful, organically grown hay. You can always get feed, but you have to pay for it. Goats are a very profitable form of farming, but they are also a labor-intensive form of farming because they are 365 days a year, twice a day.

ACRES U.S.A. Of course, all of the literature we've read about goats, usually out of USDA, is how they destroyed the Mideast because of overgrazing.

COLEBY. Yes, and that is exactly why. The mineral requirements of goats are so high—you know the old myth about them eating the washing—they will go anywhere and to any lengths to get what they need. When you get the trace elements right in a goat's feed, they do not strip the trees, and they live perfectly well on what they have got.

ACRES U.S.A. They don't go after the label off the tomato can or anything?

COLEBY. I am just resuscitating one that was virtually dead from the drought—someone left it with me. My vet said only an idiot would try to save it. I did have to let that goat into the garden to eat little bits and pieces for the first five weeks, just to get it eating again. Normally in a drought they just go straight through with whatever hay you get.

ACRES U.S.A. How do you dole out the proper feed rations?

COLEBY. I make up my feed rations. I won't let anybody that works with me—whether it is horses, cattle, even camels—buy pre-made rations. With my own rations I can make an analysis—I know what I have to put in that isn't there, and I can do it with quality materials. There is some difficulty in the states, one of my horse people wrote in and said they could not get chaff (chopped up oat or wheat hay cut in short lengths), which is a staple thing we use, adding whatever grain or minerals we need to that.

ACRES U.S.A. Stepping off your place and looking at others around you, do you run into anything like chronic wasting disease in Australia?

COLEBY. Yes, it is a deficiency of vitamin B12. We have it on what we call the swayback belt here, and it was referred to as chronic wasting disease. This is on the border of Victoria, where I live, and South Australia next door. Sheep in this area, if they do not have adequate cobalt in their food, will get chronic wasting disease. We make up a stock lick, which is discussed in my books. We add one kilogram—that's two pounds of cobalt—to every lick we put out, and that will stop the sheep from getting sick.

ACRES U.S.A. Let's go back to copper for a moment. Mark Purdey claims that the inability to assimilate copper is implicated in many diseases.

COLEBY. Mark Purdey is absolutely right. We have three so-called plagues: Johne's disease, which is evidence of a lack of zinc and copper; Foot-and-Mouth disease, also a zinc and copper deficiency; and Mad Cow disease, which is marked by copper deficiency. Mark's work is brilliant. I think he is a very clever young man.

ACRES U.S.A. The trick is getting the minerals right, then.

COLEBY. There is a country paper here called the Weekly Times, which recently ran an article by a vet. I am not very popular with a lot of vets. I work with a great many of them, but the ones I don't work with don't like me much. At any rate, this vet has written a half-page article on a disease called "string halt," where the animal lifts its hind legs very suddenly and turns. In bad cases, the front legs are also affected. You may have seen it on old, heavy horses on farms. It is in fact a magnesium deficiency, caused by a lack of magnesium to the muscle enzymes. This man had written an article about this "new" disease and recommended treatment with drugs. This vet is very highly qualified and has stature. We cure the disease by making sure the animal has the right kind of magnesium—that is, a form that can be absorbed very quickly—exercising the animals, and seeing they get their proper minerals. String halt can happen to cattle and goats, though most often we see it in horses.

ACRES U.S.A. The current thinking is to recommend drugs for diseases caused by poor nutrition.

COLEBY. They are all nutritional diseases.

ACRES U.S.A. Do you have brucellosis turning up in Australia?

COLEBY. Yes, for the same reasons you do. How many issues of Acres U.S.A. ago was it that you wrote how, as long as copper, cobalt, manganese and iodine were present, you did not get brucellosis?

ACRES U.S.A. Dr. Albrecht knew that, and back when we edited *Veterinary Medicine* we knew that, but we can't seem to get the point across to the government, which still requires that all animals be inoculated.

COLEBY. Which is ridiculous—and it applies to tuberculosis as well, that particular mixture of copper, cobalt, manganese and iodine. Tuberculosis does not happen when these elements are present. Now in Australia, I have to add calcium and magnesium to those, but in America that would very rarely happen, because you have enough of the last two.

ACRES U.S.A. In developing this feed ration that you build yourself, what is your norm?

COLEBY. The norm is equal parts oat chaff, alfalfa chaff and bran. I don't get the very purified bran because all bran is rolled wheat, and the less purified it is, the more natural vitamin E you get in it, and that is what it is for. But that is the basic ration, and it turns out about 13 to 14 percent protein.

ACRES U.S.A. Are you able to pick up the trace minerals such as zinc and copper in that ration?

COLEBY. We can add them as needed according to our analysis of the ration. We make horses take seaweed ad lib, but it is good for all animals to take it as they want it because they will never take more than they want. People cannot understand that animals know what nutrition they need better than we do—but they do.

ACRES U.S.A. So this is also applicable for goats as well?

COLEBY. For everything—particularly goats. We put out separate minerals for the goats, and we have had them eating raw copper sulfate when they were really quite deficient. People won't believe that either, but I have seen it happen.

ACRES U.S.A. Copper is one of the missing elements in the pastures down under, isn't it?

COLEBY. Copper is missing wherever they have used superphosphate or urea, that stops copper immediately.

ACRES U.S.A. The last time we went to Australia, it seems we ran into an awful lot of talk about phosphates. You got the impression that they had phosphate on the brain.

COLEBY. On the brain and everywhere else! The man who does the analysis for us here does not sell anything like that. Many of your big analysis firms in America, in fact most of them, don't monitor cobalt, while in Australia it means death if you don't know what your cobalt reading is. They don't monitor total locked-up phosphorous. Everyplace in Australia, even in the rain forest—and we've managed to do analyses in virgin rain forest—has tons of locked-up phosphorous, but without the right calcium and magnesium ratios, you cannot use it. That is the crux of the matter. If they go on putting artificial phosphorous on land, along with urea, they will literally tie up all of the important trace minerals—and magnesium and sulfur, as well.

ACRES U.S.A. When you don't have the trace minerals tied up, how many of them do you actually check to make certain you have?

COLEBY. We check for copper, zinc, iron, manganese, cobalt, molybdenum, boron, total organic matter, total phosphorous, and total aluminum nitrogen if needed. At first I had to get boron information from the USDA because no one here seemed to know what it was, but it is just as necessary as any other nutrient.

ACRES U.S.A. Does selenium figure in the equation at all?

COLEBY. No, selenium does not. It is very difficult to monitor. But the fact about selenium is that as long as your sulfur levels are right, you do not have a problem. There are very few places, even in Australia, where selenium is missing. I know you have one or two areas in the states, but some years ago the vets where I live in Southeastern Australia kept saying that the government keeps on telling us there are no selenium deficiencies, yet we find selenium-deficient animals on every farm that we look at! I told them to run an analysis and they would find they were excessively low in sulfur, which they were.

ACRES U.S.A. From what you are saying, we suspect you really don't have any vet bills. Of course, you are your own vet, too.

COLEBY. Only for mandatory testing for Johne's disease and a goat disease that is virtually goat AIDS, Caprine Arthritis Encephalitis (CAE). You do have it in the states. I have to test for those. It is money in the vet's pocket. He roars with laughter and so do I, because we both know damn well I cannot get either due to the way I feed.

ACRES U.S.A. We have a tremendous epizootic of Johne's disease in the United States.

COLEBY. Because of a lack of copper.

ACRES U.S.A. Is this why it is happening so often in these confinement dairy herds?

COLEBY. You know Mark Purdey went and looked at the deer in Colorado for BSE, and some of that soil is probably so poor that there is very little copper in it, which must mean that they are at risk. But any farmed land that is low in copper in this country, because we have copper mines, is always a result of too much artificial manure—superphosphate or urea, some form of phosphate.

ACRES U.S.A. You know they are also pouring phosphate down the spine of the animal, and this inhibits the uptake of copper even if it is out there in the pasture.

COLEBY. That's right. Copper is knocked down, for want of a better word, by high manganese, high zinc, phosphorous, urea, and in some cases by excessive calcium and magnesium. It is a very difficult mineral to keep if you are not farming or feeding right.

ACRES U.S.A. All of the time we read in the papers stories about so many tons of meat being recalled because of *E. coli* contamination. *E. coli* is made to sound like some sort of biological ogre. Are we correct in assuming that what is happening is that the *E. coli* in an acidic animal literally explodes to all parts of the animal?

COLEBY. Well, I would say that. The other thing I would say—this goes back to your comment about enzootic diseases—one of the interesting spin-offs I discovered when a vet and I did our first tests on copper in 1968, was that once I got those levels right, things like coccidiosis and quite a lot of other enzootic diseases never occurred again, and it didn't matter what animal it was. Those diseases are also due to a lack of copper.

ACRES U.S.A. As you said, they are all nutritional diseases.

COLEBY. All nutritional. Of course, Dr. Albrecht said that you don't get worms in animals whose copper intake is correct, and he could not have been more right. I don't know whether he realized that it affected enzootic diseases, because at first I did not. I just suddenly realized that I haven't had them.

ACRES U.S.A. Reading your books, we get the impression that if farmers paid attention to what you have to say, they would cut their vet bills down to absolute zero.

COLEBY. Well, they would. I don't have any vet bills. But the thing is that you don't win any marks, do you?

ACRES U.S.A. Not with the veterinary profession.

COLEBY. No. I was lucky in that I worked with vets. Because I was not allowed to be a vet, I went and worked with vets after World War II. I said to the vets I worked with that if they had any operations, anything out of my working hours, I wanted them to call me so I could come and help. This was before penicillin, before antibiotics, and I learned a great deal from them.

ACRES U.S.A. For instance?

COLEBY. They taught me that for wounds you disinfect once and never again. This is because disinfection inhibits healing, so you must get it right the first time. They taught me a lot of basic stuff that nowadays vets don't learn at college. My local vet told me that all they learn is the latest antibiotic, regrettably.

ACRES U.S.A. Your book, *Natural Goat Care,* has a chapter in it that refers to the psychology of goats. At first we thought it might be a misprint and that you meant "physiology," but no, you meant psychology. Can you explain that please?

COLEBY. Well, all animals have a pecking order, all animals have a safe distance if you are trying to handle them, whether they are horses, cattle or anything else. If a person is going to be a reasonable stock keeper, they learn those rules pretty quickly. For instance, this goat that I have been bringing back from the brink, it was virtually dead when it was dumped in my yard. It had been bullied because the person who owned it did not understand goats. The owner said, "I gave him lots of food," but he had different ages in the pen, and this goat was not getting food. They have a pecking order, they have ways of thinking through. A lot of animals have a plane of understanding that we cannot reach, I think.

ACRES U.S.A. You mean animals understand more than we give them credit for?

COLEBY. That is right. The first time I really saw this was when I was going to sell a doe. When she had her kid, I milked her out, and I didn't let her see the kid at all because I thought she would fret. I reared the kid with milk from the others, and then she went to her new owner. Five years later, I brought that kid back. I unloaded her into the paddock, and she walked around. When I came out for evening milking, that kid and her mother were lying together, and they never separated again.

ACRES U.S.A. And they had been separated for five years?

COLEBY. They were separated for five years, and they had never seen one another since being separated after the kid's birth, yet they seemed to know one another.

ACRES U.S.A. What other tips could you give animal handlers as far as the psychology of animals is concerned?

COLEBY. I think people have to think more, their personal psychology is not always so hot, is it? Perhaps that is why they find it difficult with animals. I know I'm regarded as a bit peculiar. The interesting thing is that with the CAE, you have to catch the kids when they are born. The kids cannot suck their mother's milk because that is how the disease is transferred. It is like AIDS in the way it spreads through bodily fluid. What we used to do originally, we would spend the night in the goat house so we could catch the kids to insure there would be no risk of contamination. Because I was dairying I would milk at the same time, morning and night. It took me about three months to realize that my goats never kidded except when I was milking. I would put the feed out, they'd have their feed, and then they would have their kids. They did it at the same time every time. All different goats would deliver their kids at the same time all the time.

ACRES U.S.A. Let's talk about cattle care for a moment. Do you prefer spring or fall calving?

COLEBY. I prefer a spring birth with any species because it is the natural time. If those cows were wild, that's when they would calve—just as spring was coming on. Certainly that is true in this country, and in England because that would be the time there would be decent grass for them.

ACRES U.S.A. You believe it is important for them to have fresh grass in their formative period.

COLEBY. I think you have to go back and look at what happens to animals in their natural state. I do a lot of work with dogs, and I note that people buy skinned butcher's meat for their dogs. I tell them to find a friend who hunts and get them to give the animal to the dog with the skin on. With all animals, I believe you have to go back to the natural state, and keep them as close to that state as possible.

ACRES U.S.A. That's right. The dog who catches a rabbit doesn't have someone skin it for him.

COLEBY. You always have to go back. In the cattle book I say that cows always run before the wind. And they do. When you do your paddocks you must make sure the fence where the prevailing wind occurs is strong, because cattle will run away from prevailing winds. Things like that need to be taken into account.

ACRES U.S.A. What is your opinion for line breeding?

COLEBY. If it is not an excuse for inbreeding, it is okay. Not surprisingly, in Australia we have problems with European stock because we do not have any natural ones. Go a bit further and blame the show ring with any animal. People often don't really line breed—they inbreed, producing some very unfortunate results. I got my best results with milking animals by using a good sire four generations back. I used two lines that had come away from him, and when I crossed those lines four generations on, I got what I wanted. I did not want to breed closer than that, though we sometimes had to because we had so few individuals.

ACRES U.S.A. You dislike the show ring?

COLEBY. The show ring, I think, is the cause of a huge amount of troubles in livestock.

ACRES U.S.A. They do seem to set the norm.

COLEBY. That's right. I remember an article in a European cattle magazine that said the last chance to buy really good stock has gone because Finland had decided to start having agricultural shows. Prior to that cows were judged by their milk record, their longevity record, and so on.

ACRES U.S.A. With goats the main market now is for the milk, is that correct?

COLEBY. It is really in cheese now, very much it is cheese.

ACRES U.S.A. Do you make cheese?

COLEBY. No, I don't. I give milk away around the village to people who are mostly older than I am—and I am getting on. But I do help several of the big cheese dairies here and the one in West Australia, which is totally natural. We have got them straight, and they are now producing the most beautiful cheese. We are encouraging people to go to the farmers' markets with goat cheese. Because if a person sells her top-class cheese to a shop, she can sell it to them for $2 a pound or so, but the shop then sells it for $14 a pound or more, and she gets none of that. I say, go for the farmer's markets, as they are doing in the States and in England.

ACRES U.S.A. In other words, go straight to the consumer.

COLEBY. Yes, exactly. It is disgraceful, the markup. I sold my milk for 52 cents a liter. I sold to a Greek cheese maker. He told me he loved having my milk because no matter what the quality of the other milk he got was, my milk pulled it up. Of course, when I asked for more money, that didn't apply.

ACRES U.S.A. So he could dilute the bad milk with your good milk.

COLEBY. Well, that's what they were doing. Now there is a woman who gets milk from me on a trade basis, she said that my milk was the best she had ever handled. I told her that when you get all the things right and you use no chemicals at all, ever, that's what happens.

ACRES U.S.A. Do you grow your own feed?

COLEBY. No, I grow my own food for myself, but at full stock on only six acres, I don't have room to grow feed for my animals. The farm I was telling you about that I have been working with in West Australia—I now have them growing all of their own feed. They have some really good pastures coming up with lots of mixed things in them. That is the ideal, of course. When I came to Australia I thought I could probably grow goats as they do in England—I knew they didn't heavily hand feed. But it is very difficult to do that here, you really have to hand feed almost anything you grow in this country except up on the range country in the north. The soil just isn't strong enough for really good food.

ACRES U.S.A. But you can build the soil. What have you been able to do for the organic matter?

COLEBY. Probably the most difficult client I have is the biggest hay cutter and exporter. A tremendous weight of every basic mineral is lost by that farm each year. To keep up with the amount that a hay producer needs is difficult. Now people are burning off crops. You have this burning off of stubbles in the States, don't you? Here I can remember driving to South Australia 25 or 30 years ago, and you could not even see across the highway because every single farm was burning. That burning takes an enormous amount of basic minerals out of the soil. If you can maintain those minerals, the organic matter isn't that difficult, actually. The other thing is that very high temperatures, 45° C, which is about 112° F, is very common here, which requires that you work compost in. If you just put compost on top, it will burn up.

ACRES U.S.A. It just cooks it.

COLEBY. That's right. I'm still learning. I grow my own vegetables. I try to grow everything I eat, and the goats get a little bit of that. Otherwise, the goats have their paddock, and I have to depend on other people for the actual hard feed.

ACRES U.S.A. We talked a little about your cattle, but we haven't talked a lot about it. What was the biggest lesson you picked up during your work with cattle?

COLEBY. The landlords on our first place were three old gentlemen whose combined ages were 270. What they didn't know about cattle, horse and sheep farming was very interesting. They didn't know anything about much modern stuff, but I learned a huge amount from them about handling cattle of all kinds, whether or not they had been handled before. I learned the basics of running beef cattle—mostly Black Angus and Herefords. Again, it always comes back to the same thing. On sick paddocks disease problems are very bad. People ring me up and say they have a terrible tick problem, and you only get that on acidic land. Many of these problems happen on acidic land. I don't like cell grazing. André Voisin's book, Soils, Grass and Cancer is the best thing he did, but not his book on cell grazing. I disagree with cell grazing because I think to lock an animal on a cell of land that she wouldn't choose for herself is counterproductive.

ACRES U.S.A. So you disagree with Voisin?

COLEBY. *Soils, Grass and Cancer* is an excellent book. But cell grazing is not good in Australia unless the individual cells are about 200 acres each. A lot of people go mad and do two-acre cells and things like that. I always let my cattle or goats graze free, and if there was an area on the farm that they would never, ever touch, then I took an analysis and found out why. I think it is a mistake to shut an animal on an area of land. The first cell grazing farm I saw here was on Kangaroo Island in the south, and it was covered with bent grass — it's about the worst grass there is. I asked the farmer if he had an analysis done, and he said he didn't need it because he was cell grazing. I looked at the coats on the cattle and the horses and knew there was something radically wrong. When we ran an analysis, we found that it was so lacking in minerals, it was terrible. But he had depended on cell grazing.

ACRES U.S.A. Well, that was a misapplication of the idea, whatever its merit. Do you have any alkaline areas?

COLEBY. We have a few, around Esperance in the south of West Australia. That is very low quality soil—they run one cow to every 30 acres and one sheep to every 10 acres. By American standards that is very amazing.

ACRES U.S.A. We have a lot of areas that are high pH and short of calcium.

COLEBY. That's right. The main deficiency here is magnesium, and that's why I do have trouble with American firms because they tend to say put on a lot of calcium and forget the magnesium. Here we have too many areas where magnesium is dangerously low. When I was working with Neal Kinsey, he told me—and Phil Callahan also told me this—that Florida is quite reasonable with magnesium both ways, and you can use dolomite in Florida. Neal told me that there is a strip up the middle of the state, as though you could take a ruler and run it straight up, and that area has quite a reasonable magnesium level, but the rest has high magnesium, and that is pretty dangerous.

ACRES U.S.A. On the bottom line we have to go back to the balances that Dr. Albrecht talked about.

COLEBY. You have to. I was intrigued when I spent five days with Neal when he came here and worked in the west. I just watched him. He said, "This is straight science. There is no arguing with it, and it wouldn't matter if it was Timbuktu or Australia or anywhere. You have to get the science straight." But as the days wore on, people would ask, "What about adding a little bit of this or that?" Arden Anderson says it on his tapes: "If you had listened to what I said early on and got the land in balance, I don't think you would be asking me these questions." But it seems that people want a quick or fashionable fix, they don't want to do the nitty gritty of the straight soil balancing.

ACRES U.S.A. The fix is contained between the covers of those excellent books you wrote, and we hope that more and more people avail themselves of them.

COLEBY. Well, every time I write a book I think of all the things I should have put in. I like your interviews in Acres U.S.A. and I have read it from the very first issue. I used to help and look after a dear old lady, and one day she said, "I'm going to give you a present, Pat," and it was the first copy of Acres U.S.A. That's how I came to know your company. I read it and thought, "This is incredible!" It still is. It is so good.

"They Ain't Been Paying Attention"

An Interview with Eliot Coleman

Originally Published: November 2014

Any person attempting to grow fruits and vegetables in the winter will likely come across the work of Eliot Coleman. A tireless innovator and skilled communicator, Coleman began writing about organic growing an astonishing 39 years ago—after a career of mountaineering and skiing.

Along with fellow writer and wife Barbara, he was the host of the TV series, *Gardening Naturally*, on The Learning Channel. He and Barbara operate a commercial, year-round market garden at Four Season Farm in Harborside, Maine, where he conducts the experiments he describes in this interview. He served for two years as the executive director of the International Federation of Organic Agriculture Movements and was an advisor to the U.S. Department of Agriculture during their landmark 1979-80 study, "Report and Recommendations on Organic Farming." Coleman's books include *The New Organic Grower, Four Season Harvest,* and *The Winter Harvest Handbook.*

In 2018, he headlined our Eco-Ag Conference & Trade Show under the headline of "50 years of Organic Farming." This interview was recorded four years earlier.

ACRES U.S.A. Didn't your wife, Barbara Damrosch, play a large part in your story? Not only personally but professionally?

ELIOT COLEMAN. No question, she is the best thing that ever happened to me. We've been married 23 years this December. She writes a weekly column for the Washington Post called "A Cook's Garden," and she has written two books by herself and one with me. I've written three by myself. When my first book *The New Organic Grower*, came out in 1989, my publisher told me the competition for it was something called *The Garden Primer* by Barbara Damrosch. After I moved back here in 1991, I was down at Ellen Nearing's place, helping her tie up her tomatoes in her greenhouse, and this very attractive brunette wandered in to visit Helen. I invited her to go for pizza, and we were married six months later. She had heard about my book, and obviously I'd heard about hers. She said she had always wanted to live on a farm, so I tell everybody that she stalked me.

ACRES U.S.A. Who were Helen and Scott Nearing?

COLEMAN. They wrote a book back in 1953 about their 20 years of homesteading experiences in the mountains of Vermont, *Living the Good Life*. I think 95 percent of the back-to-the-land hippies in the '60s and '70s read it. It was definitely one of the bibles of the movement. Scott was an old leftist who was thrown out of his professorship at the University of Pennsylvania in 1912 because he objected to child labor—amazing that something like that could actually have happened in the 20th century. He was put on trial by the U.S. government in 1918 for writing a pamphlet titled *The Great Madness*, which suggested that the reason for World War I was the colonial powers just trying to one-up each other, which, in retrospect, was a pretty good analysis. He served as his own lawyer during the trial and he won. He had quite an active left-wing career. He ran for Congress once in New York City on the Communist party ticket, but none of the parties really liked him because he was a maverick who thought for himself. By 1933 he couldn't get jobs anywhere teaching, so they moved to Vermont and supported themselves growing their own food and selling maple syrup. *Living the Good Life* was about how to live in the country and support yourself, and that was a very appealing idea to a lot of us. After I read the book I came to visit them. I visited again after I had been fired from my teaching job for being a contrarian, and was looking for land with my first wife. We got along very well with the Nearings, and Helen suggested we buy the back half of their place since they weren't using it. They sold us 60 acres for $33 an acre—Scott was a socialist who walked his talk. He said he hadn't done anything to it, so why should he get more than he had paid for it? He didn't believe in what he called "unearned" income. We thought that was such a wonderful gift that we subsequently sold 20 of those 60 acres to three different friends for the same $33 an acre.

ACRES U.S.A. What was it like?

COLEMAN. It was all wooded land, spruce and fir forest, very poor, about 3 inches of topsoil, initial pH of 4.3, and a glacier had left most of its rocks here. It was quite an adventure. We now have 14 acres cleared. A lot of that is in pasture, but we have an acre and a half that we have been able to clear enough rocks out of and get enough fertility in to grow the best vegetables you can imagine. It was all done, interestingly enough, with local resources. We brought seaweed from the coast. When the summer people mowed their hay fields just because they wanted them to look nice, they'd give us the hay and we composted that. We used to get clamshells from neighbors who shuck clams and crab shells from neighbors who picked the meat out of crabs. We dumped all that in and it was amazing. We now have about 12 inches because it's been continually tilled deeper and deeper and used deep-rooting green manures like sweet clover. The soil will grow anything with no pests at the moment, so that's the story in a nutshell.

ACRES U.S.A. Where is the land we're talking about?

COLEMAN. The Nearings were in Vermont from 1933 until '53. There was a mountain to the south of them which at that point was becoming Stratton Mountain, a ski area. Scott and Helen sold that farm and bought land in Maine in '53, so that's how they came to be here. We're about two-thirds of the way up the coast, about an hour south of Bangor up the Penobscot River. We're on a peninsula called Cape Rosier, and my farm is sort of in the middle at the south end. There are three beaches I can reach to get seaweed from, all about a mile away in three different directions.

ACRES U.S.A. A peninsula in Maine means a short growing season?

COLEMAN. Yes, we're in Zone 5, going to 20 below in the usual winter. For the first 10 years of my market gardening career here I more or less turned my business over to the Californians every fall and then tried to get it back every spring. They used to find this very annoying, so we started looking into ways we could inexpensively produce food throughout the frozen months. At first we had no money to buy regular greenhouses, so we built our own greenhouse. We used the old-time technology of cold frames. As we continued experimenting with that in the '80s, we put cold-frames inside a homemade greenhouse, and all of a sudden that was a great leap forward. It turned out that each layer of covering moved the covered area climatically about 500 miles to the south. So outside I'm in Maine. I walk into the greenhouse, the first layer of covering, and I'm in a climate like New Jersey. If we have a cold frame in there, I reach my hand into the second layer of covering, and my hand is in the Georgia winter climate. Obviously there are no tomatoes or peppers or eggplants in there, but there's plenty of spinach, carrots, scallions, Swiss chard, kale and Asian greens that don't mind freezing at night as long as they're safe from the dry, desiccating cold winds. That's what actually kills outdoor plants in the winter.

ACRES U.S.A. How many cuts to you get off the spinach?

COLEMAN. This is the ideal situation for spinach. As anybody who lives where it gets warm quickly knows, they're lucky to get one or two cuts off the spinach before it quickly goes to seed in the spring as the temperature warms up and the days get longer. But our overwintered spinach we plant on the 15th of September and that's ready to harvest by the middle of November, and it keeps sending up new leaves every time we cut off a leaf because the temperature is so cold it knows it isn't supposed to go to seed and the days are so short. We can keep harvesting from those beds right on through until the end of March, and by the end of March we have new crops that we planted through the winter coming up in other greenhouses, so it's a wonderful system. Basically the soil barely freezes in a double-covered greenhouse, and we can replant almost any day in the middle of the winter.

ACRES U.S.A. You're growing on how many acres?

COLEMAN. We only have an acre-and-a-half in vegetables. The soil we began with was so unpromising it's taken a good part of my life to get an acre and a half of it, but last year we sold $165,000 of produce off of that. If anybody tells you organic can't feed the world, they ain't been paying attention. We'll triple crop, even some of our outdoor fields. In the greenhouses we get six crops a year. We finish a harvest, we're pulling out the residue, we're putting down more compost and we're putting in transplants that afternoon. We keep it moving.

ACRES U.S.A. Where did cold frames come from?

COLEMAN. Cold frames—just a wooden box with a glass top, it's technology that goes back to about the 1600s in Europe, once they started being able to manufacture glass in slightly larger sheets. That's no longer what we use. That's what we used when we experimented with these ideas, but now the inner layer in the greenhouses is one of those spun-bonded fabrics that you see over huge fields on market gardens in the spring as frost protection. People use the heavier weight models to cover winter strawberries. We have wire wicket that holds that spun-bonded fabric about a foot above the ground inside the greenhouse. When you use it in the fields, you always have to figure out how to anchor it so it won't blow away. The nice thing about using it in a greenhouse is there's no wind.

ACRES U.S.A. Before all of these winter stratagems, did just about everything die back around September?

COLEMAN. Everybody dreaded the first frost and would usually be out there with sheets or blankets to put over their crops. We did that too before we decided it was worth investigating what we could do by ignoring the winter. There's a wonderful quote from Buckminster Fuller that I put up on my bulletin board — "Don't fight horses; use them." Okay, rather than fighting the force of cold by trying to put heat in the greenhouse and then go broke paying for the fuel, how can we use it? Well, we can use it by growing the crops that don't mind the cold. That's pretty much a simple capsule description of what we do. The more we investigated it, I started lecturing about it, I wrote a couple of books detailing it, and now enough people have gotten interested that I've put together a conference. It took place back on the 10th and 11th of August at an inn in Vermont, centrally located for New England growers. We got some government money to pay for plane tickets for growers from the upper Midwest. We rounded up the two dozen best grow-in-the-winter and unheated-greenhouse growers. We had a wonderful time exchanging information and trying to figure out how to do what we do better.

ACRES U.S.A. Could you describe some of the work you've done on shifting green-houses around without taking them apart?

COLEMAN. We've been experimenting for years with ways to move greenhouses, moving the whole structure from one field to the next. The advantage is that I can have a beautiful summer crop of tomatoes or peppers or eggplants growing in a greenhouse, which I need to do here on the coast because our summers are so cool. It won't get cold enough around here to freeze inside the greenhouse until well into October, but remember I need to plant my overwintered spinach crop in the middle of September. The way we make that happen is to have the spinach field right next to the tomato greenhouse, leave the tomatoes in there, and get the harvest as late as we can. The spinach is perfectly safe out of doors almost until November and then once the tomatoes freeze out, we roll the greenhouse next door and now it's covering the spinach for winter. This is a fantastic way to get far more use out of a greenhouse than you would get if you had to leave the crop that's in there until it matures and then figure out how to plant the next crop. We have some rotations where the greenhouse moves four or five different times. If we add up the number of months from seed to harvest for all the crops we're covering—in that case we're only covering them when they need protection—we find that we're getting actually 24 months of use out of that greenhouse every 12 months. As a businessman who like any other is concerned with getting a return on my capital investment, that's a pretty good deal.

ACRES U.S.A. Someone looking at a photograph of your rolling greenhouses on tracks might think they look pretty expensive and labor-intensive. Is that the case?

COLEMAN. Well, yes. There are many, many ways to move greenhouses, and we have experi-mented with all of them. Some of the early techniques we no longer use, but they were all based on the fact that we wanted the greenhouse to be movable for no more than 10 percent above the cost of the greenhouse itself. The model on pipe rails with wheels on the bottom of each hoop is one I designed for a greenhouse company but I've never been happy with it. It moves very easily, it's a wonderful model but each of those wheels is between $25 and $35, and on a 50-foot house there are 26 of them with the hoops four feet apart. That's a lot of capital to be sitting there doing nothing except when you're moving the house. We have redesigned all of them. We've had ones that used sleds on which the greenhouse just slid across the ground, but that's hard to move. You need to get a tractor in there, and sometimes you have such close spacing between fields you can't get the tractor in. We now have a simple arrangement using one set of 10 wheels and those 10 wheels, five on each side of a house, will move every greenhouse on the farm and then when you finish moving it, you take the wheels off and store them away and then bring them back out to move another house.

ACRES U.S.A. In other words, you're bringing down the price of rolling greenhouses?

COLEMAN. As part of trying to figure out how to earn a living at this game, we have been working very hard to make movable greenhouses as inexpensive as we can make them. For the benefits they give, that extra 10 percent is paid for very quickly. We actually have other models that we can make here on the farm using a pipe-bending form sold by Johnny's Selected Seeds catalog. We can build those for about $1 per square foot of covered space. We build them in modules so a 48-foot house that's 14-feet wide would be made of three modules, each 14-feet wide by 16-feet long. Those can be independently picked up by four people and carried to where you want them. You put them down and slide them in next to each other, then put in ground anchors to hold them down. Thus there are actually no wheels or almost hardly any expense with that one. We've tried out everything that seems to make sense economically as we've pursued this concept.

ACRES U.S.A. You're just constantly finding a newer, cheaper way to skin the cat.

COLEMAN. It's no different than what every farmer for the last 10,000 years has been doing—trying to find a better way to grow for a living.

ACRES U.S.A. You mentioned Buckminster Fuller. His concepts and his approach to engineering were a radical break with the past, and he seems to have influenced you quite a bit.

COLEMAN. Interestingly enough, his summer home was on an island about 10 miles off the coast from where I live, so maybe his spirit is still living on around here. I refer to it as imagination engineering or imaginative engineering—just looking at what exists and not thinking that that's the only way to do it. Looking at what exists and saying, "OK, how can we do this differently?" Sometimes it's really fascinating when you're looking for a solution—you turn the problem 180-degrees backward and wonder if you can solve it that way. An awful lot of thinking in our world today is, I think, 180-degrees backward from an intelligent way of doing things.

ACRES U.S.A. Did you figure out a way to make the wheels affordable?

COLEMAN. We still have those 10 wheels, but they cost $50 each, so I've got $500 invested, but these wheels now can move every one of the greenhouses on the farm. They are stored away so they'll last forever, rather than 26 wheels at $35 each sitting under each greenhouse forever and ever. The wheels on this new system are actually wheelbarrow wheels, so they have rubber tires on them. There is another greenhouse I helped design with easily rolling, ball bearing wheels, industrial wheels that roll on a V-track. The company that makes them is called Four Season Tools in Kansas City. They make what I think is probably the very best movable greenhouse. Because of the quality of their materials, it's slightly more expensive than others, but I think it's the top of the heap. Again, I'm designing for the small farmer, and I want the small farmer to make a living and survive. As a small farmer, I find the Four Season Tools' greenhouses more expensive than I could invest in, so we've been looking for less expensive ways of doing it.

ACRES U.S.A. If you have something you can move with only four people, you're finding it.

COLEMAN. Yes, and we move the ones with the wheelbarrow wheels using two winches. There's a winch pulling straight down each side of the greenhouse so the thing doesn't wreck. I presented a paper about it at an organic greenhouse conference in France last fall.

ACRES U.S.A. What did they think about it?

COLEMAN. I think everybody was impressed, but like all of these conferences I was the only farmer there. All these other guys were what I call organic bureaucrats or organic professors. The fascinating thing about this conference had nothing to do with greenhouses; it had to do with my sitting around, having coffee and talking to some of these people. These were some of the leading French organic researchers, and they were a lot younger than I am. None of them had ever heard of Chaboussou's book, *Healthy Crops*. It was originally written in French as *Sante des Cultures*, and he was a researcher at the French Agricultural research group. Here were all these leading organic researchers in France, and they had never read this book that now has been translated from French to English, and when I told them that we use no pesticides because we have no pests if we've done right by the soil, these organic researchers looked at me as if I was about to sell them a bridge in Brooklyn. I just couldn't believe that these guys were the best Europe had researching organics, because they didn't understand the most important aspects of it.

ACRES U.S.A. Can you tell some of his story?

COLEMAN. Chaboussou died back in the '80s. I heard him speak at a conference in Paris in 1974. I'd read all the Rodale books and I was fascinated by that line in all of the old organic books that healthy plants resist pests, but whenever I mentioned that to anybody in agriculture they would treat me like a nutcase. Then I'm at this large organic farming conference, it was very impressive. It's in a huge hall in Paris, and here's this speaker up there saying just as casually as he can that when you prepare the soil correctly there aren't any pests. I speak enough French to get by over there, and I told him it was just amazing to hear someone who is a professor in the French version of the USDA saying these things. He looked at me and said, "Doesn't everybody know that?" Apparently not. The English translation was published in 2004 by a British company called Jon Carpenter.

ACRES U.S.A. Did you enter growing from an engineering background?

COLEMAN. No, I came into growing with an adventurer's background. I was teaching; I got a master's degree after I got out of college so I could teach. I taught literature, mainly so I could have my summer, winter and spring vacations free to go adventuring. I used to be a rock climber, white water kayaker and mountaineer. The thing that got to me when I read the Nearings' *Living the Good Life*, about feeding and housing themselves as homesteaders in the wilderness, was that it sounded like a fun adventure. That's basically how I got into it. I tell people the reason I remained utterly fascinated for the past 48 years is that the dullest part of climbing a mountain was getting to the top. The fascinating part was trying to figure out how to get up the cliff or down the river or whatever you were doing. The nice thing about farming is that the mountain doesn't have a top. There's always some way to figure out how to do it better and that's been utterly entrancing.

ACRES U.S.A. Growing food is obviously a lot of hard work. It seems an intellectual or emotional shift happened deep inside, and you suddenly saw it as an adventure, and that's helped sustain you.

COLEMAN. Yes, but I remember something. It was probably during the first six months I was here. We'd built the house, and I was cutting down trees and chopping out the stumps to start clearing land so we could begin as vegetable growers. A neighbor came by. I was sweating and covered with dirt and chopping at the roots of this thing, and he said, "My gosh, Eliot, isn't that a lot of work?" I looked at him and said, "Well, you might think so, but let me tell you what I used to do for fun. I used to go on mountaineering expeditions and spend three weeks freezing my butt off in a tent with a 70-pound pack on my back all day, chopping steps in an icy cliff to pack supplies up to camp. Compared to that, this is pretty pleasant."

ACRES U.S.A. Atypical reasoning if there ever was such a thing.

COLEMAN. If I were a little kid in school today, they would have me on an intravenous drip of Ritalin. I always had a lot of energy, and this has been a great way to use it in some sort of socially redeemable way.

ACRES U.S.A. How does what you're doing compare to the way agriculture was traditionally practiced in Maine or Vermont?

COLEMAN. There are lots of new, young farmers moving in so I think that's just the tenor of the times. The young people today are fascinated by what farming has to offer. They come to Maine because land is less expensive, but once you get away from the coast land is less expensive because the farmland was never that great. Where it was good, up in northern Maine in Aroostook County where tons of potatoes are still grown, that used to be the breadbasket of New England.

That's where all the wheat was grown before those crops moved to the Champlain Valley of Vermont. Then they eventually moved out into the Midwest. Probably the main agricultural crops are wild blueberries, low-bush blueberries. Blueberries love acid soil. The rocks are no problem for them because this is a wild crop that you manage by burning the field off every other year. This acts as a form of pruning and stimulates the berries to come back thirstily so that next year they give two or three times as much crop as they might otherwise give. Mainers learned that from the Indians. Since we're right next to the ocean, you can hardly eat a lobster in the United States that didn't come from Maine, and that's a whole other industry. There have always been people making their living from nature.

ACRES U.S.A. They've depended on the sea for a long time.

COLEMAN. Take an average Maine guy—there's not much industry up here, so how are you going to get along? Well, digging clams has always been known as down-east welfare. If you didn't have a job, you could always dig clams and make money selling them to the wholesalers. So this guy might be digging clams in the winter, he might be sugaring—making maple syrup—in the spring, then probably digging some more clams, then raking blueberries in the summer. In the fall there's a whole industry of making Christmas wreaths. People have to go out in the woods and cut the tips off of fir trees for material to make the wreaths from. So this guy would be tipping fir trees in September and October and then helping make wreaths or delivering them, and then back to clamming again. There's always been a tradition of hard work up here. My neighbors are some of the most wonderful, hard-working people you'd ever want to know.

ACRES U.S.A. Does what you're doing introduce a whole new world of eating and food to life up there, where fresh vegetables were traditionally harder to find?

COLEMAN. Three of our crops have captivated people who might not otherwise eat vegetables. The first are the carrots we grow. We're very careful about soil preparation. We grow an oat and pea green manure early in the season, and then we sow carrots around the first of August, quite late actually, up until the 10th of August. We harvest some of them out of the very cold ground just before things freeze up, but for the majority of them, we slide one of our greenhouses over them and then we can harvest them right out of the cold ground all winter long and when you leave carrots in the ground, they protect themselves against the cold by changing some of their starch to sugar, sort of like antifreeze. These are known locally as candy carrots. We've been told by parents that our carrots are the trading item of choice in local grade-school lunchboxes.

ACRES U.S.A. Let's pause here to let that sink in.

COLEMAN. Then there's our spinach. Parents are absolutely fascinated that their children will eat our spinach. I've always thought that the thing that made spinach taste good is getting enough calcium into it. I think you can probably do it with gypsum, but here we have access to all these crab shells. Crab shells are made out of chitin. I think that the calcium is delivered in a different way, but anyway we grow the sweetest tasting spinach anybody ever ate. Since it's available all winter long the cold temperatures also keep it sweet. The third crop is our potatoes. When we started running our market garden we grew the old-time fingerling potatoes. Mainers who eat a lot of potatoes had never eaten anything quite as tasty as some of these yellow-fleshed fingerlings. Everybody always says their children won't eat vegetables. Well, the local children eat our vegetables. Children have good taste buds, and they know that conventional vegetables taste like crap, and they have no interest in eating them. When you read about carrots having a sort of petroleum taste, that's because the main carrot herbicide for years was called carrot oil, basically kerosene. People learned by pure chance that very few weeds would grow in kerosene-soaked soil, but carrots would.

ACRES U.S.A. Are you able to raise vegetables that are normally associated with a more Mediterranean style of eating?

COLEMAN. We grow incredibly delicious tomatoes but we grow them in greenhouses since we're right next to the coast. Everybody comes to Maine in the summer because you can sleep at night. Here a 75-degree day in August is a hot day. Thus if we want to grow tomatoes or peppers or eggplant or cucumbers, we have them in greenhouses. All of those crops really respond to plenty of manure compost. It's just amazing when you see what you can do with techniques that have been known to farmers for 10,000 years. How the chemical companies ever talked farmers out of using what they had for free and buying inputs is beyond me.

ACRES U.S.A. As long as we're taking the long view, on your video lecture you also mentioned the amazing things that went on in 19th century Paris, when the city grew much of its own food.

COLEMAN. All their vegetable growing was based on horse manure. Of all the manures I've ever used as a vegetable grower, if you can get horse manure from horses bedded on straw, not sawdust or shavings, that's almost a magic soil amendment for growing good vegetables. It was amazing what the Parisian growers were doing. They were on 6 percent of the land area of Paris. I researched this—I have a bunch of the old books these guys wrote. There was no such thing as organic farming back then, but in the books these guys are saying that they had no pests. If they had a good crop rotation and used plenty of compost, they had no pests.

ACRES U.S.A. Was this conference the first of a series?

COLEMAN. It would be nice if the extension agents who were there picked up on the growers' suggestions as to research they could do make what we do better and easier. If any of them pick up on that I could see every two years getting a group together again to review the new information, what new greenhouse covering materials or what new floating covers have come out. I'm quite sure what I do today is going to look very old-fashioned in 10 years because people will be doing it so much better.

ACRES U.S.A. How are your relations with the agriculture establishment in Maine, into greater New England, do they just ignore you?

COLEMAN. All the university types really don't dare say what they're not supposed to say. We've been here for 48 years. No entomologist from the University of Maine has ever bothered to come down and look at what we're doing. Back in 1979, I got a call from Bergland's USDA, that was when they were starting that study they did, report and recommendations on organic agriculture, the first one the USDA ever put out. They had heard I ran tours of European organic farms and they wanted me to tell them about European organics. I said, "I'll do better than that; I'll take you there." They put together a four-man study team and we flew over to Europe. I took them to all the best organic farms. One day we stood on the edge of a field in Germany on this one farm, talking to a really good guy through the translator. One of the four people on loan from the USDA was the Dean of Entomology at Michigan State. He wandered out in the vegetable field, and we could see him leaning over, sweeping his hand over the back of the crops looking for pests and pest damage, like entomologists do. Finally we noticed he was just standing there, stunned. We looked over at him and he said, "There aren't any pests. We can't even do this well with pesticides." I think the guy was never the same again. For a lot of those guys it's a pretty scary thing to go to a successful organic farm and find out that the things they've been saying, that without chemicals you can't grow anything, is pure bull.

ACRES U.S.A. It's never pleasant to see years of your career vanish; just drift away on the breeze. Futility hits people hard.

COLEMAN. I can imagine some PhD who's spent his whole career developing pesticides. Now he's about to retire, and he sure doesn't want to meet me.

ACRES U.S.A. Do you plan to hold more conferences?

COLEMAN. Oh, we'll see. If extension reacts to the suggestions we made and there's some more new information coming out, I think it could happen every other year just to keep people up to date on what's going on. We all learn so much doing it all winter, and a lot of it we share just informally by emailing each other.

ACRES U.S.A. Then you were able to attract some extension people to the event?

COLEMAN. Yes, a guy in Vermont who works with a lot of the Vermont growers, and there were two guys from the University of Michigan, an extension agent from Massachusetts and one from New Hampshire, so we did pretty well.

ACRES U.S.A. Did the difficult land of Maine come out of special conditions?

COLEMAN. Whenever there are rocks all over this part of the world, they were usually dumped here by the glacier. The middle of our farm is a slight knoll. If you understand geology, it's what is called a terminal moraine. That means that the glacier melted back at the same speed it moved forward for a whole bunch of years there, so it dumped a lot more rocks than normal. That's why there's a slight rise to the land.

ACRES U.S.A. Then Mainers have always had to remove a lot of rocks?

COLEMAN. Yes. We're on glacial till, glacial outflow. Our soil type is a sandy acid podzol, one of those soil names that came from the early Russian soil scientists.

ACRES U.S.A. Last thoughts?

COLEMAN. It's just that every day when I walk out the door and get to work, whether I'm cultivating or seeding or tilling or whatever, I just cannot believe these techniques will grow such continually magnificent crops, yet modern agriculture is determined to tell everybody it won't work. That just blows my mind. I look at the beautiful fields of stuff we've got growing—we grow some 45 different crops—and it all looks good. We work hard to cultivate and grow, but just the idea that they've been able to convince farmers that this doesn't work is beyond belief.

"They're Seeds of Freedom"

An Interview with Dr. Vandana Shiva

Originally Published: January 2016

Americans who visit India often come back more or less over-whelmed by its vast size and complexity, and if they are not stunned into silence they are at least much less willing to engage in generalities. Timeless beauty, explosive economic growth, persistent poverty and about a billion people all make for an intense experience if you're used to the rhythm of Americana.

One thing that does emerge from the ancient nation's recent history, though, is the way societies that seem chaotic and disorganized to outsiders actually offer opportunities for their citizens who are willing to act with boldness, imagination and fierce resolve. Gandhi was one such actor, and Vandana Shiva is on the path to become another. Increasingly well-known here as an author and lecturer, her popularity makes her, quite simply, a pain in the neck to proponents of industrial agriculture.

It's a whole other story back in India, however—there Shiva is a force for change not only among the commentariat but also on the ground. She agitates for legislation and political change at one end of society in while leading a movement to empower farmers at the other. Shiva is that rarity in modern life, an intellectual who sees possibilities for action in the world outside her study and moves to set them in motion, working with fellow sojourners to build and sustain a counterforce opposing the corporate status quo over the long haul. On a trip to California to speak at the Soil Not Oil conference as well as the Heirloom Expo in 2016, Shiva covered an amazing amount of ground in less than an hour on the phone. In 2017, she was the keynote speaker at the Eco-Ag Conference & Trade Show where, when asked why she had enough hope to continue her advocacy, she said, "Because not a day in my life have I worked against my conscience."

Contextual Note: The Bhopal disaster—a massive tragedy linked to agricultural chemicals—occurred in India. According to media outlets, on December 2, 1984, lethal methyl isocyanate (MIC) gas from a Union Carbide pesticide plant blanketed Bhopal, India, killing as many as 16,000 people (one of the highest estimates) and injuring 500,000 others. A 2006 government affidavit stated that the leak caused 558,125 injuries, including 38,478 temporary partial injuries and approximately 3,900 severely and permanently disabling injuries. The cause was never proven. No one was ever held accountable.

ACRES U.S.A. How should we approach the story of Indian agriculture?

SHIVA. The first thing you need to remember is that India is a land which has farmed for 10,000 years continuously and sustains more than a billion people on its agriculture. India is the land where when the British were the rulers, and in 1891 they sent John Augustus Voelcker to make a survey. He wrote a report on Indian agriculture that was published two years later. He said he could find more ways that Indian farmers could advise Great Britain about how to improve its farming than ways the British could advise India. He wrote that Indian agriculture was not backward, and that in many areas there was little or no room for improvement. Then the imperial British government sent Albert Howard to India in 1905. He arrived to find the fields were fertile. He found no pests damaging the crops, and he decided to make the study of peasant agriculture his profession. The agricultural testament that resulted from his studies became the basis of the organic movement worldwide—the soil association in the U.K., Rodale in the United States, all of them came out of Howard's information, and Howard's inspiration was ancient Indian agriculture. He so clearly distinguished between, as he said, the agriculture of the Occidental world and the agriculture of the Orient.

ACRES U.S.A. Only a few decades later, Voelcker and Howard and World War II had come and gone. How did the story change from admiration to alarm over impending starvation in the 1960s?

SHIVA. It changed to, "Oh my God, these pathetic people, they don't know how to farm, we're going to teach them, they're starving, we're going to bring them food." They first had to show that was the situation. After all, what is the narrative of the Green Revolution? The introduction of chemicals into India or into the Third World was first tried in India in the name of the Green Revolution in Punjab. The narrative of the Green Revolution is that we were starving before it came in. This was what got me into agriculture, because I was trained as a physicist in quantum theory—this was not my chosen area of work. When India erupted in violence after the Bhopal disaster on December 2, 1984, it forced me to see that something was wrong with agriculture. I decided to look into it. I worked for the United Nations University at the time, and I did a book called The Violence of the Green Revolution. The Punjab has been devastated, but not just Punjab. That was in '84 and we are now in 2015, and I was in Punjab before coming to California. Farmers are committing suicide on a very large scale. The soils are dead. There's a hospital train called "the cancer train" that leaves Punjab because of the toxic chemicals. The water is disappearing. The farmers grow commodities—wheat and rice, not for food, and those commodities then rot in storage. This is not a food system by any means. The true story of India's agriculture as a sophisticated agro-ecology system doesn't get told. The negative impacts of the Green Revolution now continue as Monsanto enters the scene with its Bt cotton seed, pushing farmers into debt by shooting the prices of seed up thousands of percent higher to collect illegal royalties—illegal both because Monsanto doesn't own the seed. In India they were not allowed to have a patent. So we have tragedy of a very severe kind—300,000 farmer suicides by official government data. All I do is go deeper to find out what the death is about. In the cotton areas, the deaths have to do with Bt cotton seed.

ACRES U.S.A. How do you and your colleagues work to fashion an alternative?

SHIVA. I love my land, I love my people, and I'm proud of India's ancient traditions, which have sustainability built into their core. These are not stagnant traditions; they're evolving all the time. Part of what we do in Navdanya, the movement I started in 1987 when I realized that these poison corporations now wanted to own our seeds, is to save thousands of seed varieties. We've trained nearly a million farmers over the years in organic farming, in awareness of seeds and why they should be using their own seed. We've built the largest domestic network of Fair Trade organic products and biodiverse organic products. These farmers are saving their seeds, doing organic farming and participating in fair trade which means they shift the market. They're sovereign in seed, they're sovereign in food, and they're sovereign in the economy. That has increased their incomes tenfold. We have increased nutrition production—we call it "health per acre, nutrition per acre"—so that a farm on half an acre can feed itself from the surpluses. Doubling food production, increasing rural incomes tenfold — that is the answer to hunger and poverty, not the myths that make superprofits for the Monsantos of the world and kill our farmers.

ACRES U.S.A. How is this movement organized?

SHIVA. Navdanya's work begins with creating community seed banks to conserve biodiversity so communities of farmers start taking care of their seed. They facilitate seed collection, they facilitate training, and then they set up community seed banks so that they have their own seed supply. That's one level. We have helped set up more than 120 community seed banks in the country. Unlike seed libraries in the United States, which are literally outside the agricultural system, our seed banks are the base of an agriculture system, a non-industrial agricultural system. For us, it's very important that we save seeds to shape another agriculture, because we can't isolate the seeds and trying to shape agriculture in isolation while the Monsantos take over agriculture. While I've been here I gave a keynote and did a panel at the Heirloom Expo. What I've been repeatedly saying is that one thing industrial agriculture did to the United States was that it not only fragmented the agriculture—it fragmented the thinking about agriculture, including the thinking of those who respond to the crisis. The pieces don't connect because of the Cartesian idea of either/or—if you don't do labeling, for example, then everything will collapse, or if you don't do this or that, there goes everything. It's all a part of one system! The seed saving is our foundation. The seed saving then leads to an agriculture that is biodiverse and ecological, and that is where the training comes into the picture.

ACRES U.S.A. What goes on at Navdanya in terms of education?

SHIVA. We do trainings locally, for coordinators at the level of different states. We work in 17 states, but we also have national trainings, which is why I have built up the Earth University. We have a farm where we grow more than 2,000 crop varieties. It's a living thing, a living university where people learn by being there. Of course they learn the science, they learn the theory—but they also learn the practice. The longest course we offer is a one-month course on the A-to-Z of agro-ecology and organic farming. We teach living seed, living soil—because industrial agriculture presents soil as dead and inert and an empty container, seed as an empty container waiting for toxic genes through genetic engineering, food as just stuff you put into your mouth, not living nourishment that brings you health and becomes what you are. We teach living economies at the time that the dominant economy is only creating catastrophe, unemployment and crashes. These courses have become very important catalysts to large groups of farmers who come from across the country to train there. When I travel to a certain region, 10 years later farmers who were trained at our farm are now running the organic movement of a state, working with the government to help declare their state as an organic state. That has happened in five states—Uttarakhand, Madhya Pradesh, Kerala, Sikkim and Jharkand. So the word from the grassroots is starting to have a trickle-up effect on policy. The production units work at the cooperative level. They are producer groups who organize themselves to do the organic production and then make sure that the amaranth in a high mountain village reaches Delhi or a very rare bal minthia (a chocolate-like fudge) from tribal areas of central India gets to market. So community seed banks coordinate at the regional level, cooperatives at the local level, and research and training back them up so we have the cutting-edge science which is needed to inform policy. We do all of that.

ACRES U.S.A. Does political activism play a role here?

SHIVA. We also do civil disobedience, which is a very important part of our work. If there's an attempt to pass a bad law, or permit a GMO, we've done civil disobedience and stopped it. One of the things I have done here at the Heirloom Expo in California is to distribute a newspaper devoted to ending seed slavery, which is available on the Seed Freedom website; seedfreedom. in. California passed an insane law that took effect on the first of January 2015 saying that people cannot exchange seed beyond three miles and only Sacramento can write any law related to seed—no county, no local governments—and that only corporations can do research and breeding, that all other breeding is unreliable. At a time when corporate breeding gives us toxic food whereas public breeding, community breeding and farm breeding give us robust, resilient seed, full of nutrition, the final straw in that seed law which must be challenged is the part that says corporations are persons. Corporations are not persons. We give entities a legal right to exist. It stops with that. It stops with a legal personality, and that legal personality should be revocable by society when the corporations become antisocial and commit crimes against nature, against humanity. The destruction is everywhere. Whatever we have been able to do in the last 30 years in India, I want to share it with others who have the same problems, including California and its very silly seed law. There should be civil disobedience against it, and not just in California. If it stays on the books in California it will be tried in other states.

ACRES U.S.A. Do you think Americans could use some education in Gandhi's principles of nonviolent resistance?

SHIVA. Of course. We have a course in Gandhi and Globalization set up for next year at Earth University, followed by one month of A-to-Z. Your readers are welcome to come visit India and learn from Ghandi.

ACRES U.S.A. How do Ghandi's principles inform or influence your struggle?

SHIVA. Every dimension of Navdanya is inspired by Gandhi. Let me take it one step back. In 1984 when violence erupted following the assassination of Prime Minister Indira Gandhi, and then the tragedy took place in Bhopal, I read in those happenings signs of violence of the highest level. It was Gandhi's teachings of nonviolence that forced me to look at the system of violence and start shaping a nonviolent agriculture, for which I turned back to Albert Howard and whoever else had done this work. I thought of Gandhi back when the British Empire controlled 85 percent of the territories of the world. What he did was take out a spinning wheel and start to spin cloth. He said the empire is based on textiles. He said we will only be free when we start making our own clothing again. He called it khadi—self-made cloth. He said the spinning wheel is not primitive compared to a factory in Manchester—the spinning wheel brings you liberation.

For us, it's exactly the same. Our seeds are not primitive; they're seeds of freedom and seeds of hope. Monsanto's GMO seeds are seeds of death and seeds of soil destruction. We've just done a soil analysis in the Bt cotton areas, and the bacteria have dropped 250 percent compared to organic farms! In four years of planting, 22 percent of beneficial soil organisms were killed by the release of the toxins. Fungi have dropped. The soil organisms have dropped in the Bt cotton area. The idea of nonviolent farming comes from Gandhi. We call it bija swaraj—bija is seed, swaraj is serenity and self-governance. I wrote a piece called "Bija Swaraj, Not Bt Raj." We attempt to make farmers sovereign in seed and food, to make communities sovereign in seed and food. Then it comes to Gandhi's most important lesson, not cooperating with unjust law. He first experimented with it when he was in South Africa and they were laying the legal framework for the apartheid regime. The British passed a law that Indians had to wear a badge showing they were Indians and different from the whites. Gandhi said we are equal citizens. After all, it was in South Africa when they pushed him out of a train's first-class compartment—he was a lawyer and traveling first class, and they pushed him out. That's when he woke up to racism. That's when he realized how cruel the world can be. Then apartheid started, and he said, "We will not obey." He did a march, he went to jail. He organized the Indian community. That took place, interestingly, on September 11, 1906. We call it, "The other September 11."

ACRES U.S.A. Didn't his first campaign in India involve poor farmers?

SHIVA. When he came back to India and the first thing he did was go to the area where the British were forcing our peasants to grow indigo. The peasants were starving so that the mills in the U.K. could depend on a supply of the blue dye. He spent time talking to farmers. His hut in Champaran was burned twice by the planters, but at the end of it the peasants organized with him what is called the Indigo Satyagraha. Imperialistic economic systems—which are what we have now except we have corporate empires—are organized around asking, "What do people need, and how can we monopolize it?"

ACRES U.S.A. Some time later another crucial satyagraha, or nonviolent rebellion, involved another food staple.

SHIVA. Yes. When they tried to make a monopoly of salt with the salt tax in 1930, Gandhi walked to the beach—the famous Gandhi march—picked up salt from the sea, and announced that since nature gives it for free and we need it for our survival, we will continue to make salt, and we will not obey your laws. The Salt Satyagraha triggered a whole new urge for breaking free of imperialism in India. In my region—we're up in the mountains near the Himalayas—there's no salt, but we had forests that were being enclosed. The people said these forests are ours and they walked into the forests and did a forest satyagraha. Hundreds died in the process. That became the basis of the later movement which was my first ecological movement when I was still a young student, the Chipko movement where women came out to embrace the trees and say you can't kill these trees, they're our mothers, you'll have to kill us before you kill them. And that stopped the logging in the high Himalaya. Gandhi's tradition of civil disobedience has been the basis of us having seeds that we can use after cyclones and after drought. Tree fruits that are delicious and tasty, seeds that have more nutrition like our wheat, seeds like the bashma seed which were patented by an American company and we had to fight that case. All the old wheat that was patented by Monsanto—everything that exists is being defined as an invention.

ACRES U.S.A. Do you believe the spirit behind the Green Revolution was deeply paternalistic?

SHIVA. I don't think it was just an issue of being paternalistic. I think the issue was that it was a system based on chemical warfare. Norman Borlaug came out of a defense lab viewpoint. After World War II the corporations deployed these chemicals into agriculture. From 1952 onward, Rockefeller and the Ford Foundation tried to push this on India. It wasn't working because native seeds and chemicals don't go together. Norman Borlaug was sent from the DuPont defense lab to Mexico to work on dwarf varieties of wheat to adapt them to chemicals, and he got that done. Then in 1965 we had a drought, and the drought raised food prices. There was no starvation, there wasn't a famine. There was a rise in food prices, and India asked for additional shipments of wheat. The United States government said, "Sorry, we won't send wheat unless you take the chemicals and shift your agriculture to chemical farming." The term "green revolution" came much later. The prime minister of that time, Lal Bahadur Shastri, said he would not experiment with an entire nation of farmers. He told them, "We can try it on a small scale. If it works, we'll adopt it, if it doesn't, we won't. But you can't force us to use these chemicals." He died mysteriously a year later in Tashkent and the pressure continued. The conditionality that if we did not adopt chemical farming we were not going to get that additional bit of wheat for that year to bring relief and stabilize prices—that's more than paternalizing, it is forcing.

ACRES U.S.A. Then it was actually more a form of aggression?

SHIVA. It was aggressive, yes, very aggressive, inhumanly aggressive. Our first agriculture minister, Panjabrao Deshmukh, was a close friend of my parents who visited us often. He said agriculture is based on two cycles—the nutrition cycle and the water cycle. These cycles have been broken because of the war and because of colonization. We have to fix the cycles to fix the nutrition cycle—which is your composting, exactly what Howard had done. To fix the water cycle we have to conserve water and do water harvesting. We knew what we had to do. We had entire programs. I once met an 80-year-old former civil servant after giving a talk who said, "You know, Dr. Shiva, when we were training in the civil service, we would learn how to compost." That is the moment in which the Green Revolution was introduced. There was so much alternative potential growing.

ACRES U.S.A. And the alternative was growing as a way of correcting the distortions of agriculture imposed by the British Raj—the colonial authority—over the previous century?

SHIVA. Yes. Over the previous century, and due to the destructions of war. Don't forget, we had the great Bengal famine during the war, 1943-44.

ACRES U.S.A. That's not widely known over here.

SHIVA. Oh! Two million people died. There was more rice than we needed but the British were extracting all the rice from the peasants to profit in trade during the war after the Japanese seized Burma. During the famine, women started a movement called "nari bahini," fighting the police who came to stop them from threshing rice and saying, "We will give our lives, we will not give our rice." 1942 was the year Gandhi called on the British to quit India. It was called the Quit India movement. But the famine occurred not because India didn't know how to produce food, but because every drop was being extracted. The combination was cash crops and trading.

ACRES U.S.A. And there is a clear parallel with the present day, as many of the world's farmers grow cash crops for the commodity markets and can barely feed their families.

SHIVA. Farmers are half of the hungry in the world because they're made to grow food and raise it in debt. They sell what they grow, and they don't have enough to eat. Yes, the Green Revolution was very badly intended, but it wasn't long after that came what is called the second green revolution, the combination of GMOs and global free trade, written on the basis of rules made by Cargill and ConAgra. What we have is a recipe for disaster. The two million who died in 1943 and 1944 are nothing compared to the famines that will hit the world when only profits guide how we do agriculture, rather than the care of the land, the dignity and justice of the farmer and our concern and awareness about our health that comes from food.

ACRES U.S.A. Where is Earth University is located?

SHIVA. Earth University is located in Doon Valley, a beautiful valley up in the Lower Himalayas. Many Americans will know it as the valley of Rishikesh, with the ashram, the yoga, the meditation—that's where the Beatles came to study with the Maharishi. Our valley is a beautiful valley. It's the first valley of the Himalayas, with the Ganges on one side and the Yamuna on the other, and the higher mountains on the north and the foothills on the south. It's not a very big land, but it's where I was born and where I returned to do this work. Because it was about taking care of the earth, I thought it had better be the place which had given me birth.

ACRES U.S.A. The average age of farmers in the United States is high, they tend to be middle-aged or older. Does India have a similar problem with the farmers growing older and not being replaced fast enough?

SHIVA. You know, farmers have always gotten old. Everyone gets old. I'm old now. But when a system is stable, younger generations take on the options that are available to them. It is only in recent times when industrial agriculture came to the United States and the Green Revolution came to India that younger generations have started to abandon farming, and it is the result of two factors. One is that these systems of agriculture are design to extract profits from the farmer. They sell costly inputs and buy cheap commodities, which means the farmer's economic livelihood becomes non-sustainable. Farmers get into debt—in this country look at the number of family farms that have disappeared, look at the number of farmers having to rent their own land back from the banks and the mortgages that made them lose land ownership. Now if a young man of 20 is seeing an indebted farmer, or in India a young person has seen their father commit suicide because Monsanto extracted so much royalty that they pushed the farmer into debt, that young person will not want to go into farming. But there is another aspect to it, and that has to do with the fact that industrial farming is not just a war against the earth and a war against farmers. It is a war against farming as an occupation. It can only work with the repeated propaganda—farming is backwards, working on the land is primitive, you only do it if you have no options. Move to the city and find a specific job; that is progress. Our leaders say that we should wipe out our villages and put everyone in the city. This very artificial narrative that farming, villages, farmers, are part of an obsolete existence that should be wiped out, that progress is destroying agriculture for industry, destroying industry for services—it got internalized, this forced narrative of progress. So young people will leave the village and toil in cities and join gangs because nobody is really providing them with alternatives. Our work in Navdanya is to make young people feel that farming is exciting, it is full of knowledge, that through seed sovereignty and ecological agriculture and your own participation in markets on your terms you can actually make a dignified living. I really feel happy that in the last 30 years so many young people have moved back to villages after the trainings.

ACRES U.S.A. Can you give us an example?

SHIVA. My colleague Darwan Singh Negi is now a master trainer at the national level. He's the one who goes to Bhutan to train the Bhutanese farmers to grow organic. As you know, the government of Bhutan asked us to help them go 100 percent organic. I go there once in a while, but Darwan goes there all the time. He was a young man who had run away from the village, came to work in Delhi and worked with me in the office. Two years later, sitting in the office, he said, "Veve, this work is good work. If you let me go back to my village and support me, I will become a Navdanya activist." In the first year he made 200 villages pesticide-free! He inspired people in his valley so much that traders of pesticides gave up selling pesticides and became organic farmers. And this is one person—there are hundreds like this. Young people who come to train with us because of the work we've done go away knowing plenty. This is the food system that needs to be put in place so that life can be full of dignity, meaning and justice.

ACRES U.S.A. How were you involved in the struggle to save the neem tree from getting patented? What has happened since then?

SHIVA. The story of the neem goes back to the Bhopal disaster of 1984. It took about three days before we were allowed to go there because of the poison in the air, but I went with a bucket full of neem saplings, and I planted them. We drew a poster, I made it myself, and the poster's title read, "No more Bhopals, plant a neem," because neem has been used in India from ancient times as a pest control agent. We use the neem leaves in our grain and our seeds, we use the neem leaves in our silk and our woolens. Everyone—my grandmother, my mother—knew that neem controls pests. And yet we were making toxic pesticides that killed 3,000 people in one night, 30,000 since then, and hundreds of thousands of babies who are still born crippled and with birth defects. We were already training farmers in organic farming, because before I started seed saving in 1987 we were already creating awareness of organic farming. I had always wanted to do a neem workshop on the side. And neem is all over the country. You'll always find neem in the school building if you're doing a school workshop, or in a village commons. That was 1984. In 1994 I read in a journal about the world's first invention of the use of neem as a pesticide. It was a company called W.R. Grace, the one that poisoned the water outside Boston, resulting in a big lawsuit and a book called *A Civil Action*.

ACRES U.S.A. Right, them. That was a big case from the early 1980s when a leukemia cluster appeared in a small town in Massachusetts, and the lawsuit against Grace, Beatrice Foods and Unifirst went on for years. Even 30 years later that is still an infamous pollution case here in the USA—it even became a popular film.

SHIVA. The interesting thing was that they had no idea what you do with neem. Nor did the person who first patented it, an American called Larson. He was traveling through India, saw women putting neem leaves into the grain, and asked why. "It controls pests," they told him. He was smart; he visited public institutions, and he collected all the publications he could find on the pest control properties of neem. He took those publications and wrote a patent application, got a patent and then sold it to Grace. Interestingly, the one we challenged was listed to Grace and the U.S. Department of Agriculture. When I saw this patent, I immediately organized a neem campaign. Over the years we must have collected more than 100,000 signatures. We held beautiful neem rallies with neem branches all over the place. We wrote books on how useful the neem was, brought it into popularity again, and initiated a legal challenge with those 100,000 signatures. The neem patent was registered in the United States and in Europe. Our laws in India did not allow these patents. When we pulled the application for the challenge in the United States—I did the one in the United States with Jeremy Rifkin—the U.S. Patent Office asked, "What is your commercial interest?" We answered, "We don't have a commercial interest, we have a public interest. We care for our knowledge and we want to stop biopiracy." The term "biopiracy" gained currency because of the neem challenge. The U.S. Patent office wrote back and said, "Sorry, we only entertain challenges from commercial interests." European patent law includes a public interest clause, and they admitted our challenge; in Europe I did it with the international federation of organic movements, IFOAM. Its president at the time was Linda Bullard, and the head of the Greens in the European Parliament was Magda Aelvoet, both of whom I knew well. I told them I wanted to bring this challenge and asked if they'd join me because I wasn't going to be able to fly over from India all the time. We did all the groundwork, we did all the evidence, we did all the finding of scientists and bringing the farmers. But it really it was three wonderful women, and a wonderful lawyer called Dr. Dolder who gave his time pro bono.

ACRES U.S.A. If memory serves, you won it on appeal?

SHIVA. We fought the case, and we won the case. Grace and the U.S. Department of Agriculture appealed, and it went for an appellate hearing. I remember so clearly—it was the 8th of March in 2005—and the judge of the patent court said to come back after lunch. We came back after lunch not knowing whether the patent would be struck down or our challenge would be struck down, and the judge said only one line: "Happy Women's Day, you've won." So even the appeal was defeated. I think it was a historic case because 100,000 people from an ancient Indian culture protected their knowledge and their biodiversity. It was global solidarity that won it. Since then we've continued to popularize the use of neem. I'm encouraging young people to set up small oil extraction units in their villages. The tree is growing, they use the leaves and so on, and the oil comes from the kernel of the seed. You can harvest it every year. The neem grows in every climate in India so it helps promote an agriculture without chemical pesticides. Neem is a very important friend in that.

ACRES U.S.A. The struggle to keep it in the public domain pointed the way to a viable enterprise?

SHIVA. I really feel the only way we're going to solve the unemployment problem is by turning to businesses related to protecting the Earth. Small-scale, decentralized economies are the key. We call them living economies.

ACRES U.S.A. How bad a problem is biopiracy?

SHIVA. Biopiracy is an epidemic. The most serious piracy involves plundering the innovation by farmers of the Third World who have evolved climate-resilient crops. Today the Monsantos and the Bill Gateses of the world are presenting the pirated climate-resilient crops as their inventions. Bill Gates wrote an article about it: "Oh, Melinda and I were visiting a farmer who is using seeds we introduced"—seeds that tolerate flooding. Well, it didn't come out of Bill Gates' labs, it came from Indian farmers. They pirate the seeds and take a patent. Monsanto, Bayer and Syngenta have 1,500 patents on climate-resilient crops! They are looking toward the climate crisis as a way to deepen their monopoly. If you look at the last few years, every time there has been a disaster—an earthquake, a tsunami, a cyclone—they have arrived with their GMO seeds. After the earthquake damaged Nepal so badly in April we kept getting calls—half of their seed banks had been damaged in the earthquake, buried under homes. The earthquake happened in April, by May we had to get the seeds there. We put the seeds together. At the border, the customs officer saw a very strange circular saying, "No seed except ..." and there was a list of companies, Monsanto and a Monsanto subsidiary. Only those seeds could enter. We checked with Nepal's agriculture minister and he said, "I never passed this order." They'll even exploit an earthquake to make a monopoly!

ACRES U.S.A. You mentioned earlier that historic resources such as the neem are protected by federal law in India.

SHIVA. We managed to get into law clauses stating that biopiracy is illegal. Of course I've been working a lot with my government to have it declared illegal at the international level in the World Trade Organization, but that's been blocked repeatedly by the U.S. government. Our work of course is to make sure this heritage is protected, and we believe it's the communities that are the biggest force of being guardians of biodiversity. In 2000 we started a very beautiful movement which then led to my writing my book Earth Democracy—it was the living democracy movement against the chemicals and all the pushing of GMOs. It became big and communities began to come together discussing how to protect their biodiversity. They wrote this most beautiful declaration, just drafted it at the local level. About 200 villages were the first to do it, and it spread to about 6,000 villages. The text basically said, "We are part of the earth family. The tigers and the wolves in the forest and the trees in the forest and the seeds in our farms are all part of our earth community; therefore we do not accept the destruction, the privatization through patenting, the pollution through chemicals, and we will protect our family, the earth family, as we protect our own family." Not only did they make these declarations, but the villages wrote postcards to the WTO chief and to the corporations who were pirating. To the WTO chief they said, "You're supposed to look after trade. Creating ownership of our family members is not about trade. This is about ethics. You are overstepping your jurisdiction. The head of WTO actually came to India as a result of these letters. But the most wonderful one was when they wrote to Monsanto and Syngenta and said, "It's not that people don't steal, people steal, but usually people steal in desperation. A child will steal food if they've been hungry. Someone else will steal something because the mother is ill and they need money to buy medicine. If you are stealing, you must be experiencing desperation. Come and sit under our people's tree, the sacred ficus, in our village commons where we resolve all issues of importance in the village. Come and explain to us what is your desperation. We will try to understand and help you out!" They never came, but oddly it gave the villagers such confidence.

"Impact is Always More Important Than Intention"

An Interview with Leah Penniman

Originally Published: April 2017

As a creative educator, regenerative farmer, writer and activist, Leah Penniman is an exceptional leader for a developing side of sustainable agriculture—food justice. After all, if the very system in which an organic farm thrives in is toxic, how healthy can the farm actually be?

Penniman, the author of the essential read *Farming While Black,* is best known for implementing the change she seeks at Soul Fire Farm in New York state, which she and husband Jonah Vitale-Wolff started as an organic family farm committed to "the dismantling of oppressive structures that misguide our food system."

Soul Fire Farm serves the couple's former urban neighbors in Albany and Troy, and since coming to the land over a decade ago, they have transformed a patch of marginal mountain ground into rich topsoil. They faithfully provisioned a sliding-scale CSA whose members often lack access to fresh produce and created a vibrant, welcoming community of learning and admirable influence. In this interview she spoke to the efforts to educate an agriculture industry plagued by its blood-stained history, and how progress can be made once a common understanding is reached.

ACRES U.S.A. At Soul Fire Farm your commitment to social justice goes far beyond making the food you grow accessible to people with limited resources. Could you start with an overview of what you do?

LEAH PENNIMAN. First off, we're restoring degraded hillsides. The land here ranks as the worst in the USDA Agricultural Soils Classification. By using regenerative and ancestral farming practices, we've brought our soil back into full health and production. Our Farm Share, a version of a CSA, predominantly serves low-income people in inner-city neighborhoods. We use a sliding scale system where people pay according to their income and wealth. Every year we host hundreds of youth at Youth Food Justice Empowerment programs. By teaching farming, cooking and leadership skills we are helping youth find a home here on the land and in the wilderness and a sense of belonging in the food system. We also train farmers—mostly Black, Latinx and indigenous—from beginners up to folks getting ready to manage their own farms.

Finally, we support activists and movement builders. Activists come here for strategic planning retreats and use the land as a basis for organizing and activism, and we support movements for food sovereignty internationally.

ACRES U.S.A. You describe Soul Fire Farm as a family farm committed to dismantling the oppressive structures that misguide our food system. What are a few of these oppressive structures and how do they negatively impact people?

PENNIMAN. From the early colonization of the Americas our food system has been based on exploited labor and stolen land. The long history of oppressive systems goes back to the genocide of Native Americans to the enslavement of people to the takeover of Mexican territory and murder of Mexicans by the Texas Rangers on to black codes, convict leasing and sharecropping. Without looking at that history, it's hard to understand why huge disparities in food access exist right now. These disparities aren't so much in the quantity of food or number of calories as food quality. Diet-related illnesses like diabetes, obesity, heart disease, some forms of cancer, asthma and poor eyesight disproportionately affect people of color due to lack of access to fresh food. There are similar disparities with land. The 2012 Agricultural Census found about 95 percent of farmland is in the hands of European heritage people, a figure higher than 20 years after slavery ended. In some ways we're sliding backward.

ACRES U.S.A. You've said, "If we're not acting to change the system, we are complicit, casting our vote for the status quo." Could you elaborate?

PENNIMAN. Americans really appreciate cheap food. According to the latest statistics, as a percentage of our income people in the United States spend less money on food than any other country in the world. If we only look at the dollars, but ignore the story of what it took to get the food on our plates, then we're participating in pushing down commodity prices, putting rural dairy farmers out of business and so on. The food system has the largest environmental impact of any industry, and it's the number one driver of climate change. And labor isn't fairly paid. All of us are complicit, and we also all have the potential to be agents of change.

ACRES U.S.A. I'm not aware of any other family-run farms as infused with community connections as Soul Fire Farm. Why and how did these relationships develop?

PENNIMAN. I really don't think of the community as separate from the farm. We've been living in this area for about 12 years. When we first moved to the south end of Albany we immediately got involved in service of community. We helped start the Harriet Tubman Democratic School. Most of the youth that go there had dropped out of Albany High. We got involved with urban gardens. There was clearly a community need for fresh food delivery and for safe rural spaces for people of color. In starting this farm we were responding to what our community, friends and comrades were asking for. It continues to be based on those relationships. We're deeply involved with the Albany, Troy and Schenectady activist and Black and brown organizing communities and also regionally and even internationally with Freedom Food Alliance, the National Black Farmers Conference and La Via Campesina. It's an illusion to act as an independent agent without those community connections. We really need each other.

ACRES U.S.A. Participating in The Food Project when you were 16 seems to have opened up a world of possibilities for you. What about that experience made it so important and formative?

PENNIMAN. As a 16-year-old—like many teens—I was confused about my identity and my place in the world. I took the The Food Project job because I needed to save money for college. It turned out that farming could be the intersection of the two things I loved most, social justice and the earth. We grew food that we sold in farmers' markets in low-income neighborhoods. We worked in homeless shelters, did leadership development and learned about fairness in the food system. I found so much richness in this work. The simplicity of planting the seed, harvesting a carrot and making sure it got to people who needed it was very compelling.

ACRES U.S.A. After that you worked on organic farms. What was that like for you?

PENNIMAN. From The Food Project I went to work at the Farm School in Athol, Massachusetts, for a year. From there I went to Many Hands Organic Farm in Barre, Massachusetts. I learned so many farming skills, and I was able to move into a co-managerial position in Barre. The challenge was that after The Food Project, organic farming was a white-dominated world. At NOFA (Northeast Organic Farming Association) conferences the speakers that I saw and the books that I read presented the white narrative of organic farming. I began to feel that I had to make a choice between organic farming and being connected to the Black community. That false dichotomy caused me quite a bit of doubt and internal strife. It wasn't until I got back to urban farming in college that I found a home place that engaged both worlds.

ACRES U.S.A. While you and your future husband Jonah were in college you started YouthGROW. Describe the program and how it came about.

PENNIMAN. Jonah and I were going to college in Worcester, Massachusetts, and the Regional Environmental Council hired him as the community gardens coordinator. He had a critique of the organization as being focused on the environment separate from human need. We brainstormed ideas to change that and decided to create a version of The Food Project in Worcester. YouthGROW would engage young people in reclaiming vacant lots and growing food for the community while they earned money. What I feel so proud of is that YouthGROW actually became their flagship program. Some of the young people who started out with us are now leaders in the program.

ACRES U.S.A. You've had the opportunity to spend time in farming communities in Ghana, Mexico and Haiti. What was the occasion for these travels, and how has this exposure influenced you?

PENNIMAN. Closest to my heart at the moment is Haiti because my family and friends there are reeling from and working hard to recover from the recent hurricane. We've been doing quite a bit of organizing to get support to them. Since 2010 Soul Fire Farm has been collaborating with Ayiti Resurrect. It's a collaboration of Haitian and Caribbean heritage folks working in solidarity with farmers, artists and healers in Komye, Leogane, Haiti. We're responding with a number of exciting projects—reforestation, compost development, small-scale solar, a well, planning out irrigation, herbalism and healing clinics and work with children in the schools. The farmers in Haiti are incredible. I've learned a lot from them. Kombit is a collective work model where all the people support each farmer in turn to get their hoeing done and their fields prepped for planting beans and maize. These work parties are infused with rhythm and song. They use a very powerful primary tillage tool—an ancient long-handled hoe invented in Africa. I've also learned about the restoration of severely degraded lands. Haitian farmers have been able to stabilize hillsides and start to trap some organic matter by using strong grass crops. After that they plant bushes and trees. We shared this strategy in Oaxaca, Mexico, where we were to serve indigenous farmers on a Fulbright fellowship last year. The Mixteca there also suffer from land degradation. It was beautiful to be able to exchange indigenous knowledge between these different cultures. We explained the use of these plants for degraded hillsides, and farmers in Oaxaca explained their intercropping to us. They intercrop three to 12 diverse crops. Each plant provides a service to others. These intercropping systems are amazing technologies that release nutrients throughout the season. They go beyond the Three Sisters. Natural chemicals that wash off the leaf of the squash actually prevent pest infestation in the beans, for example. The Mixteca also taught us to use nixtamalization to make the niacin and protein in maize more available for human digestion. That's not a common practice in Haiti so we shared it there. The peasant farmers of the world still grow over 70 percent of the world's food. They have many of the solutions we need to feed the world sustainably. These conversations and exchanges and mutual support have been a great and humbling honor.

ACRES U.S.A. Does your family have Haitian roots?

PENNIMAN. My maternal lineage is Haitian.

ACRES U.S.A. I understand that you also went to Ghana.

PENNIMAN. I've been to Ghana a few times. Right after college I lived there for five months. I worked with farmers to create income-generating projects for people who were marginalized in society—the disabled, orphans and people living with HIV. There was a small but strong movement led by Mr. Kwabla in the village of Ojepa Djerkiti to stop using agrochemicals and treated seed which some of the foreign organizations were pushing. His farm was very productive so people would visit. His model would get them thinking that maybe they didn't need the agrochemicals. I spent a lot of time with Mr. Kwabla on his farm. Their gender roles were different from what I was accustomed to, but I won their favor and got to use a machete to help clear the fields without burning or spraying herbicides.

ACRES U.S.A. After seeing your sister Naima perform her poetry before your keynote at last year's NOFA Summer Conference, I became curious about your family background. Can you identify things in your growing up that contributed to your creativity and courage as a passionate and original leader?

PENNIMAN. That's very generous. I also need to shout out to our little brother, Allen Penniman, the urban planner for the City of Providence, Rhode Island, and quite a progressive leader and justice advocate in his own right. In some ways I think adversity contributed to our ability to be leaders and to see injustice and act. We have experienced poverty and different types of abuse. We have been surrounded by folks who struggle with mental illness and have experienced the fallout. From a very young age we had to come up with survival skills and mechanisms and develop fortitude. I think having compassion for those who are suffering came out of our own experiences. Certainly our parents have strong social justice consciousness. My mother was very active in the Civil Rights movement, as was my father, though he was more active on land protection and environmental stewardship. Growing up mostly in rural Ashburnham, Massachusetts, shaped the heartfelt connection to the earth that informs our work.

ACRES U.S.A. What sorts of personal growth does it take to be an agent of change in the movements toward liberation and justice?

PENNIMAN. It's probably different for every person. I come at this from a spiritual activist perspective. I consider it a sacred duty to heal and repair the world, and doing this fills my soul and gives me purpose and joy. But I need to maintain personal self-care practices. I run every morning at 6, even in the dark, and I do my weekly co-counseling and have a prayer relationship with the divine. Those things have allowed me to do the work. Besides rootedness and spirituality, I think a sense of humility and a learner's mind are important. Yesterday and today I've been having conversations with people who have participated in our programs. I'm asking them,

"How can we do better? Where does it hurt? What are the things I'm not seeing that I need to see?" so that we can keep our integrity in moving toward justice.

ACRES U.S.A. You've worked as a high school biology teacher for a number of years. Could you talk about that aspect of your life?

PENNIMAN. In 2002, right out of college, I started teaching biology and environmental science full-time in a public high school. I'm now teaching environmental science and agro-ecology two and a half days a week at an independent school. That has been a really important shift. Farming is a full-time job, but we needed that supplemental income and community connection. Now that we're moving toward a more financially sustainable model we don't need to be off farm as much.

ACRES U.S.A. Turning to another subject, what does the phrase food justice mean to you?

PENNIMAN. I defer to La Via Compesina's *Declaration of Food Sovereignty* as the authority on this, but in simple terms there are four ingredients to food justice. The first is everyone having access to nutritious, culturally appropriate food regardless of race, income, class, gender and geography. The second is fair wages and working conditions and dignity for everyone involved in the production of that food. Third, in the process of producing the food we improve the quality of the natural environment, rather than diminishing it. And fourth, democratic rather than corporate control of the food system at all levels.

ACRES U.S.A. Going back to that first element, talk about food apartheid and the seriousness of its impact.

PENNIMAN. Food apartheid is a term I learned from Karen Washington, a black farmer and food justice activist, friend and mentor. It refers to geographical areas and communities where people live in poverty and do not have access to fresh, affordable, culturally appropriate food grown with sustainable methods. The USDA maps these neighborhoods and calls them food deserts. We prefer "food apartheid" because this is a human-created system of segregation, not a natural biome. We can measure the impact of food apartheid on communities of color in the incidence of diet-related diseases. The number of children who go to bed hungry is also measurable. Less measurable is what happens to a hungry and sick people in terms of ability to organize, resist, and live free from anxiety and distraction. A hungry population is less likely to be able to analyze and resist the structures of oppression.

ACRES U.S.A. An important part of your mission is making real food available to everyone. How do you overcome economic and other barriers to serving marginalized communities?

PENNIMAN. We don't have all the answers, but we're committed to doing our best. In our model right now—the farm share model—some people pay more than market value for their food, while others pay substantially less. That washes out to keep the farm afloat in terms of income. We try to reduce the transportation barrier by offering doorstep delivery to people who live in food apartheid neighborhoods. We also do what we can to grow culturally familiar foods. We grow a lot of tomatoes, peppers and alliums, collard greens and okra—foods familiar to new Americans and Black Americans. With less familiar foods we offer culturally familiar recipes with substitutions. Using other greens to make callaloo is an example of that. Additionally, a foundation of authentic and caring relationships with community members is crucial. We spend a lot of time in the winter at organizing meetings and at events talking with folks in the neighborhood about their needs and how as a farm we can better serve them. It's out of those relationships that almost all of our members have come.

ACRES U.S.A. You have interesting ways of explaining the CSA farm share to make it more culturally relevant.

PENNIMAN. Our friend Masai coined the term "Netflix for vegetables," which breaks down some of the immediate misunderstandings about how our farm share works. We also credit the clientele membership club developed by Booker T. Whatley, in blessed memory, a Black man who was an agricultural professor at Tuskegee Institute in Alabama. We talk about him and Tuskegee, and we talk about the Kwanzaa principle of Ujamaa or cooperative economics. The Swahili word is about moving beyond the casual relationship between producer and consumer encouraged by capitalism into a community-based system of mutual commitment.

ACRES U.S.A. Do many of the people in your CSA eventually come to the farm?

PENNIMAN. Oh yes, most.

ACRES U.S.A. How is the low price of food in the United States related to race?

PENNIMAN. In this country we still very much rely on exploited labor—which is predominantly done by people of color—to keep the price of food low. Many of the farm workers that grow 75 percent of our food are here through the H-2A guest-worker program. They're invited to the United States to labor on the farms, but they don't have a pathway to citizenship or land ownership or the right to collectively bargain, or in most states the right to overtime, sick days, health insurance, high-quality housing or safe transportation.

ACRES U.S.A. Or the right to change farms if the farm they're on is not working out for them.

PENNIMAN. Exactly. One of the arguments that people use for capitalism is that labor has the freedom to move, but in the guest worker program workers don't have this freedom.

ACRES U.S.A. As a small farmer do you experience the contradiction of trying to pay your workers fairly, while also compensating yourselves as family farmers?

PENNIMAN. Absolutely. For the first years of the farm we were all volunteers. That included us as the leaders of the project and everyone who came to work here. We are committed to working toward a living wage and have ended the practice of volunteers performing any of the crucial work on the farm. That said, I don't think that figuring out how to fairly pay folks on the farm is solely the responsibility of farmers. It's a systemic issue. The imbalance between the percentage of the USDA budget going to support commodity crops and large-scale industrial Ag compared to small-scale farmers is completely out of line. We need to subsidize small farmers for the environmental services they provide like carbon sequestration, pollinator protection and purifying the air. Those things have a real dollar value. At the same time polluters should pay for the externalities like water pollution and the Dead Zone in the Gulf of Mexico.

ACRES U.S.A. The USDA has had a strong role in the decline in the number of black farmers and in black land ownership. There is a long history, but maybe you could give some highlights.

PENNIMAN. It's important to know about the class action lawsuit Pigford v. Glickman, which was settled in 1999. Black farmers suing the USDA for decades of discrimination proved that discrimination existed and the case was settled out of court. Farmers in the suit received up to $50,000 each. Unfortunately, by then almost everyone had lost their land, and as farmers we know that $50,000 is not going to get your land back. It really was too little, too late. By that point black farmers only owned 1 percent of farms, down from a high of around 14 percent in 1910. That's 16 million acres of land lost. Many push factors are behind why people left their land. Some of it was just straight up racist violence and harassment as black land ownership was seen as a threat to white supremacy in the south. White folks wanted black people to stay in their 'place' as sharecroppers and tenant farmers and they would lynch and harass black farmers and burn crosses in their yards and burn their homes. When the option of moving to the north opened up, it became more attractive than staying and facing the risks. Of course, county USDA offices controlled the disbursement of credit and emergency relief, and at every turn black folks were denied. When the cotton boll weevil struck in the early 1900s and decimated the cotton crop, white farmers got relief and black farmers didn't. And white farmers could attend land grant universities to get training, but black farmers couldn't.

ACRES U.S.A. Wasn't there a system of separate universities?

PENNIMAN. Yes, but the resources allotted to them were unequal so over time extreme disparities developed. Many black farmers lost their land to foreclosure and legal trickery, and they didn't generate enough income to sustain it.

ACRES U.S.A. You say that giving credit where credit is due is an essential step in ending racism. What are some of the contributions people of African descent made to American agriculture?

PENNIMAN. The very concept of regenerative farming—farming where soil health is of primary importance—came out of Dr. George Washington Carver's work at Tuskegee. He helped a generation of black farmers move away from cotton monocropping and into diversified horticultural systems. He was the first person to put these practices together in writing, give them a name, and start promoting them through the university system so many people consider him the founder of U.S. organic agriculture. I have already mentioned Booker T. Whatley and CSA. African-Americans also made significant contributions to the farmer cooperative movement through the Colored Farmers' National Alliance and Cooperative Union, and later through the Federation of Southern Cooperatives. Shirley and Charles Sherrod started the first U.S. land trust on one of the largest black-owned pieces of property in the history of the country, which was collectively owned by 500 families in Georgia. The Community Land Trust movement was born out of the Sherrod's pioneering work.

ACRES U.S.A. What about all the things contributed by the people forcibly brought here from Africa?

PENNIMAN. They brought seeds and technologies. They weren't just bodies kidnapped from African shores. They were experts in agriculture—rice growers, cattle herders and tobacco farmers. Folks brought their skills and taught European Americans how to cultivate land in regions where the climate resembled West Africa's more than Europe's.

ACRES U.S.A. You've said we fail to notice and honor resistance movements of people of color, such as the refusal of Haitian peasants to adopt GMOs.

PENNIMAN. They are under such pressure to accept paternalistic donor aid, but they have maintained courage in the face of real scarcities that exist in their country. Monsanto has been dumping GMO products on the Haitian market, trying to get people hooked on them and displace Creole native seed. The Haitian Peasant Movement rose up and said, "We're not doing that.

We need our seed. That's our sovereignty. That's our life." At a demonstration they burned Monsanto seed. For their organizing, they won the Global Food Sovereignty prize a couple of years ago. We have to be aware of a U.S. strategy in the farm bill called Tied Aid, which is designed to create dependence. Whenever we give food aid to other countries, it has to come from U.S. farmers and U.S. corporations and be shipped by U.S. companies. The United States has strategically dumped this food aid onto Haitian markets at harvest time. Rice arrives during the rice harvest and peanuts during the peanut harvest. No farmer can compete with free food. This undermines farmers who don't have savings, causing migration to the cities. The president of Haiti has been calling for the cessation of this dumping, but the United States continues to dump. We have to be vigilant. Things that look like helping gestures can actually be undermining people's autonomy.

ACRES U.S.A. How did the Black Farmers and Urban Gardeners Conference come about?

PENNIMAN. My part of the story begins at a NOFA Summer Conference. There were very few black and brown folks there. I wanted to have a safer space for people of color to talk about our place in the movement so I asked people to meet at a certain place and time. Karen Washington and many others came. We felt we needed a conference focused on issues of importance to us. In 2011 Karen organized the first conference. It's been held every year since at different locations around the country, bringing together many hundred urban and rural black farmers.

ACRES U.S.A. What obstacles still make it harder today for people of color to become farmers in the United States?

PENNIMAN. One is access to land. Like I mentioned, European-Americans own almost all the farmland so without some sort of land reparations or land reform, that's an obstacle. Land is prohibitively expensive, especially land close enough to the urban communities that people are connected to. Training is another. Land grant institutions tend to be in isolated, rural white areas. Most professors are white, and they may not be culturally sensitive or know our history. And black and brown farmers don't have the resources and support to offer trainings. We need to think about putting some federal and state agricultural dollars toward supporting black and brown farmers. One recommendation is hiring black and brown farmers as adjuncts at a distance from land grant universities to provide training and support. Another policy suggestion is creating business start-up clinics that offer legal and financial services and low-cost subscriptions to software for farm businesses. A lot of paperwork goes into running a business. Offering these free and low-cost supports would let farmers focus on farming.

ACRES U.S.A. Explain how historic trauma prevents people of color from becoming farmers and how the land-based healing that Soul Fire Farm does help people overcome it?

PENNIMAN. It's impossible to have your ancestors go through hundreds of years of brutal enslavement followed by tenant farming and sharecropping and forced expulsion from your family land and not inherit some type of trauma. For a lot of black folks, when they think about farming, the first word that comes up is slavery. To undo that requires land-based spaces for healing justice. Our farming programs don't just teach how to propagate plants or make sauerkraut. We explore the history of our people on land through our hearts. Our history of dignity on the land is much longer than our history of land-based oppression, so tapping back into that pre-slavery cellular memory is part of the healing justice work. We use our ancestors as healing tools through song, dance, herbal bathing, communal meals and talk therapy. By processing that grief and trauma we can find our way home to the land.

ACRES U.S.A. What did you set out to accomplish with the Black and Latinx Farmer Immersion?

PENNIMAN. We created this program because there were many black and brown folks interested in farming, but they didn't have the experience necessary to engage in a year-long apprenticeship or take a job on a farm. We took all that we had learned from training apprentices and new farmers and created a week that offers essential gardening skills and enough knowledge that people would know what questions to ask if they decided to further their learning. We also wanted to help them decide if they wanted to pursue farming as a career, in a way that was culturally relevant, historically rooted and committed to healing justice. We try to accomplish a lot. People walk away with tangible knowledge. They learn things like how to interpret a soil test, how to transplant and how to judge ripeness for harvest and also where to find more resources. We do early business planning and teach how to cook, and we process chickens and go through animal processing safety. Our days are packed. We go from 6 in the morning to 10 at night for the entire week. In the evenings we watch documentaries, do role playing and explore our history. We make art and do massage and body movement. We want folks to be able to taste, touch, feel and breathe what is possible if we were truly liberated and to have that vision so cemented in their hearts that they're not willing to settle for anything less! We've had hundreds of people come through the program and we've heard that this is what happens. Many have gone on to become farmers or food justice leaders, folks doing farm-based education for youth or running urban gardens. It's been very powerful.

ACRES U.S.A. How many times do you hold that each summer?

PENNIMAN. About three times.

ACRES U.S.A. Can you give an example or two of the types of assumptions, attitudes, or behaviors that your Uprooting Racism Farmers Immersion helps people understand and overcome?

PENNIMAN. You know the saying, "Fish can't see water?" People living with white privilege and positional power might not even notice that food is cheap, let alone ask why that's the case. It may just seem normal, so we uncover some of those realities that we've discussed. One of the biggest blind spots for well-meaning people is how much Euro-centric culture can permeate an organization, so we spend time together unpacking that invisible knapsack. For example, there's language inequality. Relying heavily on the written word and on the English language for communications might exclude people who are more accustomed to an oral tradition and face-to-face communication, or people who don't have English as their first language. Some of the other European values that people imagine to be normal are the drive toward perfectionism, always rushing and quantity over quality in relationships.

ACRES U.S.A. How do you respond when you're asked to serve on a board or advisory committee?

PENNIMAN. It's really not possible. I'm working on a farm 20 hours a week. I run all of the educational programming, which in 2016 served 3,650 human beings. I raise all of the money for the organization, and next year we're going to have to pull in over $250,000 to pay everyone and run all these programs. And I'm a mother of two and very active in our local Black Lives Matter community. I have to say no to taking a leadership role in other organizations at this point, but I do notice that folks are excited about these issues and invite me to give talks and advise. I do the best that I can to respond, but we need to uplift the many voices of people of color and women and other folks from marginalized communities who are doing leadership work. We have a Speakers Collective and we're trying to come up with a list of organizations, but it's also important for folks to take a minute and look around their own communities. You may realize that around the block there's a new Burmese American community of farmers, but you've never talked to them to see what they need or want to say.

ACRES U.S.A. What is your philosophy of education?

PENNIMAN. We're very rooted in the popular education model of Paulo Freire and Myles Horton. Everyone should know who these men are. Myles Horton was one of the most powerful white anti-racist organizers in the history of the United States. We also use the African Griot tradition of dynamic storytelling to convey information and raise consciousness. That's a very participatory form that gets people rooted in their own experience, in their hearts and their ancestors. We use a lot of theater, movements, art, storytelling and other connected ways of learning. With all the things to touch, feel, smell and experience, the farm is a perfect container for that. It would be ridiculous to keep your nose in a book or eyes on a screen in that context.

ACRES U.S.A. You talk a lot about nature as teacher. What are some examples of nature as a model for human behavior?

PENNIMAN. We like to think about nature as not just providing food, but also providing lessons for us. I'll give you two examples. Working in the orchard, we were talking about how it's best not to harvest tree fruit for the first several years. You pick off blossoms so the tree concentrates its energy in developing the roots, trunk and branches. That's also a metaphor for how we like to think about our projects. Early on you need to be investing in the foundation and the scaffolding. Grabbing a harvest then would undermine your long-term stability. Only after you develop your leadership and organizational ideals and practices are you ready to pick fruit. Another metaphor that was really powerful for us came up when we were pruning tomatoes. Again folks were really tempted. You see a sucker coming off with little tomatoes on it so you want to keep it. But no, we have to concentrate on the apical meristem. If you have that singular focus, rather than getting distracted by all these sidelines, in the long-term you'll have a healthier plant and more abundant harvest. Often in modern life we get strung too thin. We try to do everything all at once, and nothing is done with quality.

ACRES U.S.A. There was a description in your newsletter of making rock picking an emotionally meaningful activity.

PENNIMAN. All credit belongs to Julie Rawson of Many Hands Organic Farm for inventing this activity. We call it Rock Therapy. We are getting ready to put our east field in production so we've got to pick rocks. On our rocky land young folks gather up rocks. We go to the edge of the field and I tell them to think of something that makes them angry or sad, and shout that thing as loud as they can while tossing the rocks into the forest. It becomes a beautiful catharsis. Many young folks need to release anger around their parents, teachers, law enforcement, politics, or just slow Internet and bad cafeteria food. That makes the activity joyful and connected to who they are beyond the work of the farm.

ACRES U.S.A. Readers would be interested in the diversity of groups that come to the farm.

PENNIMAN. This summer 25 or 30 different groups of youth came to the farm. They range from court-adjudicated youth that we're working with through the county to youth in foster care,

Boys and Girls Club and Jack and Jill clubs to school groups. These youth are predominantly, but not exclusively, black, Latino and urban. Most come for a six-hour program or a day and a half overnight. Some come for longer programs. Adults come to our Black and Latinx Farmer's Immersion and Uprooting Racism programs from 20 states around the nation. About half are from our region. That's similar to our one-off programs, like the Seed-keeping weekend or Healing Justice weekend. Those folks come from everywhere, too.

ACRES U.S.A. Given that many young people coming to the farm have no particular knowledge, interest, or familiarity with agriculture, are you able to reach most of these kids?

PENNIMAN. We've never had anyone not eat the food and not leave with a smile, but definitely there's some skepticism. Young people show up who don't want to get out of the van, or their hoods are drawn or earphones are in, but I think having a lot of people of color on the farm makes a big difference. When we start to talk about issues of importance to them, like mass incarceration and hunger, that also makes some connection. And the land is beautiful and healing. I can't say that we reach everyone, but we always have a closing circle where folks say what they're grateful for and what they're taking with them. Almost universally we hear they're taking a sense of belonging to the land, of a renewed caring about food quality, hunger or farms so it's worth it.

ACRES U.S.A. What makes the way you farm at Soul Fire Farm regenerative?

PENNIMAN. First of all, we farm on soils considered so low quality that people say you can't grow food on them. As you can see, even in late fall, we're growing a lot of beautiful bountiful food. With our methods we can feed about 70 families per acre with a bushel and an eighth per week. Our practices include semi-permanent raised beds, minimal tillage, heavy straw mulch and no plasticulture. Our beds are on contour. We use drip irrigation and keep our soils very moist. We pay attention to micronutrients and add them when needed. We fertigate with mixtures of beneficial fungi and bacteria. We use a lot of clovers and bell beans as understory cover crops, and we cover crop at the end of the season. We rotate through and have perennials integrated into many parts of the farm—kiwis, grapes, fruit trees, hazelnuts, herbs—and pollinator-attracting flowers. We raise a couple hundred birds per year for meat on pasture using movable chicken tractors. I created a lightweight design that works for the female body to move by herself. We also have a small flock of laying hens on pasture. We use organic seed when we can source it. For animals, we get our feed from a farmer and prioritize local over certified organic.

ACRES U.S.A. We love this statement, "Farming involves as much intricacy, complexity, and attention to detail as medicine and law. Imagine if society valued it as highly." What are some of the dimensions of farming that tend to go unnoticed?

PENNIMAN. If you counted the number of sub-skills you need there would be thousands, from being able to determine the peak ripeness of all kinds of produce to identifying pests and diseases and the right remedy to all that behind-the-scenes accounting, sourcing, ordering and marketing.

ACRES U.S.A. What is the role of intention on your farm and in your work and community?

PENNIMAN. Impact is always more important than intention. We certainly have an intention to dismantle racism and oppression in the food system, but more important are the actual results. It's been great after a couple of years to start to hear back from people who have come here and to follow alumni and see what they're up to. Understanding that there is a ripple effect in the world is what's most meaningful to me.

"We Have to Change the Whole Mindset"

An Interview with Gabe Brown

Originally Published: January 2018

Gabe Brown is one of the great bridge builders in farming. No matter what corner of agriculture you come from, or even if you don't work in agriculture at all, Brown explains how regenerative farming can restore our ravaged soils to vitality. Moreover, he does it with a plainspoken, pragmatic aplomb that always captivates and never alienates, instead drawing listeners into the pleasure and excitement he gets from trying out new ideas. He explains techniques with a clarity that eludes many professional educators, and when the moment requires he can drive straight to the core of an issue with one clean stroke. At an Acres U.S.A. conference some years ago, an audience member said it all sounded great, but asked why he should he put in the extra work. Brown simply asked him if he cared about his grandchildren.

After many years of explaining his soil-building wizardry in person, Brown somehow found time to write a book, *Dirt to Soil,* which tells his story and explains what he does and why it works. The book includes farming practices, a philosophy of nature and the story of how Brown and his family survived several years of natural disasters in the mid-1990s, an ordeal that proved pivotal.

We first interviewed Brown in our October 2013 issue. We reached him for another in-depth talk a little more than five years later at his farm in North Dakota.

ACRES U.S.A. How did people such as Ray Archuleta, Dr. Kris Nichols, Dr. Christine Jones and others impact your effort to reinvent your whole way of working?

GABE BROWN. In my book, *Dirt to Soil,* I tried to tell the story in chronological order as to the people I met along the way and how they influenced me. I learned bits and pieces from many individuals, organizations and nature herself, and it was up to me to take that information and apply it on my ranch. I wanted to show other producers that you're not alone. You can glean information from many places, and it's up to you to take that information and apply it as best you can in the stewardship of your own operation.

ACRES U.S.A. Are you confident that as some of the original thinkers grow old and leave the scene, we have enough thinkers who can advise young farmers as you were advised? The personal touch seems to make all the difference.

BROWN. You're exactly right, and that's why I put that in the book. As the regenerative agriculture movement continues to grow, we are already seeing a new generation of leaders develop, spreading the word about what they are doing on their operation. I can go to all 50 states now and name younger producers who have really grasped regenerative agriculture and are moving these principles forward. That's exciting to see. It's time for us to allow these young leaders the opportunity to share their stories. I often tell people that what took me 25 years to learn and achieve on my operation is now being achieved in five years by some of these young regenerative agriculturalists.

ACRES U.S.A. Would it be right to say that you went from farming as a set of procedures to farming as a continuing experiment?

BROWN. Well, the way I look at it is, the current industrialized, commoditized production model is one of the recipe cards. You're following a recipe card, whether you're a livestock producer, a cash grain producer or a vegetable producer. We get these prescriptions more or less spelled out on a recipe card, and then it's just a matter of doing those practices. Regenerative agriculture is one of observation. It's one of real stewardship where you have to be adaptive; you see what's going on in the ecosystem, and what the ecosystem is trying to tell you. And then you just use these tools, whether it be livestock, or cover crops, or no-till drills. You use those tools to massage the ecosystem, so to speak, and advance soil health and ecosystem function. That's the way I look at it as a recipe card versus observation, experimentation and adaptation.

ACRES U.S.A. You write in the book that you had a running contest with a couple of fellow farmers to see who can come up with the most interesting way to advance soil health.

BROWN. They were David Brandt in Ohio and Gail Fuller down in Kansas. The three of us liked to challenge each other to see who could try the craziest things as far as cover crop mixes and different polyculture cash crops. One year I did the 70-plus species chaos garden as my entry into the challenge, so to speak.

ACRES U.S.A. How did that work out?

BROWN. Very well. One of the principles of a healthy soil ecosystem is diversity, so 70-plus species growing together—it was a jungle. Now, saying that, it's not economically feasible to do that for vegetable production or flower production because it was too hard to harvest, but it was sure fun once.

ACRES U.S.A. Do you think a playful spirit is one of the things missing in a lot of farming education now? Would it be a good thing to encourage?

BROWN. Absolutely. When I tell people about my ranch, I tell them that we try to fail at something every year. If we don't fail at something, we're not trying enough new things. Failures are simply learning experiences. It's what you do with that failure and how you change that dictates your path in life. We take these failures — make sure they are small, you don't do an experiment on the whole operation, you just do it on a small percentage of it—but you learn from and you grow from it. It also makes it fun just to see what you can do.

ACRES U.S.A. There is a saying in politics that you never want to waste a good crisis, implying that a crisis always comes with opportunities for change. When you look around your community and your state, do you think the crisis of the mid-1990s—the disastrous drought—was exploited well?

BROWN. I certainly do not think people learned from it. Last year was a good example. Here in North Dakota we had a major drought, and the answer to that drought by the government was simply, let's give financial assistance, rather than one of education. I tell people that last year we didn't have a drought on our farm. It didn't rain any more than it did anywhere around here, but we have built resiliency into our operation. By advancing soil health, increasing infiltration rates and increasing the water-holding capacity of our soils we more or less drought-proof our operation. Now don't get me wrong, if we have an extended drought, we're going to feel that too. But the fact of the matter is we did not have to sell a single animal because of the drought. And we had enough forage to weather it. That's because we've built resiliency into our ecosystem. I think that we do a travesty to producers when we give them this financial assistance because there is no incentive for them to change their management. What we should be doing is educating producers on how to build resiliency into their operations.

ACRES U.S.A. How do you think the crop insurance idea, borne of good intentions way back when, went in the wrong direction?

BROWN. You look at federal crop insurance. First of all, it's heavily subsidized. It's also geared toward certain commodity crops. With revenue insurance, 95 percent-plus of planting decisions in the United States are based on which crop is going to guarantee the most income. What this does is guarantee the over-production of those crops. The crop insurance program really makes sure we are in this over-production mode of certain crops. My family and I haven't taken part in crop insurance or any government program now for many years. This allows us a great deal of flexibility. We can change our cropping plans based on weather conditions, moisture conditions, etc. We can change based on prices if we so desire—now, we're trying to direct-market most everything we grow, but it allows us a lot of flexibility, and that's a good feeling. We're not bound in any way to that program. The other thing is, I really don't think the citizens of the United States should be subsidizing the insurance of producers. I say it this way—we're not subsidizing or allowing revenue insurance for Ma & Pa's Restaurant on Main Street, are we? Then why are we doing that for farmers? That to me is just not good business. Plus, I tell my fellow producers, "If you can't make it in this business without that subsidized insurance, perhaps you need to be in a different business." I catch a lot of flak for that, but it doesn't bother me.

ACRES U.S.A. The over-production of commodities never ceases to amaze.

BROWN. You have to look at another thing that's happening. Last year there was enough food produced in this world to feed about 10.2 billion people. There are less than 8 billion people in the world, so we're already over-producing. Plus, you take into account that 70 percent of the food in the world is being produced by peasant farmers. So this mantra that's touted all the time, that we have to feed the world—we're not feeding the world now. What we're producing is mainly going for things like ethanol or grain to feed livestock—ruminants — that shouldn't be fed grain. We have to change the whole mind-set of production agriculture.

ACRES U.S.A. We try to demolish that feed-the-world canard every day, and it's a never-ending job. It keeps coming back like the villain in a horror movie.

BROWN. More people are producing soybeans outside of "normal" soybean-producing areas. Because of federal crop insurance, we are now seeing an expansion in the acreage of these crops that are guaranteed the most income due to revenue insurance.

ACRES U.S.A. Going back to the crisis period that began in 1995, when you were farming conventionally, how did this period that changed everything about how you approach your work change your feelings about farming as a way of life?

BROWN. I tell people that those four years of natural disasters were hell to go through, but in the end they were the best thing that could have happened to myself and my family. Those years really made us realize what is important, and that is faith and family. If we got through that we can get through anything. It also made us realize that we can be profitable in a way that is not degrading our natural resources and in a way that is leaving our resources better for future generations. We're also producing food higher in nutrient density. To us, that's what it is all about. It's about faith and family, being good stewards of our natural resources, and producing healthy, nutrient-dense food. So, the crisis totally changed our mind-set from a bushel-yield-pound mentality to one of stewardship, faith and doing what's right for the resource.

ACRES U.S.A. Did the crisis give you a whole new definition of what constitutes a rough time?

BROWN. We realized that we're pretty resilient, and we also learned to enjoy the simple things. I tell people, when you're dead broke—we were so broke the bank knew when we bought toilet paper—it makes you realize that you can enjoy life without these materialistic things.

ACRES U.S.A. North America is having a brutal summer as we speak here in early August. How are water conditions in North Dakota and southern Canada?

BROWN. Large parts of North Dakota and southern Canada have been receiving timely rains. There is a portion in the middle part of the state that is a bit dry right now, but most of the area is having an average growing season.

ACRES U.S.A. How did learning that the fungal component of soil is the most important in a plant's early life change your approach to soil health?

BROWN. When I saw the results of Dr. David Johnson's work, which showed that early in most plant's life cycles, their association, or lack thereof, was critical to their development, I knew I had been on the right track. I had paid close attention to mycorrhizal fungi for years, proliferating it on my ranch. Dr. Johnson's work validated my work.

ACRES U.S.A. You mention your Australian colleague Colin Seis' idea of pasture cropping in your book. How is this practice significant, and does it work well outside the antipodes?

BROWN. Colin is from New South Wales, Australia. He and a friend, Daryl Cluff, developed what they call pasture cropping. Over there they have perennial pastures comprised of predominantly warm-season species. Colin and Daryl came up with the idea that when the warm-season perennials go dormant, they would seed a cool-season cash crop such as oats, canola, or barley into it. That annual cash crop will then grow, and as the weather starts to warm up and the warm-season perennial understory starts to grow, the cash crop is combined off, leaving the perennials to grow. Where I farm, on the northern plains, we only have about 120 frost-free days; there just isn't the growing season to allow us to pasture crop, but I know there is time in the southern United States, and I know people who are starting to move down that path. You need a long growing season, and you need predominantly warm-season perennials so you can seed a cool-season cash crop into them. I think there is real potential in the southern United States to do just that. Now, interestingly enough, a spinoff we are seeing and using on our operation is what we call polyculture cash crops. That's where you seed a mix of cash crops together, combine them together, and if you like you could separate the seed. An example of that would be canola, peas and barley—grow those three together. Their growing seasons match, you combine them together and separate the seed. I know in southern Canada they're doing a lot of work with this, and they are seeing 20 to 60 percent higher net returns per acre from that type of management scenario. I really think we'll see a lot more of that in the future.

ACRES U.S.A. Why do you recommend avoiding conventional soil testing?

BROWN. The vast majority of soil tests being performed today do not give a true indication of the fertility levels of the soil being tested. Nearly all soil tests today only measure the inorganic fraction of the nutrients in a sample. These tests do not tell you the organic fraction of those nutrients in the sample and how much of that will be cycled to plant-available forms via biology. Biology is key and critical to nutrient cycling. I recommend the Haney soil test. It is, in my opinion, the best test available to help producers begin to understand the level of biology in their soil and how that biology will cycle nutrients.

ACRES U.S.A. You have a section in the book called "Planning For The Long Haul," and since we hear a lot of grim predictions about the future these days, I wonder if you could turn that around and imagine a better future for the upper Midwest as the regenerative model gains adherents.

BROWN. I think one of the things we need to look at is more of a perennial type of system. Let's face it, much of the land in the upper Midwest is marginal land as far as rainfall and soil type are concerned. Let's put it into a perennial-type system with stacked enterprises of livestock, grazing on perennial pastures. We can integrate many species on those pastures. Stacking enterprises adds a lot more potential income per acre, and also reduces the risk because you have more income streams. I see a world of potential for that type of production model.

ACRES U.S.A. What do you mean by stacking enterprises?

BROWN. I'll just explain what we do on our operation. We grow not only cash grain crops and cover crops on some of those acres, but we raise grass-finished beef, grass-finished lamb and pastured pork. Some of the grain we grow feeds our pastured hogs. We have 1,400 laying hens out on pasture, and those hens are fed grain screenings from the cash grain operation. We have bees, we have vegetable garden, we have orchards—we have all these different enterprises and income streams. It makes us very resilient to fluctuations in commodity prices.

ACRES U.S.A. How did you build your direct marketing business? Does not living near a huge population center, like an upstate New York farmer for instance, pose a challenge?

BROWN. One of the things people need to realize is that we're in the 21st century. Let's look at the fact that it's very easy with the internet and social media to sell your wares to a broader market. We market the majority of our products over the internet. We sell them in North Dakota, but even in North Dakota there is a delivery service that will drop products right to people's doors for a very low price. We can take a product and ship it a hundreds miles away for $15 to $20. That makes buying very easy for our customers because they can order online; they pay with a credit card online; we pack up the product and take it to the delivery service; and there it is. It's delivered to somebody's door the very next day at a very reasonable price. I don't think producers should use their location as an excuse, not in this day and age.

ACRES U.S.A. What is an example of a successful deliverable product for you?

BROWN. We offer over 120 different cuts of beef, lamb and pork—anything from ground to steaks to roasts, to jerky to hot dogs—they're all frozen so we're able to ship them anywhere in the state. All you do is put them in a cooler bag, put it in a box and the next day it's on somebody's doorstep.

ACRES U.S.A. Do you regard livestock diversity as a critical or crucial component of regenerative agriculture?

BROWN. I tell people this: "You have to do what you enjoy doing. Otherwise it's going to be a chore, and it's not going to be something you will want to do every day." I know there are people out there who only want to cash-grain farm. That's fine, you can do that, but realize the more enterprises you stack, the more profitable it becomes, the further you can advance soil health, the more resilient your operation becomes. Now, if you don't want to run livestock on your operation, maybe there is a young person who does. Why not allow them to bring livestock onto your operation? It's going to advance the health of your soil, and it will give a young person a chance. Now, are livestock critical? We can advance soil health without livestock. We cannot advance soil health without soil biology and without insects. Grazing animals add another dimension. I tell people it's just like climbing up a set of stairs. Every one of the practices you do—whether it's no-till, cover crops, diverse cash crops, animal integration—they all take you farther up the stairs. How high you go is up to you.

Books from Acres U.S.A.

Most of the books mentioned in these interviews are available on Amazon, but you don't have to shop there. By supporting small, independent bookstores and publishers, including Acres U.S.A., you are helping support the advancement of unique, contrarian ideas that may never else get published.

To order call 1-800-355-5313
or order online at Bookstore.AcresUSA.com

Find more interviews in every Acres U.S.A. magazine.
Subscribe for a digital or print subscription
at www.AcresUSA.com/subscribe

www.ingramcontent.com/pod-product-compliance
Lightning Source LLC
Chambersburg PA
CBHW080642270326
41928CB00017B/3164